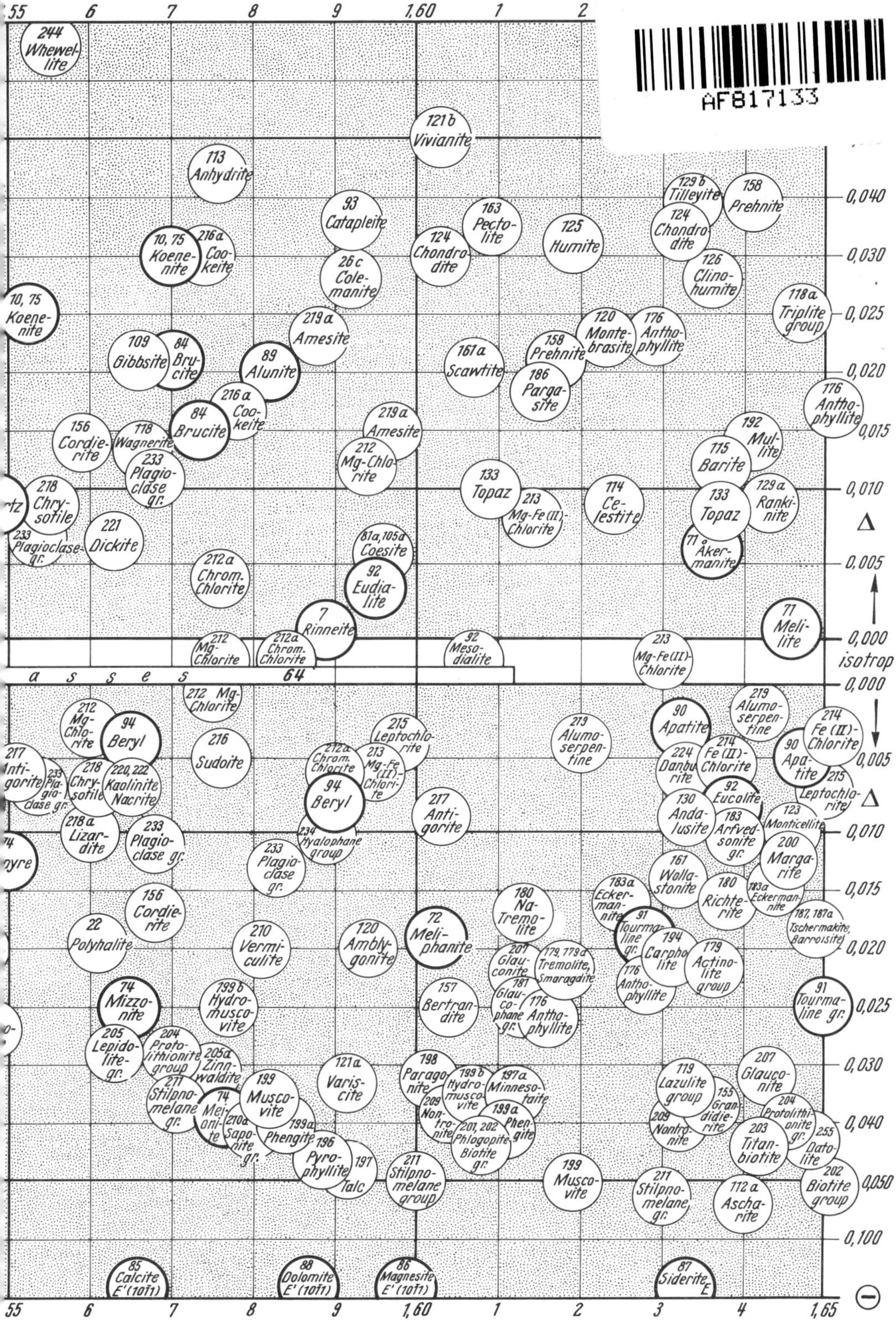


/ $n_E$ are given separately)

refractive index $n_Y$ / birefringence $n_Z - n_X$

$n_Y = 1.44 - 1.65$

rcles enclose names of uniaxial minerals

continued on back cover

W. E. Tröger

# Optical Determination of Rock-Forming Minerals

## Part 1 Determinative Tables

English Edition of the Fourth German Edition

by

H. U. Bambauer, F. Taborszky, H. D. Trochim

Translated by Chr. Hoffmann
in conjunction with the authors

With 1 chart of interference colors, 2 stereograms (as appendices) and
112 diagrams and 264 figures in the text

E. SCHWEIZERBART'SCHE VERLAGSBUCHHANDLUNG
(NÄGELE u. OBERMILLER) STUTTGART 1979

First edition (in German) 1952



Printed in Germany

ISBN 3 510 65311 4

Offsetdruck Julius Beltz, 6944 Hemsbach üb. Weinheim

# Preface to the English Edition

The following is a freely translated and very much condensed version of the preface heading the German editions.

This translation of the fourth German edition of the determinative tables adheres to the widely acknowledged arrangement of the late W. E. TRÖGER. Each mineral is given an index number which is included with each relevant diagram, crystal sketch, so that related information can be easily found.

An unknown mineral is usually first examined for refractive indices and birefringence, and so these parameters are provided in a reference chart on the inside covers of this book (ore minerals are not included). The use of this reference chart is recommended (especially for the beginner) even if a determination "at first sight" seems feasable. One will be on guard against similar minerals after looking at this reference chart.

Important changes of the fourth German edition as compared to earlier German editions are:

1. International mineral names (after STRUNZ) are used throughout in the first column. Common synonyms are given in brackets.
   The chemical formulas are given as structural formulas (after STRUNZ) as these allow better recognition of the mineral's chemical variability than the previously used oxide formulas.
2. In order to avoid confusion, more references about the crystallographic setting are given in the second column and with the crystal sketches. However some "old" crystallographic settings have been retained for internal consistency with previous editions.
3. The space in column ten ("occurrences") of the determinative tables only allows the printing of some key words; more accurate paragenetic descriptions are to be found in the German text volume or in comparable English textbooks.
4. To maintain consistency between all German and English editions, when a new mineral is included it is given a letter prefix; thus 122 a and 122 b are minerals inserted after the first edition, and are not related to mineral 122. For the same reason the numbering system does not follow all recent advances of mineral systematics.
5. Towards the end of the book is a determinative table with the d-spacings of the four strongest X-ray reflections of all listed minerals. X-ray data have also been added to some determinative diagrams.

For some determinative diagrams it was no longer possible to quote a reference as the data came from several papers. The full quotation can always be found in the German text volume W. E. TRÖGER: „Optische Bestimmung der gesteinsbildenden Minerale, Teil 2 Textband", Schweizerbart'sche Verlagsbuchhandlung, Stuttgart, 1967. Also, on p. 7 a list of English compilations is provided in which most of the literature cited can be found.

The variation-diagrams were deliberately printed without a "graph-paper" background to avoid an exaggerated impression of precision. These diagrams either were reproduced from the literature or were constructed from data taken from the literature. In only a few cases were there enough data for statistical treatment. Also, in many papers, estimates of the accuracy of chemical analysis etc. were not provided. Inclusion of all individual points would result in an overcrowded and confusing diagram, and so we have shown only the mean variation curve. It should be stressed that even the "best" determinative diagram rarely allows a determination better than $\pm$ 2–5 mol%. With most diagrams a determination with reasonably good approximation should be possible. In many cases the optical determination is complicated because the minerals concerned are not truly binary solid solution series. In such cases additional methods must be employed.

The determinative tables for plagioclases were taken from BURRI, PARKER & WENK: „Die optische Orientierung der Plagioklase", Birkhäuser Verlag, Basel, 1967; additional diagrams can be found in this book.

The cooperation of the Schweizerbart'sche Verlagsbuchhandlung is gratefully acknowledged; in particular we want to thank Miss H. SCHRÖDER for her constant help with the great number of corrections which occurred during different stages of translation. In connection with this we are also grateful to Dr. A. BLAXLAND † and Prof. P. H. RIBBE for their occasional advice on translation problems.

We would be grateful to all readers for any information concerning faults, misprints, omissions etc. Also, as a review of the current literature becomes increasingly difficult, we would be glad to receive reprints of pertinent papers from our colleagues.

November 1977 H. U. BAMBAUER, F. K. TABORSZKY, H. D. TROCHIM

Prof. Dr. H. U. BAMBAUER
Institut für Mineralogie
Universität Münster
Gievenbecker Weg 61
D-4400 Münster

Doz. Dr. F. K. TABORSZKY
Mineralogisches Institut
Universität Freiburg
Albertstr. 23 b
D-7800 Freiburg

Prof. Dr. H. D. TROCHIM
Institut für Mineralogie
Ruhr-Universität
Universitätsstr. 150
D-4630 Bochum

# Translators Remarks

This English edition is essentially an unchanged translation of the fourth German edition of the Optische Bestimmung der gesteinsbildenden Minerale, updated by BAMBAUER, TABORSZKY & TROCHIM. Some corrections were made by the authors and the translator.

The German work consists of two volumes:

1. The determinative tables (presented here in English)
2. The descriptive text volume (so far available only in German)

Despite the recommendations of the I.M.A., the symbols $n_x$, $n_y$ and $n_z$ are used for the refractive indices rather than $\alpha$, $\beta$ and $\gamma$. This is to avoid confusion in columns two through five where $\alpha$, $\beta$ and $\gamma$ are used as symbols for angles; also this is consistant with the symbols X, Y and Z for the axes of the indicatrix. The usage adopted here is the same as in WINCHELL's book and also in the earlier German editions of these tables.

The descriptions of colors (column 7) are sometimes a bit lengthy in the German edition (e.g.: "pale yellow – light yellowish green – light green"). In some such cases the descriptions were condensed (e.g.: "pale yellow – pale green") leaving it to the user to imagine all intermediary shadings.

The limited space in column ten ("occurrences") allows only a few key words. The prefixes "epi-", "meso-" and "cata-", stemming from the GRUBENMANN & NIGGLI concept of metamorphism, are not in common use among English-speaking workers, and this concept has now largely fallen out of favor. Therefore, grades of metamorphism were described by the terms "low-", "medium-" and "high-grade", sometimes indicating additional features (e.g.: "high pressure").

In the light of recent knowledge some information given in column ten of the German edition were updated (e.g. No. 54, Koppite "occurring in contact metamorphic marbles", now given as "occurring in carbonatites").

Freiburg, November 1977 CHR. HOFFMANN

# Contents

# Explanation of Nomenclature, Abbreviations and Symbols used in the Tables

**Column 1**

Crystal system: cub = cubic; tetr = tetragonal; trig = trigonal; hex = hexagonal; orh = orthorhombic; mcl = monoclinic; trcl = triclinic.

Chemical formulas are given as in "Mineralogische Tabellen" by STRUNZ. In solid solutions the predominant element is underlined, e.g. (Mg, Fe).

**Column 2**

The listing of the crystal forms approximately corresponds to their frequency and importance. If for a given mineral different crystallographic settings are used in the literature, the author whose setting has been chosen here will be named.

◇[uvw] = twinning axis, ◇(hkl) = twinning plane, comp.pl. = composition plane; lam. = lamellar twinning frequent or predominant, pseudom. = pseudomorph; ps. = pseudo-; param. = paramorph (isochemical pseudom.), e.g. microcline after sanidine.

**Column 3**

# (hkl) = cleavage. If necessary, cleavage angles will be given. $K_1$ = plane of slip in deformation twins; $K_2$ = plane of the corresponding second circular section; $\sigma_1$ = direction of slip; T = plane of translation with t = direction of translation. H = hardness (MOHS-scale); D = average density. The full range of densities is given in the diagram on pages 154–157.

**Column 4**

Compare the corresponding crystal drawings. In practical microscopy the sign of the extinction angle is of minor importance, as the crystallographic orientation of the section has to be known. Generally, a positive (negative) sign means that the optical direction measured is located in the obtuse (acute) angle between two crystallographic axes. In the case of feldspars an exception is made following the rules of SCHUSTER and MICHEL-LÉVY; see figs. p. 125, and vol. 2, p. 742 f.

a, b, c = crystallographic axes
X, Y, Z = axes of the optical indicatrix
X', Z' = minor and major axes of the elliptical random sections of the indicatrix
A, B = optic axes, especially of feldspars and pyroxenes
O. A. P. = optic axial plane
E, O = vibration directions of the extraordinary and ordinary wave, respectively
bi = indicates X or Z being acute bisectrix
r $\lessgtr$ v = dispersion of the extinction angle
1 = (+) means that in crystals of prismatic or tabular habit Z is parallel to the prism axis or plate -normal respectively. Lath-like sections therefore show "positive elongation". Similarly I = (–) means "negative elongation". If Z or X deviate slightly from the crystallographic directions, I' = (+) or I' = (–) is used instead.
Z∧c, X∧a etc. = extinction angles

**Column 5**

$n_x < n_y < n_z$ = the three main refractive indices to three places of decimals, generally for white light (identical with $n_\alpha$, $n_\beta$ and $n_\gamma$). In cases of strong dispersion the wavelength or color of light used or simply "Na-light" (= monochromatic light of sodium vapor lamp) is given. Two limiting values indicate the range of variation; the chemical and/or structural reasons of such variations are mentioned if they are known. Extreme values are neglected. Four decimal places are given only for the invariably pure minerals (e.g. quartz). $n_E$ and $n_O$ are the refractive indices for the extraordinary and ordinary waves, respectively (identical with $n_\varepsilon$ and $n_\omega$).

⊕ or ⊖ = optical character of the mineral;

$\Delta$ = value of birefringence $n_z - n_x$.

**Column 6**

In all cases the optical angle of the acute bisectrix is given. Reasons of variability are given as far as known.

$2V_X = (2v_\alpha)$ = optic axial angle about the (negative) axis X

$2V_Z = (2V_\gamma)$ = optic axial angle about the (positive) axis Z

$r \lessgtr v$ = dispersion of the optical axes.

**Column 7**

Colors given correspond to thin sections of normal thickness only. Pleochroism is related to the vibration directions, or in a few cases to crystallographic directions.

$X \lessgtr Z$ or $O \lessgtr E$ shows the relative intensity of color in the directions of absorption concerned.

**Column 8**

It should be kept in mind that distinctness and intensity of a chemical reaction varies with the variability of the mineral. If the optical determination does not lead to an unequivocal result, hints for the use of additional methods may be found in this column: chem. anal. = chemical analysis; X-ray = X-ray powder diffraction pattern or other X-ray methods; DTA = differential thermal analysis.

**Column 9**

sm = smaller; gr = greater. Otherwise the same abbreviations are used as in other columns, e.g. # = cleavage etc.

**Column 10**

Only the most important modes of occurrence are given as key words. For a more complete description see vol. 2.

d-spacings for the four strongest X-ray powder reflections are listed in a special table on pages 164 through 174. It should be pointed out that for many minerals, particularly the feldspars, these four d-spacings do not permit a definite determination. In case of doubt, the complete powder diagram should be considered in connection with the chemical and structural variation of the mineral.

# Standard Texts Recommended for Use in Conjunction with the Tables

BURRI, C., PARKER, R. L., WENK, E. (1967): Die optische Orientierung der Plagioklase. Birkhäuser, Basel. – DEER, W. A., HOWIE, R. A., ZUSSMAN, J. (1962): Rock-forming Minerals. Longman, London. – FREUND, H.: Handbuch der Mikroskopie in der Technik, Band IV, Teil 2. Umschau-Verlag, Frankfurt/Main, 437 pp. – PARFENOFF, A., POMEROL, C., TOURENQ, J. (1970): Les Minéraux en Grains. Masson, Paris. – SMITH, J. V. (1974): Feldspar Minerals. Springer, Berlin, Heidelberg, New York. – TRÖGER, W. E. (1969): Optische Bestimmung der gesteinsbildenden Minerale, Teil 2, Textband. Schweizerbart, Stuttgart. – WINCHELL, A. N. (1951): Elements of Optical Mineralogy, Part II. Wiley, New York.

# Technische Anmerkungen sowie Erklärung der in den Tabellen gebrauchten Abkürzungen, Bezeichnungen und Zeichen

**1. Kolonne**

Kristallsysteme: cub = kubisch; tetr = tetragonal; trig = trigonal; hex = hexagonal; orh = orthorhombisch; mcl = monoklin; trcl = triklin.

Als chemische Formel wird konsequent die Strukturformel nach STRUNZ angegeben. Bei Mischkristallen wird gegebenenfalls das mengenmäßig vorherrschende Element unterstrichen, z. B. (Mg, Fe).

**2. Kolonne**

Die Reihenfolge der aufgeführten Flächenformen entspricht etwa ihrer Bedeutung für den Habitus. Falls in der neueren Literatur mehrere kristallographische Aufstellungen gebräuchlich sind, wird der Autor genannt, dessen Vorschlag hier benützt wird.

◇ [uvw] = Zwillingsachse, ◇(hkl) = Zwillingsebene, compl. pl. = Verwachsungsebene, lam. bedeutet, daß die Zwillingsbildung vorwiegend lamellar ist. Pseudom. = Pseudomorphose; ps. = pseudo (z. B. ps. mcl. = pseudomonoklin); param. = Paramorphose (isochemische Pseudomorphose).

**3. Kolonne**

⋕ (hkl) = Spaltbarkeit; bei Bedarf Angabe des Spaltwinkels.

$K_1$ = Gleitfläche für einfache Schiebungen mit $K_2$ = zweiter Kreisschnittebene. $\sigma_1$ = Schiebungsrichtung. T = Translationsebene mit t = Translationsrichtung. H = Mohs'sche Härte.

D = Dichte-Mittelwert. Grenzwerte hierzu siehe im Diagramm auf S. 154–157 oder evtl. in den Bestimmungsdiagrammen.

**4. Kolonne**

Man vergleiche hierzu stets die zugehörigen Kristallfiguren! Das Vorzeichen der Auslöschungsschiefe hat in der mikroskopischen Praxis nur eingeschränkte Bedeutung, da man dazu ja die kristallographische Orientierung des Schnittes kennen muß, z. B. aus der Lage mehrerer Systeme von Spaltrissen. Im allgemeinen bedeutet ein positives Vorzeichen, daß die gemessene optische Richtung sich im stumpfen Winkel zwischen zwei kristallographischen Achsen befindet, negatives Vorzeichen entsprechend im spitzen Winkel. Eine Ausnahme machen nur die Feldspäte, bei denen die Regeln von SCHUSTER und von MICHEL-LÉVY gelten, die durch Figuren auf S. 125 erläutert sind (siehe auch Teil 2, p. 742 f.).

a, b, c = kristallographische Achsenrichtungen.
X, Y, Z = Richtungen der Indikatrixachsen
X', Z' = Richtungen der kleinen und großen Achse der in einer beliebigen Schnittfläche liegenden Schnittellipse der Indikatrix.
A, B = Optische Achsen, spez. bei Feldspäten und Pyroxenen.
E, O = Schwingungsrichtungen des außerordentlichen, bzw. des ordentlichen Strahls.
O. A. P. = Ebene der optischen Achsen.

bi kennzeichnet die auf derselben Zeile aufgeführte Achse X oder Z als spitze Bisektrix.

$r \lesseqgtr v$ = Dispersion der Auslöschungsschiefe

I = (+) bedeutet, daß bei säulig oder tafelig ausgebildeten Kristallen Z genau parallel zur Säulenachse, bzw. X senkrecht zur Tafel liegt. Leistenförmige Schnitte zeigen dann positiven „Charakter der Hauptzone". Analog I = (–). Wenn die optischen Hauptrichtungen von den kristallographischen etwas abweichen, gilt I' = (+) bzw. I' = (–).

$Z \wedge c$, $X \wedge a$ u.s.w. = Auslöschungsschiefe

**5. Kolonne**

$n_x < n_y < n_z$ = die drei Hauptbrechungsindizes (d. h. identisch mit $n_\alpha < n_\beta < n_\gamma$); im allg. für weißes Licht, abgerundet auf 3 Dezimalen. Bei starker Dispersion ist die benützte Wellenlänge oder Lichtfarbe angegeben. Zwei Grenzwerte kennzeichnen meist den Variationsbereich (wobei ausgefallene Extremwerte nicht erfaßt sein müssen), möglichst unter Nennung der chemischen Ursachen. Angaben auf 4 Dezimalen für Na-Licht nur bei invariabel reinen Mineralien.

$n_E$ und $n_O$ sind die Brechungsindizes für die außerordentliche bzw. ordentliche Welle (identisch mit $n_\varepsilon$ und $n_\omega$).

$\oplus$ oder $\ominus$ = Optischer Charakter des Minerals („optische Vorzeichen"). $\Delta$ = Betrag der Doppelbrechung $n_z - n_x$ für die o. g. Wellenlänge.

**6. Kolonne**

Es wird stets der Winkel der optischen Achsen um die spitze Bisektrix angegeben. Grenzen der Variabilität möglichst unter Angabe der chemischen Ursachen.

$2V_x$ = Achsenwinkel um die (negative) Mittellinie X (identisch $2V_\alpha$)

$2V_z$ = Achsenwinkel um die (positive) Mittellinie Z (identisch $2V_\gamma$).

$r \lesseqgtr v$ = Dispersion der optischen Achsen um die angegebene Mittellinie.

**7. Kolonne**

Alle Farbangaben gelten nur für normale Dünnschliffdicke!

Der Pleochroismus bezieht sich auf die angegebenen Schwingungsrichtungen, in Ausnahmefällen auf angegebene kristallographische Richtungen.

Die Absorptionsformel $X \lesseqgtr Y \lesseqgtr Z$ bzw. $O \lesseqgtr E$ gibt die relative Farbintensität in drei bzw. zwei Absoptionsrichtungen an.

**8. Kolonne**

X-ray = Röntgendiagnose mit Pulverdiagramm empfehlenswert.

**9. Kolonne**

D = Dichte; $\Delta$ = Doppelbrechung; gr. = größer; sm. = kleiner.

**10. Kolonne**

Als Begleiter werden nur charakteristische Minerale genannt. Die so wichtigen paragenetischen Angaben mußten aus Platzmangel extrem knapp gehalten werden und sind daher oft nicht vollständig.

Auf Seite 164–174 befinden sich Tabellen mit den d-Werten der vier stärksten Röntgeninterferenz-Linien. Die dort angegebene Literatur befindet sich in den entsprechenden Abschnitten des Textbandes (Teil 2). Es wird darauf hingewiesen, daß diese vier d-Werte bei vielen Mineralen, insbesondere bei den Feldspäten, nur einen ersten Hinweis, aber keine eindeutige Bestimmung erlauben. Im Zweifelsfall sollte stets das vollständige Pulverdiagramm unter Berücksichtigung der chemischen und strukturellen Variationsbreite der betreffenden Mineralart betrachtet werden.

# Annotations techniques et explication des abréviations, désignations et signes employés dans les „Tables“

**1ère colonne**

systèmes cristallins: cub = cubique; tetr = quadratique; trig = trigonal; hex = hexagonal; orh = orthorhombique; mcl = monoclinique; trcl = triclinique.

La notation chimique utilisée sans exception est la notation structurale d'après STRUNZ. Pour les séries isomorphes on a souligné le cas échéant l'élément dominant, p. ex. (Mg, Fe).

**2ème colonne**

L'ordre de citation des faces cristallines correspond à peu près à leur importance pour l'habitus. Lorsque plusieurs orientations cristallographiques sont admises dans la littérature récente, on a indiqué le nom de l'auteur de l'orientation choisie.

[uvw] = axe de macle; ◇ (hkl) = plan de macle; Comp. pl. = plan d'association; lam. veut dire que les cristaux maclès se présentent de préférence sous formes de lamelles multiples; pseudom. = pseudomorphose; ps. = pseudo (p. ex. ps. mcl. = pseudo-monoclinique); param. = paramorphose (= pseudomorphose isochimique).

**3ème colonne**

# (hkl) = plan de clivage; L'angle entre deux plans de clivage est quelquefois indiqué.

$K_1$ = plan de glissement des macles mécaniques avec $K_2$ = deuxième plan de section circulaire; $\sigma_1$ = direction de glissement dans $K_1$; T = plan de translation avec t = direction de translation. H = dureté suivant l'échelle de MOHS; D = valeur moyenne de la densité. Les valeurs extrêmes pour une espèce minérale donnée se trouvent dans le diagramme des pages 154 à 157 ou, dans plusieurs cas, parmi les diagrammes de détermination.

**4ème colonne**

Comparez toujours les figures des formes cristallines correspondantes! Le signe de l'angle d'extinction n'a dans la pratique du microscope polarisant qu'une valeur réduite, puisqu'il suppose l'orientation cristallographique de la section étudiée.

En général, un signe positif veut dire que la direction optique observée se trouve dans l'angle obtus entre deux axes cristallographiques; un signe négatif indique une position dans l'angle aigu. Seuls les feldspaths font exception: pour eux sont valables les règles de SCHUSTER et de MICHEL-LÉVY qui sont expliquées par les figures de la page 125 (voir aussi tome 2, p. 742 ff.)

a, b, c = directions des axes cristallographiques
X, Y, Z = directions des axes de l'indicatrice
X', Z' = directions du petit axe et du grand axe de l'ellipse découpée dans l'indicatrice par un plan d'orientation quelconque.
E, O = direction de vibration respectivement du rayon extraordinaire et du rayon ordinaire.
O. A. P. = plan des axes optiques
bi = 1ère bissectrice (p. ex.: Z = Bi)
r $\lessgtr$ v = dispersion de l'angle d'extinction
I (+) = extinction droite et allongement positif
I' (+) = extinction oblique et allongement positif
I (–) et I'.(–) = signification analogue dans le cas d'un allongement négatif
Z ∧ c, X ∧ a etc = angles d'extinction

**5ème colonne**

$n_x < n_y < n_z$ = les trois indices principaux de réfraction (identiques à $n_\alpha < n_\beta < n_\gamma$), mesurés en général en lumière blanche et arrondis à trois décimales. En cas de forte dispersion, on a indiqué la longueur d'onde ou la couleur de la lumière utilisée. Les deux valeurs données limitent le domaine de variation des indices d'une espèce. On a indiqué, dans la mesure du possible, les caractères chimiques et structurels influant sur les indices. Les valeurs exceptionnelles n'ont pas été prises en considération. Seuls les indices de minéraux de compositions constantes ont été donnés avec quatre décimales.

$n_E$ et $n_O$ = indice de réfraction du rayon extraordinaire et du rayon ordinaire (ils sont identiques à $n_\varepsilon$ et $n_\omega$).
⊕ et ⊖ = signe optique
$\Delta$ = biréfringence ($n_z - n_x$).

**6ème colonne**

L'angle des axes optiques donné est toujours l'angle aigu. Les valeurs limites sont indiquées et corrélées, dans la mesure du possible, avec les variations de composition chimique.
$2V_x$ = biaxe négatif (identique à $2V_\alpha$)
$2V_z$ = biaxe positif (identique à $2v_\gamma$)
$r \lessgtr r$ = dispersion des axes optiques autour de la 1ère bissectrice.

**7ème colonne**

Toutes les couleurs citées ne sont valables que pour une épaisseur normale de la plaque mince. Le pléochroisme est à rapporter aux directions de vibration indiquées, ou, pour quelques exceptions, aux directions cristallographiques, indiquées elles aussi.
Les formules d'absorption $X \lessgtr Y \lessgtr Z$ et $O \lessgtr E$ précisent l'intensité relative des couleurs dans les trois ou les deux directions d'absorption.

**8ème colonne**

Pour le cas où la détermination optique s'avérerait insuffisante, on pourra employer les méthodes supplémentaires indiquées dans cette colonne.
chem. anal.: analyse chimique;
X-ray: analyse par rayons X (poudre ou autre)
DTA: analyse thermique différentielle

**9ème colonne**

sm = plus petit, gr = plus grand;
Sinon, mêmes abréviations et signes que dans les autres colonnes.
p. ex.: # = plan de clivage.

**10ème colonne**

Seuls les minéraux les plus caractéristiques coexistant avec l'espèce en question sont cités. Les informations, pourtant importantes, sur les paragenèses, ont du être réduites, par manque de place.
Les valeurs $d_{hkl}$ des quatre raies les plus intenses sur les diagrammes de poudre sont données, sous forme de tableaux, pp. 164–174. On trouvera les citations bibliographiques complètes dans les chapîtres correspondants du tome 2. Dans le cas de beaucoup de minéraux, feldspaths en particulier, ces quatre valeurs $d_{hkl}$ ne permettent qu'une détermination très approximative. Dans ces cas douteux, il faut toujours interpréter le diagramme de poudre complet, en tenant compte des variations chimiques et structurelles possibles de l'espèce minèrale à dèterminer.

| No. | Name<br>Symmetry<br>Chem. Composition | Common Crystal Forms, Twinning<br>Crystal Habit<br>Aggregates | Cleavage<br>Translation Gliding<br>Deform. Twinning<br>Hardness, Density | Optic Orientation | Refractive Indices<br>Birefringence |
|---|---|---|---|---|---|
| 1 | Ice I hex<br>$H_2O$ | {0001}, {10$\bar{1}$0};<br>skeletal crystals,<br>needles, grains, radiated | # (0001) ?<br>T = (0001), plast.;<br>H = 1½(−5°C)<br>D = 0.918 | Z ∥ c E = Z | $n_O$ = 1.3091<br>$n_E$ = 1.3105 (−3°C)<br>⊕Δ = 0.0014<br>$r < v$ weak |
| 2 | Calciumoxide<br>(Lime) cub<br>CaO | {100}<br>granular | # {100} perfect;<br>H = 3 – 4 D = 3.40 | isotropic | n = 1.837 |
| 3 | Villiaumite cub<br>NaF | xenomorphic in<br>interstices | # {100} good;<br>H ~ 3 D = 2.79 | if abnormally tetr.:<br>X ∥ c E = X | $n_O$ ~ $n_E$ ~ 1.328<br>⊖Δ = very weak |
| 4 | Halite cub<br>(Rock Salt)<br>NaCl | {100}; granular, may be<br>fibrous ∥ [110] and<br>[100], hopper crystals | # {100} perfect;<br>T = (110),<br>t = [1$\bar{1}$0], plastic;<br>H = 2½ D = 2.168 | isotropic;<br>may be abnormally<br>birefringent<br>lamellae ∥ (110) | n = 1.544 |
| 5 | Sylvine cub<br>KCl | {100}, {111};<br>granular, spathic, rarely<br>prisms, often acicular | # {100};<br>T = (100)<br>t = [1$\bar{1}$0], plastic;<br>H = 2 D = 1.99 | isotropic;<br>may be abnormally<br>birefringent | n = 1.490 |
| 6 | Bischofite mcl<br>$MgCl_2 \cdot 6H_2O$ | {100}, {111}, {110};<br>β = 94°; pseudohex.<br>◈[1$\bar{1}$2], composition<br>plane ~ (110), lam.;<br>granular, spathic,<br>fibrous ∥ c | perfect parting<br>∥ {110}, 108°;<br>$K_1$ ~ (1$\bar{1}$0) irrat.,<br>$\sigma_1$ = [1$\bar{1}$2], $K_2$ = (1$\bar{1}$1),<br>T = (110), very plastic;<br>H = 1½ D = 1.60 | X ∥ b<br>Y ∧ c = + 10°<br>Z ∧ a = − 6°, bi<br>one opt. axis ~⊥(110);<br>see Vol. 2, p. 15 | $n_X$ = 1.494<br>$n_Y$ = 1.507<br>$n_Z$ = 1.528<br>⊕Δ = 0.034 |
| 7 | Rinneite trig<br>$K_3Na[FeCl_6]$ | {11$\bar{2}$0}, {10$\bar{1}$1}, {0001};<br>◈[0001];<br>tabular (0001)<br>short prisms [0001],<br>coarse grains | # {11$\bar{2}$0} distinct;<br>H = 3 D = 2.35<br>(2.55) | Z ∥ c E = Z<br>I = (+) | $n_O$ = 1.5886<br>$n_E$ = 1.5894<br>⊕Δ = 0.0008<br>disp. $r < v$ strong |
| 8 | Carnallite orh<br>$KMgCl_3 \cdot 6H_2O$ | {111}, {001}, {0Kl}<br>pseudotrig.;<br>◈{110} lam.;<br>granular, fibrous | fracture conchoidal<br>$K_1$ = (110), $K_2$ = (1$\bar{3}$0),<br>plastic;<br>H = 2½ D = 1.60 | X ∥ c<br>Y ∥ b O.A.P. (010)<br>Z ∥ a, bi | $n_X$ = 1.467<br>$n_Y$ = 1.475<br>$n_Z$ = 1.494<br>⊕Δ = 0.027 |
| 9 | Tachydrite trig<br>$Mg_2CaCl_6 \cdot 12H_2O$ | {10$\bar{1}$1}, {0001};<br>◈ lam.;<br>granular, spathic | # {10$\bar{1}$1} perfect<br>106½°;<br>parting ∥ ◈;<br>H = 2 D = 1.66 | X ∥ c E = X | $n_O$ = 1.522<br>$n_E$ = 1.513<br>⊖Δ = 0.009 |
| 10 | Koenenite trig<br>$4NaCl \cdot 4(Mg,Ca)Cl_2 \cdot$<br>$5Mg(OH)_2 \cdot 4Al(OH)_3$ | {0001}<br>barrel-shaped form,<br>scaly | # (0001) perfect<br>T = (0001), flexible;<br>H < 1 D = 1.98 | Z ∥ c E = Z<br>I = (−)<br>Fig. Vol. 2, p. 32 | $n_O$ = 1.540 – 1.555<br>$n_E$ = 1.565 – 1.585<br>⊕Δ = 0.025 – 0.030 |
| 11 | Thenardite orh<br>$Na_2SO_4$ | {111}, {010}, {101};<br>◈{110} and {011};<br>dipyram., tabular (010),<br>rarely acicular [100],<br>powdery | # (010) perfect,<br># {101} distinct,<br># {100} poor;<br>H = 2½ D = 2.67 | X ∥ c<br>Y ∥ b, O.A.P. (010)<br>Z ∥ a, bi | $n_X$ = 1.468<br>$n_Y$ = 1.473<br>$n_Z$ = 1.483<br>⊕Δ = 0.015 |
| 12 | Glaserite trig<br>(Aphthitalite)<br>$(K,Na)_2[SO_4]$<br>K : Na ~ 3 : 1 | (0001), {10$\bar{1}$2};<br>◈{11$\bar{2}$0} ps.-hex. triplets<br>◈(0001);<br>tabular, flaky, sometimes<br>rhombohedra | # {10$\bar{1}$0} distinct,<br># (0001) poor;<br>H = 3 D = 2.66 – 2.70 | Z ∥ c E = Z | $n_O$ = 1.487 – 1.491<br>$n_E$ = 1.492 – 1.499<br>⊕Δ = 0.005 – 0.008 |
| 13 | Vanthoffite mcl<br>$Na_6Mg[SO_4]_4$ | granular, fine-grained<br>masses | fracture uneven<br>H = 3½ D = 2.69 | | $n_X$ = 1.485<br>$n_Y$ = 1.488<br>$n_Z$ = 1.489<br>⊖Δ = 0.004 |
| 14 | Langbeinite cub<br>$K_2Mg_2[SO_4]_3$ | {111}, {100};<br>tetrahedra,<br>usually granular | conchoidal fracture,<br>brittle<br>H = 3 – 4 D = 2.83 | isotropic | n = 1.535 |
| 15 | Glauberite mcl<br>$CaNa_2[SO_4]_2$ | {001}, {111}, striated<br>{110}; β = 112°;<br>tabular (001)<br>columnar [$\bar{1}$01] or<br>[001], granular | # (001) perf.,<br># {110} distinct,<br>97°;<br>conchoidal fract.; brittle;<br>H = 2½ – 3 D = 2.86 | X ∧ c = + 31°, bi;<br>< 18°C: Z ∥ b,<br>O.A.P. ⊥ (010)<br>Y ∧ a ~ − 90°;<br>> 52°C: Y ∥ b,<br>O.A.P. ∥ (010)<br>Z ∧ a ~ − 9°<br>(for all λ)<br>see Vol. 2, p. 27 | $n_X$ = 1.507 – 1.515<br>$n_Y$ = 1.527 – 1.532<br>$n_Z$ = 1.529 – 1.536<br>⊖Δ = 0.022 – 0.021<br>strongly dep. on<br>temperature |

| Angle and Dispersion of Optic Axes | Color in Thin Section Pleochroism | Chemical Reactions Special Remarks | Distinguishing Features of Similar Minerals | Occurrences | |
|---|---|---|---|---|---|
| | colorless | | | | 1 |
| | colorless | reacts with $H_2O$ to form portlandite; easily soluble in acids | | in volcanic ejecta | 2 |
| | abnormal:<br>E golden-yellow<br>O carmine-red | | fluorite, cryolite: no pleochroism | in alkaline rocks with alkali feldspar, zeolites, foids | 3 |
| | colorless; in thick sheets: blue-violet; yellow: α-rays | almost insoluble in ethanol; percussion figure on {100} with rays ‖ to [011], [0$\bar{1}$1] etc. | sylvine: sm. n | primary or metam. in all salt rocks, with anhydrite, gypsum, kieserite, polyhalite, sylvine, carnallite; prod. of sublim. | 4 |
| | colorless or very lightly colored, whitish, opalescent, reddish due to inclusions; yellow: α-rays | almost insoluble in ethanol; percussion figure on {100} unsymmetr. relative to [001], [010] etc. | halite and langbeinite: gr. n | primary with halite, anhydrite; metam. with halite, kieserite ± anhydrite, langbeinite | 5 |
| $2V_Z = 80°$<br>r > v weak<br>crossed dispersion | colorless | very hygroscopic, soluble in ethanol, deformable, ◈ lamellae | carnallite: sm. n., sm. $2V_Z$; decomposed by ethanol with sylvine-precipit.; kainite: sm. Δ, gr. $2V_X$, not hygroscopic, soluble in ethanol | subordinate in carnallite rocks; secondary from metam. brines | 6 |
| | colorless, yellowish to violet; quickly attaining a brown tinge in open air | hygroscopic, decomposed by ethanol, forming NaCl and KCl; interference colors abnormal: N < 2.7 | | metam. with halite, sylvine, anhydrite, and ± kieserite | 7 |
| $2V_Z = 66°$<br>r < v weak | colorless, red due to inclusions of hematite | hygroscopic, decomposed by ethanol, forming KCl | kainite, bischofite: larger n; schoenite: smaller Δ, no ◈ thenardite: # perf., larger D, diff. paragen. | frequent prim. and metam. mineral in oceanic salt deposits | 8 |
| | yellowish, colorless | very hygroscopic, decomposed by humid air, soluble in ethanol and water | loeweite: sm. n, gr. Δ, gr. D | metam. in carnallitic potassium-salt deposits | 9 |
| | colorless, yellowish<br>O reddish brown<br>E colorless | decomposed slowly by boiling $H_2O$; hematite inclusions | brucite: distinct r > v, colorless ascharite: gr. n, fibrous; talc: X ⊥ #, opt. ⊖ | secondary in anhydrite and salt-cont. pelites | 10 |
| $2V_Z = 83°$<br>r > v weak | colorless | easily soluble in $H_2O$ | vanthoffite: gr. n, sm. Δ; $2V_X$; bloedite and leonite: sm. Δ, sm. D; $2V_X$; schoenite: sm. D, sm. $2V_Z$; carnallite: sm $2V_Z$, sm. D, gr. Δ | prim. or diagen. in terrestrial salt deposits; product of fumarolic activity | 11 |
| if abnormal:<br>$2V_Z \sim 0°$ | colorless | soluble in $H_2O$ | koenite: gr. Δ, gr. n; loeweite: uniaxial ⊖, gr. Δ; tachydrite: uniaxial ⊖, gr. n | secondary in the cap rocks of sulfatic potassium-salt deposits; product of volcanic exhalations | 12 |
| $2V_X \sim 84°$<br>r < v very weak | colorless | soluble in $H_2O$; stays clear upon heating | bloedite: sm. D; leonite: sm. D; thenardite: sm. n, biaxial ⊕ | locally and always metam. in oceanic sulfate-rich salt deposits | 13 |
| | colorless | hygroscopic, soluble in $H_2O$ | halite: gr. n, #; sylvine: sm. n, # | locally as metam. product in sulfate-rich salt deposits | 14 |
| $2V_X = 9°$ at 5°C<br>horiz. disp. r ≫ v<br>$2V_X = 0°$ at 43°C<br>$2V_X = 8°$ at 85°C<br>inclined disp. r ≫ v<br>($2V_X$ for $Na_D$!) | colorless | decomposed by $H_2O$ with formation of gypsum, soluble in HCl; interference colors abnormal | sygenite: sm. n; gypsum: sm. Δ, gr. $2V_Z$; polyhalite: gr. n, gr. $2V_X$ | in salt lakes, metam. in oceanic salt deposits | 15 |

| | Name<br>Symmetry<br>Chem. Composition | Common Crystal Forms, Twinning<br>Crystal Habit<br>Aggregates | Cleavage<br>Translation Gliding<br>Deform. Twinning<br>Hardness, Density | Optic Orientation | Refractive Indices<br>Birefringence |
|---|---|---|---|---|---|
| 16 | Kieserite mcl<br>$MgSO_4 \cdot H_2O$ | {110}, {111}, $\{\bar{1}12\}$;<br>β = 116°; ◈ (001)<br>granular, nodules,<br>rarely massive | # {110}, {111} perf.<br># $\{\bar{1}11\}$, $\{\bar{1}01\}$, {011}<br>poor;<br>H = 3½ D = 2.573 | X ∧ c = + 41<br>Y ∥ b, O.A.P.(010)<br>Z ∧ c = − 49.2, bi | $n_x$ = 1.523 – 1.520<br>$n_y$ = 1.535 – 1.533<br>$n_z$ = 1.586 – 1.584<br>⊕Δ = 0.063 – 0.064 |
| 17 | Epsomite orh<br>(bitter salt)<br>$MgSO_4 \cdot 7H_2O$ | {110}, {111}, {010}<br>◈ (110);<br>fibrous-columnar<br>∥ c, granular | # (010) perf.,<br># {011} poor;<br>H = 2 – 2½ D = 1.68 | X ∥ b, bi;<br>Y ∥ c, O.A.P.(001)<br>Z ∥ a | $n_x$ = 1.4325<br>$n_y$ = 1.4554<br>$n_z$ = 1.4608<br>⊖Δ = 0.0283 |
| 18 | Bloedite mcl<br>(Astrakanite)<br>$Na_2Mg(SO_4)_2 \cdot 4H_2O$ | {001}, {110}, {010}<br>{hkl}; β = 100½°;<br>granular, prismatic,<br>nodules | no cleavage;<br>conchoidal fract.;<br>H = 3 D = 2.23 | X ∧ a = 37°, bi;<br>Y ∥ b, O.A.P. (010)<br>Z ∧ c = + 47° | $n_x$ = 1.483 – 1.486<br>$n_y$ = 1.486 – 1.488<br>$n_z$ = 1.487 – 1.489<br>⊖Δ = 0.004 – 0.003 |
| 19 | Leonite mcl<br>$K_2Mg(SO_4)_2 \cdot 4H_2O$ | {100}, {010}, {hkl};<br>β = 95°;<br>◈ (100) lam.;<br>massive | no cleavage<br>H = 3 D = 2.20 | Z ∧ a ~ 0°, bi;<br>Y ∥ b, O.A.P.(010)<br>X ∧ c ~ + 5° | $n_x$ = 1.478<br>$n_y$ = 1.481<br>$n_z$ = 1.486<br>⊕Δ = 0.008 |
| 20 | Loeweite trig<br>$Na_{12}Mg_7(SO_4)_{13} \cdot 5H_2O$ | columnar, granular | # (001) poor;<br>H = 3 D = 2.37 | X ∥ c E = X | $n_O$ = 1.490<br>$n_E$ = 1.471<br>⊖Δ = 0.019 |
| 21 | Schoenite mcl<br>(Picromerite)<br>$K_2Mg(SO_4)_4 \cdot 2H_2O$ | {001}, {110}, $\{\bar{2}01\}$<br>and other;<br>β = 104¾°<br>crusts, massive, powdery | # $\{\bar{2}01\}$ perf.;<br>H = 2½ D = 2.03 | X ∧c = + 14°<br>Y ∥ b, O.A.P.(010),<br>Z ∧ a = − 1°, bi | $n_x$ = 1.4607<br>$n_y$ = 1.4629<br>$n_z$ = 1.4755<br>⊕Δ = 0.0148 |
| 22 | Polyhalite trcl<br>$K_2MgCa_2(SO_4)_4\ 2H_2O$ | always ◈ (100) and<br>(010); may be lamellar,<br>granular,<br>tabular (010),<br>fibrous ∥ b,<br>radiated, massive | # $(10\bar{1})$ perf.,<br># (010) parting;<br>H = 3½ D = 2.78 | In $(10\bar{1})$: X ∧ c = −12°<br>Z' ∧ ◈ = 28°<br>In (010) X' ∧ ◈ = 28° | $n_x$ = 1.548 – 1.547<br>$n_y$ = 1.562 – 1.560<br>$n_z$ = 1.567 – 1.567<br>⊖Δ = 0.019 – 0.020 |
| 23 | Syngenite mcl<br>$K_2Ca[SO_4]_2 \cdot H_2O$ | {100}, {hk0}, {h0l};<br>β = 104°;<br>◈ (100);<br>tabular (100),<br>prismatic [001],<br>granular, reticulated | # (100) perf. ⟩ 106°,<br># (110) perf.<br># (010) distinct;<br>H = 2½ D = 2.58 | X ∧ c = +88°, bi;<br>Y ∧ c = −2°, O.A.P.<br>⊥(010) (< 127°C)<br>Z ∥ b; X ≈ ⊥(001); | $n_x$ = 1.5010 }<br>$n_y$ = 1.5166 } at 29°C<br>$n_z$ = 1.5176 }<br>⊖Δ = 0.0166<br>optics strongly |
| 24 | Kainite mcl<br>$K_4Mg_4[Cl/SO_4]_4 \cdot 11H_2O$ | {100}, {421}, $\{42\bar{1}\}$,<br>{010}, {001}, {201};<br>β = 95°;<br>irregular penetr. ◈;<br>polysynth. ◈;<br>fine-grained, massive,<br>thick-tabular (100),<br>rarely fibrous | # (001) perf. ⟩ 54°;<br># {111} dist.<br>H = 2½ – 3 D = 2.13 | X ∧ a = −8°, bi;<br>Y ∥ b, O.A.P. (010)<br>Z ∧ c = +13° | $n_x$ = 1.494 – 1.495<br>$n_y$ = 1.505 – 1.506<br>$n_z$ = 1.516 – 1.520<br>⊖Δ = 0.022 – 0.025 |
| 25 | Trona mcl<br>$Na_3H[CO_3]_2 \cdot 2H_2O$ | β = 106⅓°;<br>tabular (001),<br>columnar, elongated,<br>fibrous ∥ b, crusts | # (100) perf.,<br># $\{10\bar{1}\}$ good;<br>H = 2½ D = 2.17 | X ∥ b, bi;<br>Y ∧ c = −7°, O.A.P.<br>≈ (001)<br>Z ∧ c = +83°;<br>Z ≈ ⊥(100) | $n_x$ = 1.412<br>$n_y$ = 1.492<br>$n_z$ = 1.540<br>⊖Δ = 0.128 |
| 26 | Borax mcl<br>$Na_2[B_4O_5(OH)_4] \cdot 8H_2O$ | {100}, {110}, {010},<br>{001}, $\{1\bar{1}1\}$;<br>β = 106⅔°;<br>◈ (100), rare;<br>short prisms ∥ c,<br>tabular ∥ (100) | # (100) perf.,<br># {110} distinct,<br># {010} poor;<br>H = 2 – 2½ D = 1.73 | X ∥ b, bi;<br>Y ∧ c = +34°, O.A.P.<br>⊥(010)<br>Z ∧ c = −56° | $n_x$ = 1.447<br>$n_y$ = 1.469<br>$n_z$ = 1.472<br>⊖Δ = 0.025 |
| 26a | Kernite mcl<br>$Na_2[B_4O_6(OH)_2] \cdot 3H_2O$ | β = 109°; large grains,<br>coarsely spathic,<br>acicular to fibrous ∥ a | # (100) perf. ⟩ 109°;<br># (001) perf.<br>H = 2½ D = 1.95 | X ∧ c = +38°, bi;<br>Y ∧ c = −52°, O.A.P.<br>⊥(010)<br>Z ∥ b | $n_x$ = 1.454<br>$n_y$ = 1.472<br>$n_z$ = 1.488<br>⊖Δ = 0.034 |
| 26b | Ulexite trcl<br>$NaCa[B_5O_6(OH)_6] \cdot 5H_2O$ | β = 109°; finely fibrous,<br>reticulated, radiated,<br>crusts | # (010) perf.,<br># $(1\bar{1}0)$ good,<br># (110) poor;<br>H = 2½ D = 2.00 | X ∧ c = −69°<br>Y ∧ c = +21°, O.A.P.<br>≉ ⊥(010)<br>Z ≈ ∥ b, bi | $n_x$ = 1.493<br>$n_y$ = 1.506<br>$n_z$ = 1.519<br>⊕Δ = 0.026 |

| Angle and Dispersion of Optic Axes | Color in Thin Section Pleochroism | Chemical Reactions Special Remarks | Distinguishing Features of Similar Minerals | Occurrences | |
|---|---|---|---|---|---|
| $2V_z \sim 57^o$<br>r > v<br>oblique disp. | colorless | may be isolated by means of $MgCl_2$-solution; weathering to epsomite; ◈ lamellae; hardness, high Δ | carbonates: uniaxial ⊖;<br>anhydrite: gr. n, sm. Δ | usually metam. in oceanic potassium-salt deposits | 16 |
| $2V_x = 51^o$<br>r > v | colorless | tastes bitter; low D; weathering in dry air | hexahydrite: X-ray<br>pentahydrite, leonhardite, schoenite : gr. n | in salt pans, in cap rocks of oceanic salt domes, forming from kieserite | 17 |
| $2V_x = 71^o$<br>oblique disp.<br>r < v strong | colorless | decrepitates upon heating, forms white crust; no weathering; sm. n and sm. Δ | vanthoffite stays clear upon heating;<br>leonite: gr. Δ;<br>loeweite: gr. Δ, uniaxial | prim. in oceanic salt deposits and in salt lakes; in nitrous deserts; sec. in kainitic caps | 18 |
| $2V_x \sim 90^o$<br>r < v | colorless | white coating in open air; ◈ lamellae frequent | vanthoffite gives off no water; bloedite: sm. Δ;<br>kainite: g. n | second. in sulfatic potassium-salt deposits with kainite, polyhalite | 19 |
| | colorless, yellow, reddish | white coating in humid air | tachydrite: gr. n; other similar salt minerals: biaxial | locally and always metam. in oceanic salt deposits; together with vanthoffite, langbeinite | 20 |
| $2V_z \sim 48^o$<br>r > v | colorless | easily soluble in $H_2O$, gives off water upon heating;<br>very small n | similar minerals have gr. n; X-ray | in kainitic cap rocks of salt domes and secondary in sulfate-rich potassium-salt deposits | 21 |
| $2V_x = 64^o$<br>r < v | colorless | decomposed by $H_2O$ to form gypsum;<br>◈ lam. | anhydrite: #, gr. n, Δ;<br>goergeyite: no ◈;<br>koenenite: uniaxial | with anhydrite, usually metam. in kieseritic and langbeinitic salt deposits | 22 |
| $2V_x = 29^o$ (21½°C)<br>r < v strong;<br>$2V_{red} = 0^o$ (127°C)<br>$2V_{blue} = 0^o$ (178°C)<br>temperature-dependent | colorless | easily decomposed by $H_2O$ to form gypsum; changes of optics with changing temperature | kainite: only one # perf., gr. $2V_x$; bischofite: gr. Δ, gr. $2V_z$;<br>glauberite: no ◈, gr. n | in oceanic salt deposits metam. from anhydritic sylvine-halite salts; rarely from volcanic vapors and in clastic sediments | 23 |
| $2V_x = 84^o$<br>r > v,<br>oblique disp. | colorless | soluble in $H_2O$; decomposes at temp. beyond 72° to form kieserite + sylvine | carnallite: sm. n, no #;<br>bischofite: gr. Δ, more easily sol. in $H_2O$;<br>schoenite: sm. n, sm. $2V_z$, no ◈ | in oceanic salt deposits; kainite in caps forming from kieserite-halite-carnallite or kieserite-sylvine-halite salts (< 72°C) | 24 |
| $2V_x = 76^o$<br>r < v distinct | colorless | easily soluble in $H_2O$; effervescent in HCl; white coating of thermonatrite in open air; high Δ | thermonatrite: sm. 2V, no # | monomineralic layers in terrestrial salt pans; authigenic in continental salt-containing pelites | 25 |
| $2V_x = 39^o$<br>r < v strong,<br>crossed disp. | colorless | solubility in $H_2O$ increases strongly with temp.; white coating of tincalconite in open air; reversible inversion to kernite at 53° – 63°C | epsomite: sm. Δ, weak dispersion;<br>trona: gr. Δ, gr. 2V;<br>kernite: gr. 2V | in continental evaporites and salt pans; efflorescence of desert soil; may replace metasom. silicates during diagenesis | 26 |
| $2V_x = 80^o$<br>r > v | colorless | very slowly soluble in cold $H_2O$ | borax: sm. 2V, strong dispersion;<br>trona: gr. Δ | in nonmarine evaporites and salt pans; metam. from borax | 26a |
| $2V_z = 78^o$ | colorless | slightly soluble in $H_2O$; soluble in acids | inyoite: ⊖Δ;<br>karnakovite: gr. ⊖Δ | frequently in continental salt lakes and salt pans; in extremely arid sediments together with nitrates and glauberite | 26b |

| No. | Name<br>Symmetry<br>Chem. Composition | Common Crystal Forms, Twinning<br>Crystal Habit<br>Aggregates | Cleavage<br>Translation Gliding<br>Deform. Twinning<br>Hardness, Density | Optic Orientation | Refractive Indices<br>Birefringence |
|---|---|---|---|---|---|
| 26c | Colemanite mcl<br>$Ca[B_3O_4(OH)_3]\cdot H_2O$ | {100}, {110}, {010}, {001}, {20$\bar{1}$}, {111}, {hkl}; β = 110°<br>short prisms ∥ c, elongated, granular, massive | # (010) perf.,<br># (001) distinct;<br>H = 4½ D = 2.4 | X ∥ b,<br>Y ∧ c = –7°,<br>O.A.P. ⊥ (010);<br>Z ∧ a = +27°, bi | $n_X$ = 1.586<br>$n_Y$ = 1.592<br>$n_Z$ = 1.614<br>⊕Δ = 0.028 |
| 27 | Soda niter trig<br>$NaNO_3$ | {10$\bar{1}$1};<br>◈ = [01$\bar{1}$2]<br>granular to mass. crusts | # (10$\bar{1}$1) perf., 73½°<br>$K_1$ = (01$\bar{1}$2), $K_2$ = (0$\bar{1}$11);<br>H = 1½ – 2 D = 2.27 | X ∥ c E = X | $n_O$ = 1.587<br>$n_E$ = 1.336<br>⊖Δ = 0.251 |

**Opaque (or Nearly**

| No. | Name<br>Symmetry<br>Chem. Composition | Common Crystal Forms, Twinning<br>Crystal Habit<br>Aggregates | Cleavage<br>Translation Gliding<br>Deform. Twinning<br>Hardness, Density | Optic Orientation | Refractive Indices<br>Birefringence |
|---|---|---|---|---|---|
| 28 | Coal amorphous<br>coaly substances | colloidal forms<br>dust-particles, flakes | H = ½ – 2½<br>D = 1.0 – 1.8 | isotropic | n = 1.6 – 1.9 |
| 29 | Graphite hex<br>C | {001};<br>flakes, grains | # (0001) perf.;<br>T = (0001) flexible;<br>H = 1 – 2 D = 2.23 | X ∥ c E = X | $n_O$ = 1.93 – 2.07 (red)<br>⊖Δ = . . . |
| 30 | Pyrrhotine hex<br>$Fe_{1-x}S$ | {0001};<br>tabular, massive | # {11$\bar{2}$0} poor,<br>(0001) parting;<br>H = 3½ – 4½ D = 4.65 | opaque | |
| 31 | Pyrite cub<br>$FeS_2$ | {100}, {210};<br>◈ (110);<br>granular | # (100) poor, brittle;<br>H = 6 – 6½ D = 5.0 | opaque | |
| 32 | Magnetite cub<br>$FeFe_2O_4$ | {111}, {110};<br>◈ {111};<br>granular | # {111} (parting);<br>$K_1$ = (111), $K_2$ = (11$\bar{1}$);<br>H = 5½ – 6½ D = 4.9 | opaque | n = 2.42 Na<br>diagr. p. 22 |
| 33 | Chromite cub<br>(<u>Fe</u>, Mg) (<u>Cr</u>, Al)$_2O_4$ | {111};<br>◈ {111};<br>granular | H = 5½ D = 4.9 | isotropic | n = 2.16 (Fe)<br>n ~ 2.05 (Fe, Mg)<br>diagr. p. 22 |
| 34 | Hematite trig<br>$\alpha - Fe_2O_3$ | {0001}, {10$\bar{1}$1}, {11$\bar{2}$0};<br>◈ {10$\bar{1}$1} lam.,<br>◈ {0001};<br>tabular, foliated, scaly, granular, fibrous, dust particles, oolithic | parting {10$\bar{1}$1} 94°,<br>parting {0001};<br>T = (0001)<br>$K_1$ = (10$\bar{1}$1), $K_2$ = (01$\bar{1}$2),<br>$K_1'$ = (0001), $K_2'$ = (02$\bar{2}$1);<br>H ~ 6 D ~ 5.26 | X ∥ c E = X<br>usually 1 = (+)<br>fig. p. 32 | λ = 5890 Å λ = 7590 Å<br>$n_O$ = 3.22 – 2.90<br>$n_E$ = 2.94 – 2.69<br>⊖Δ = 0.28 – 0.21<br>strong disp. |
| 35 | Ilmenite trig<br>$FeTiO_3$ | {0001}, {10$\bar{1}$1}, {02$\bar{2}$1};<br>◈ {10$\bar{1}$1} lam.,<br>◈ {0001}; flaky, gran., skeletal crystals | parting {10$\bar{1}$1} 94½°<br>parting {0001}<br>H = 5 – 6 D = 4.72 | X ∥ c E = X<br>fig. p. 32 | n ≫ 2.7<br>⊖Δ = very strong |
| 36 | Pyrophanite trig<br>(<u>Mn</u>, Fe) $TiO_3$ | {0001};<br>thin flakes, scales | # {02$\bar{2}$1} perf.,<br># {10$\bar{1}$2} good;<br>H = 5 – 6 D = 4.54 | X ∥ c E = X<br>1 = (+)<br>fig. p. 32 | $n_O$ = 2.48<br>$n_E$ = 2.21<br>⊖Δ = 0.27 |
| 37 | Geikielite trig<br>(<u>Mg</u>, Fe) $TiO_3$ | {0001}, {10$\bar{1}$1};<br>◈ {10$\bar{1}$1}, lam.,<br>granular | parting {10$\bar{1}$1},<br>parting {0001};<br>H = 5 – 6 D ~ 4.05 | X ∥ c E = X<br>fig. p. 32 | $n_O$ = 2.31<br>$n_E$ = 1.95<br>⊖Δ = 0.36 |
| *<br>39 | Columbite group orh<br>(Niobite-Tantalite)<br>(Fe,Mn) (Nb,Ta)$_2O_6$ | {100}, {h0l}, {001}, {hkl};<br>◈ {021} and {023};<br>Nb: tabular ∥ (100), prismatic ∥ b;<br>Ta: columnar ∥ c<br>(100) striated ∥ c, divergent, massive | # (100) distinct;<br>T = (001)? ;<br>H = 6 – 6½<br>Nb: D = 5.0<br>Ta: D = 7.9<br>diagr. p. 60 | X ∥ b;<br>Y ∥ a, O.A.P. (100)<br>Z ∥ c, bi;<br>fig. p. 46 | Nb Ta<br>$n_X$ = 2.40 – 2.24<br>$n_Y$ = 2.45 – 2.30<br>$n_Z$ = 2.53 – 2.41<br>⊕Δ = 0.13 – 0.17 |
| 40 | Ludwigite orh<br>(<u>Mg</u>, $Fe^{2+})_2$ $Fe^{3+}$<br>$[O_2/BO_3]$ | {110}, {410}, {310}, {210};<br>divergent ∥ c, stellate;<br>fig. p. 47 | mechanically resistant;<br>H = 5 D = 3.6 – 4.3 | X ∥ a<br>Y ∥ b, O.A.P. (010)<br>Z ∥ c, bi;<br>1 = (+)<br>fig. p. 47 | Fe-rich Fe-poor<br>$n_X$ = 1.85 – 1.84<br>$n_Y$ = 1.85 – 1.85<br>$n_Z$ = 2.02 – 1.98<br>⊕Δ = 0.17 – 0.14 |
| 41 | Ilvaite orh<br>(Lievrite)<br>$CaFe^{2+}{}_2Fe^{3+}$<br>$[OH/O/Si_2O_7]$ | {110}, {120}, {101}, {010};<br>columnar ∥ c,<br>fibrous, massive | # (010) distinct<br># (001), (100) poor<br>H = 5½ – 6 D = 4.03 | X ∥ b<br>Y ∥ a, O.A.P. (100)<br>Z ∥ c, bi;<br>1 = (+)<br>fig. p. 59 | $n_X$ ~ 1.878<br>$n_Y$ ~ 1.890<br>$n_Z$ ~ 1.926<br>⊕Δ ~ 0.048 |

*see No. 66

| Angle and Dispersion of Optic Axes | Color in Thin Section Pleochroism | Chemical Reactions Special Remarks | Distinguishing Features of Similar Minerals | Occurrences | |
|---|---|---|---|---|---|
| $2V_Z = 55^o$<br>$r < v$ | colorless, yellow | not soluble in $H_2O$; soluble in hot HCl | | in terrestrial salt deposits and salt pans, filling joints in borax; geodes in associated pelites | 26c |
| $n_{E'}(10\bar{1}1) = 1.467$ | colorless | easily soluble in $H_2O$; slightly hygroscopic; very large Δ, but sm.n | niter: sm.n, biaxial; carbonates: gr.n, gr. H, slightly soluble in $H_2O$ | in extremely arid deserts, in salt pans, as efflorescence of soils, together with halite,thenardite etc. | 27 |

## Opaque) Minerals

| | | | | | |
|---|---|---|---|---|---|
| | reflection dull black to brown, in thin lamellae dark brown to grey translucent | combustible; may be soluble in KOH-solution, benzol, xylol | magnetite: idiom. graphite: idiom. flakes; goethite: red with crossed nicols | in sediments with organogenic components, also in low grade metam. rocks | 28 |
| | refl. shiny grey, in extremely thin lamellae greenish grey translucent | barely combustible; produces black marks; insoluble in HF | molybdenite: loses $SO_2$ upon heating; coal: more easily combustible | in metam. sediments as organogenic component | 29 |
| | reflection brownish, bronze-colored | $SO_2$ upon heating; with HCl: $H_2S$-formation; magnetic up to 348° | chalcopyrite and pyrite: brass-colored, not magnetic | in bas. magmatites with chalcopyrite, magnetite, ilmenite ± pyrite in skarns | 30 |
| | reflection bright brass-colored | blowpipe test: $SO_2$-formation and bluish flame; fine powder soluble in $HNO_3$ | marcasite: refl. birefringent; pyrrhotite: bronze-colored | in magmatic, metamorphic and sedimentary rocks | 31 |
| | reflection black | soluble in HCl, insoluble in HF; strongly magnetic | ilmenite, chromite: insoluble in HCl, not magnetic | with other iron-ores, pyrogenic, metam. hydroth., metasom. | 32 |
| | dark brown, almost opaque | decomposed by $KNO_3$-melt | magnetite: soluble in HCl, strongly magnetic; picotite: sm. D, more translucent; melanite: {110},zonal coloration | magmatic in peridotites, rarely in dolomitic marbles, in sands, placer deposits | 33 |
| | O: brownish red, E: yellowish red, O > E, in extremely thin scales greyish yellow | slightly soluble in HCl; streak: red | goethite, lepidocrocite: lighter colored, more brownish; ilmenite: black streak, ore microscope | with other iron-ores; magmat., pneumat., or contact-metam.; in low and medium grade metamorphites; diagenet. in sediments | 34 |
| | opaque, in extremely fine scales dark brown E > O | slightly soluble in HCl; decomposed by $KHSO_4$-melt; alteration to leucoxene | hematite: more red; pyrophanite, geikielite: X-ray diffraction-pattern | accessory of nearly all magmatites; in sands and placers | 35 |
| | opaque, in extremely fine scales dark red to yellowish red, O = E | slightly soluble in hot HCl | hematite, ilmenite, geikielite: X-ray, ore microscope | in Mn-skarn deposits | 36 |
| | opaque, in extremely fine scales: brownish red or purple | slightly soluble in hot HCl | see pyrophanite | in contact-metam. dolomitic marbles, with rutile in placers | 37 |
| $2V_Z \approx 75^o$<br>$r < v$ | Nb Ta<br>X = reddish brown to yellow<br>Y = dark reddish brown to reddish brown<br>Z = opaque to dark red-brown<br>X < Y < Z | decomposed by $KHSO_4$-melt, insoluble in acids; if Fe-Nb-content high: almost opaque in all sectional planes | hematite: uniaxial ⊖; rutile: uniaxial, gr. Δ; cassiterite: uniaxial, sm. Δ; magnetite, uraninite, wolframite: X-ray, ore microscope | in granitic and syenitic pegmatites with cleavelandite, cassiterite, F-, Be-, B-, Li- and W-minerals | 39 |
| $2V_Z$ very small<br>$r > v$ very strong | X = Y = dark green, Z = dark brown, X = Y < Z; Fe-rich specimen nearly opaque | soluble in acids; red upon heating; strongly paramagnetic; mechanically resistant | ilvaite: sm. Δ; columbite: gr. n; tourmaline: uniaxial | pneumatolytic; contact-metasomatic in limestones and skarns | 40 |
| $2V_Z \sim 40^o$ (– ~ 90°)<br>$r \ll v$ very strong | X = yellowish brown to brown,<br>Y = dark brown to opaque,<br>Z = dark green to opaque,<br>X<Y<Z | gelatinizes with HCl; strong pleochroism; nearly opaque | pseudobrookite: gr. n and Y = red; goethite: red-brown; ludwigite: gr. Δ, sm. 2V; kaersutite: sm. absorption, aenigmatite and rhoenite: oblique extinction | contact-metasomatic in skarns, with hedenbergite, hornblende, epidote etc. in diabase-hornfelses; deuteric in gabbros | 41 |

| No. | Name<br>Symmetry<br>Chem. Composition | Common Crystal Forms, Twinning<br>Crystal Habit<br>Aggregates | Cleavage<br>Translation Gliding<br>Deform. Twinning<br>Hardness, Density | Optic Orientation | Refractive Indices<br>Birefringence |
|---|---|---|---|---|---|
| 42 | Aenigmatite trcl<br>(Cossyrite)<br>$Na_4 Fe^{2+}_{10} Ti_2$<br>$[O_4/(Si_2O_6)_6]$ | $\{010\}$, $\{100\}$, $\{110\}$, $\{001\}$,$\{1\bar{1}0\}$; $\beta = 96.5°$; ps.-mcl., similar to hornblende, columnar ‖ c, always ⟐ (010) lam. | # $\{110\}$ # $\{1\bar{1}0\}$ > 66°<br>H = 5½ D = 3.79±<br>fig. p. 99 | O.A.P. ~ (010)<br>Z Λ c in (100) = 4°–5°; (010) ~ 45°<br>fig. p. 99 | $n_x$ = 1.81<br>$n_y$ ~ 1.82<br>$n_z$ = 1.88<br>⊕Δ = 0.07 |
| 43 | Rhoenite trcl<br>pseudomonocl.<br>$Ca_4(\underline{Mg},Fe^{2+})_8 Fe^{3+}_2 Ti_2$<br>$[O_4 \mid (AlSiO_6)_6]$ | $\{100\}$,$\{011\}$,$\{0\bar{1}1\}$; $\beta = 109°$; ⟐ (010), lam.; columnar ‖ c, tabular ‖ $(1\bar{1}0)$ | # $\{201\}$ and $\{\bar{2}01\}$ good<br># $\{110\}$ and $\{1\bar{1}0\}$ distinct;<br>H < 5½ D ~ 3.58 | Z'Λc in (010) ~ 28°–40°<br>Y'Λa in (010) ~ 9°–21°<br>O.A.P. ~ ⊥ (010)<br>X ~ b (0° – 6°)<br>fig. p. 99 | $n_x$ = 1.808<br>$n_y$ = 1.824<br>$n_z$ = 1.840<br>⊕Δ = 0.032 |

## Optically Isotropic (or Weakly

| No. | Name<br>Symmetry<br>Chem. Composition | Common Crystal Forms, Twinning<br>Crystal Habit<br>Aggregates | Cleavage<br>Translation Gliding<br>Deform. Twinning<br>Hardness, Density | Optic Orientation | Refractive Indices<br>Birefringence |
|---|---|---|---|---|---|
| 44 | Sphalerite cub<br>ZnS | $\{110\}$, $\{311\}$;<br>⟐ $\{111\}$ and $\{211\}$, also lamellar; granular | # $\{110\}$ perf.;<br>D = 4.09 (pure); 3.94 (28 mol % FeS);<br>H = 3½–4 | isotropic, may have tensional birefringence | n = 2.396 (pure); 2.47 (28 mol % FeS) |
| 45 | Fluorite cub<br>$CaF_2$ | $\{001\}$, $\{111\}$;<br>⟐ $\{111\}$;<br>granular, xenomorphous in interstices | # $\{111\}$ perf.,<br>T = (001), t = [110]?<br>brittle if strained for short time;<br>H = 4 D = 3.18 | in thin section: almost always isotropic | n = 1.434 |
| 46 | Cryolite mcl<br>$Na_3[AlF_6]$ | paramorphous (cub): $\{001\}$, $\{110\}$<br>⟐ $\{001\}$,$\{110\}$ and $\{112\}$ cub., lam.; mcl $\beta = 90.2$; granular | # $\{001\}$ cub. perf.,<br># $\{110\}$ cub. good,<br>parting?<br>H = 2½–3 D = 2.98 | X ‖ b, O.A.P. ⊥ (010)<br>Y Λ a = +44°<br>Z Λ c = –44°, bi | $n_x$ = 1.3385<br>$n_y$ = 1.3389<br>$n_z$ = 1.3396<br>⊕Δ = 0.0011 |
| 47 | Periclase cub<br>MgO (frequently admixtures of Fe, Mn, Zn) | $\{111\}$, $\{100\}$;<br>granular | # $\{001\}$ perf.,<br># $\{110\}$ poor;<br>T = (110), t = $[1\bar{1}0]$;<br>H = 5½–6 D ~ 3.56 | rarely a weak abnormal birefringence: lam. ‖ (110) | n = 1.730 – 1.739 |
| 48 | Manganosite cub<br>MnO (frequently admixtures of Mg, Fe) | $\{111\}$, $\{100\}$;<br>granular | # $\{100\}$, may have parting $\{111\}$;<br>H = 5½ D ~ 5.4 | isotropic | n (green) = 2.19<br>n (Li) = 2.16 |
| 49 | Calciumoxide cub<br>CaO (Lime) | $\{100\}$;<br>granular | # $\{100\}$ perf.;<br>H = 3 – 4 D = 3.40 | isotropic | n = 1.837 |
| 50 | Spinel group cub<br>(except Nr. 32,51)<br>$(\underline{Mg},\underline{Fe},Mn,\underline{Zn})(\underline{Al},Fe,Cr)_2O_4$ | $\{111\}$, rarely $\{110\}$;<br>⟐ $\{111\}$;<br>granular | # $\{111\}$ distinct (parting)?<br># $\{100\}$ distinct, rare;<br>H = 7½–8 D = 3.6 – 4.6<br>diagr. p. 22, 23 | isotropic | usually<br>n = 1.718 – 2.05<br>diagr. p. 22, 23 |
| 51 | Chromite cub<br>$(\underline{Fe},Mg)(\underline{Cr},Al)_2O_4$ | $\{111\}$;<br>⟐ $\{111\}$;<br>granular | H = 5½ D ~ 4.9 | isotropic | n = 2.16 (Fe)<br>n ~ 2.05 (Fe, Mg)<br>diagr. p. 22 |
| 52 | Perovskite mcl?<br>$CaTiO_3$ | paramorph. (cub): $\{001\}$ and $\{111\}$;<br>⟐ $\{111\}$ cub.,<br>⟐ $\{110\}$ and $\{100\}$ cub., complex lam.; isometric, skeletal crystals | # $\{100\}$ cub distinct<br>H = 5½ D ~ 4.0 | opt. mcl. lam. ‖ (100) cub. = ‖ (110) mcl<br>Z ‖ a cub<br>Y ‖ b, O.A.P. (010) cub<br>Z Λ a mcl = 45°; small crystals apparently isotropic | n = 2.30 – 2.34 (pure)<br>n = 2.38 (Ce-rich) |
| 53 | Dysanalyte mcl?<br>$(\underline{Ca},Na,Ce)(\underline{Ti},Nb,Fe)O_3$ | paramorph. (cub): $\{001\}$;<br>granular | # $\{001\}$ cub distinct;<br>H = 5 – 6 D ~ 4.1 | isotropic;<br>may have abnormally birefringent domains | n ~ 2.33<br>may have ⊕Δ small |
| 54 | Pyrochlore cub<br>hydroxyl-bearing niobate of (Ca,Na) with Ce,Y,U,Ti,Fe,Ta | $\{111\}$: pyrochlore and pyrrhite, $\{110\}$: koppite;<br>⟐ $\{111\}$; granular | # $\{111\}$ poor;<br>H = 5 – 6<br>pyrrhite: D = 4.20<br>pyrochl.: D = 4.30<br>koppite: D = 4.50 | isotropic;<br>may have abnormal birefringence | pyrrhite, koppite:<br>n = 2.12 – 2.18<br>pyrochlore:<br>n = 1.96 – 2.02 |
| 55 | Boracite orh<br>$(\underline{Mg},Fe,Mn)_3(\underline{Cl},Br)B_7O_{13}$<br>>265°C: cub. | paramorph.: $\{100\}$,$\{111\}$, $\{1\bar{1}1\}$ and $\{110\}$ cub;<br>⟐ $\{111\}$ cub rarely,<br>⟐ $\{110\}$ cub lam.;<br>nodules, lenses; "Stassfurtite": fibrous | conchoidal fracture<br>$K_1$ = (110) cub,<br>$K_2$ = $(1\bar{1}0)$ cub?<br>H = 7 D = 2.95<br>"stassfurtite" D = 2.90 | <265°C sect. orh. lam.<br>Z ⊥ (110) cub<br>O.A.P. ‖ (001) cub<br>"stassfurtite": l = (+) | $n_x$ = 1.658 – 1.662<br>$n_y$ = 1.662 – 1.667<br>$n_z$ = 1.668 – 1.673<br>⊕Δ = 0.010 – 0.011 |
| 56 | Garnet group cub<br>$R^{2+}_3 Z^{3+}_2 [SiO_4]_3$<br>R = (Mg,Fe,Mn,Ca)<br>Z = (Al,Fe,Cr) | $\{110\}$, $\{112\}$;<br>granular | # $\{110\}$ poor; brittle<br>H = 6½– 7½<br>D = 3.5 – 4.3<br>diagr. p. 24, 25 | isotropic; grossularite and andradite frequently with abnormally birefringent sectors | n = 1.71 – 1.89<br>melanite (Ti-bearing andradite)<br>n ~ 1.86 – 2.00<br>diagr. p. 24, 25 |

| Angle and Dispersion of Optic Axes | Color in Thin Section Pleochroism | Chemical Reactions Special Remarks | Distinguishing Features of Similar Minerals | Occurrences | |
|---|---|---|---|---|---|
| $2V_Z = 32^o$<br>$r < v$ | X = light red-brown<br>Y = dark chestnut-brown<br>Z = dark black-brown<br>X < Y < Z | partly decomposed by HCl; O.A.P. approx. halves the obtuse angle of # (010) to (100) | basalt. hbl: diff. position of ◈ to (110) and gr. Δ; rhoenite: X = greenish brown | magm. with aegirine and arfvedsonite in sodic syenites, Na-trachytes and phonolites | 42 |
| $2V_{Z,X}$ = large, > 70° fig. p. 99 | X = brown<br>Y = red-brown<br>Z = dark red-brown to opaque<br>X ⩽ Y < Z | O.A.P. approx. halves the angle of #; abnormal interference colors | ilmenite: no pleochr., no #; aenigmatite: gr. $n_Z$, gr. Δ, sm. 2V; titaniferous hbl: sm. n | in nepheline-basanites, alkali-basalts and nephelinites; only in volcanic rocks | 43 |

## Birefringent) Minerals

| | | | | | |
|---|---|---|---|---|---|
| | colorless, yellowish, brown with increasing Fe-content; zonal structure | soluble in $HNO_3$<br>blow pipe test: Zn crust.<br>semi-metallic luster in reflected light | dysanalyte, pyrochlor: X-ray, microchem. test; rutile, brookite, cassiterite: gr. Δ | with other sulfides in hydrothermal veins and contact-metam. limestones | 44 |
| | colorless, may be zonal blue or purple; radioactive haloes | decomposed by $H_2SO_4$ with formation of HF, insoluble in HF | opal: no #; sodalite: gr. n, different #; cryolite: sm. n, different #; villiaumite: sm. n | rarely primary magmatic; pneumatolytic together with other F-minerals; usually hydrothermal | 45 |
| $2V_Z = 43^o$<br>$r < v$,<br>horiz. disp. | colorless | decomposed by conc. $H_2SO_4$ with formation of HF; only partly soluble in HCl | villiaumite: pleochr., sm. n; fluorite: gr. n, different #; opal: isotropic, no # | pegmatitic with sulfides, oxidic ores and F-minerals | 46 |
| | colorless, brownish | soluble in HCl, $HNO_3$, precipitates $Ag_2O$ from $AgNO_3$-solution; brucite-rims | light col. spinels have different chem. characteristics | with monticellite and spinel in contact-metam. limestones and dolomites; rare | 47 |
| | emerald-green, commonly black stained | soluble with effervescence in conc. $H_2SO_4 + H_2O_2$; black with $H_2O_2$ | hercynite, gahnite, uvarovite: sm. n, different chem. characteristics | only in Mn-(Zn-) skarn deposits in metam. limestones | 48 |
| | colorless | reacts with $H_2O$ to form portlandite, easily soluble in acids | | in volcanic ejecta | 49 |
| | spinel: colorless, pink, bluish; chlorospinel: pale green; pleonast: grey, greenish grey; hercynite, gahnite: dark green; picotite: yellow, brown | only Fe-rich members soluble in acids; soluble in $KHSO_4$-melt; {001} not observed | garnet: no {111}; periclase, manganosite and calciumoxide: different chem. characteristics, X-ray | high grade contact and regional metam.; in basic magmatites; in sands and placers | 50 |
| | dark brown, nearly opaque | soluble in $KNO_3$-melt | magnetite: sol. in HCl, magnetic; piocotite: more translucent, sm. D; melanite: {110}, zonal coloring | magmat. in peridotites; rarely in dolomitic marbles, in sands and placers | 51 |
| $2V_Z \sim 90^o$<br>$r > v$ approx. Z | colorless, violet-grey, red-brown, zonal<br>Z > X weak | decomposed by hot conc. $H_2SO_4$; may be radio-active; alteration to leucoxene | picotite, melanite sm. n; dysanalyte, pyrochlore: X-ray, microchem. test | with melilite in Ca-rich basic volcanics and in contact-metam. limestones; in chlorite- and talc schists | 52 |
| | brownish yellow to grey-green, partly zonal {001};<br>(X < Z) | decomposed by hot conc. $H_2SO_4$; may be radioactive; alteration to leucoxene | perovskite: X-ray | in carbonatites | 53 |
| | yellow to red; frequently patches and zonal structure | decomposed by hot conc. $H_2SO_4$; may be metamict, especially the edges with lower n and D | perovskite, dysanalyte: X-ray, microchem. test; spinel: harder, metamict gadolinite: different paragenesis | pyrochlore: in foyaites; pyrrhite: in pegmatites; koppite: in carbonatites | 54 |
| $2V_Z = 83^o$ | colorless, greenish | slowly soluble in HCl | | diagen. and metam. in rock salt layers, especially in the gypsum cap | 55 |
| | pyrope: pink; almandine: pale pink; spess.: yellowish; grossularite: colorless, pale greenish to yellow; andradite: greenish to yellow; melanite: yellowish to brown, zonal | almost insoluble in acids; zonal structure | spinel: {111} | pyrope: serpentinites, eclogites; alm.: granulites, eclogites, granites, pegmatites; spess.: low grade metam., pegmatites; andrat.: contact-metam. skarns; melanite: foyaites, phonolites; grossularite: contact-metam. marbles | 56 |

| No. | Name<br>Symmetry<br>Chem. Composition | Common Crystal Forms, Twinning<br>Crystal Habit<br>Aggregates | Cleavage<br>Translation Gliding<br>Deform. Twinning<br>Hardness, Density | Optic Orientation | Refractive Indices<br>Birefringence |
|---|---|---|---|---|---|
| 57 | Hibschite cub<br>$Ca_3Al_2$<br>$[(OH)_4/(SiO_4)_2]$ | {111}, rarely {110};<br>granular | no #;<br>H = 6½ D ~ 3.1 | isotropic; abnormally birefringent sectors like grossularite | n = 1.671 – 1.682 |
| 58 | Opal amorphous<br>$SiO_2 \cdot nH_2O$ | colloidal forms, in interstices and vesicles | brittle:<br>H = 5½ – 6½<br>D = (1.9) – 2.0 – 2.2<br>diagr. p. 28 | isotropic, may have weak abnormal birefringence | n = (1.40)–1.44–1.46<br>diagr. p. 28 |
| 58a | Lechatelerite amorph.<br>$SiO_2$-glass | xenomorphic | H < 7 D ~ 2.2 | isotropic | n = 1.45 |
| 59 | Melanophlogite cub<br>$SiO_2$, containing hydrocarbons | {100} | D = 2.0 (after heating) | isotropic after heating; abnormally birefringent patches before heating | n = 1.47 (before heating)<br>n = 1.425 (after heating) |
| 60 | Cristobalite tetr<br>$SiO_2$<br>> 267 (220)°C<br>High Cristobolite cub | paramorph. cub.:<br>{111}; ◈ {111} (cub).<br>tabular; ◈ complex lam.: [001] tetr ‖ {100} cub (penetration-triplets); grains, skeletal crystals, “balls”; chalcedony-like: lussatite, lussatine | H = 6 – 7 D = 2.33 | < 220 – 275°: complex pattern of tetr. lamellae<br>lussatite: l = (+)<br>lussatine: l = (–) | $n_E$ = 1.484<br>$n_O$ = 1.487<br>⊖Δ = 0.003 |
| 61 | Analcite cub<br>$Na[AlSi_2O_6] \cdot H_2O$ | {211}, {100};<br>no ◈ visible<br>in interstices and vesicles | # (001) poor<br>H = 5½ D = 2.25± | as groundmass usually isotropic, crystals in cavities often abnormal: weakly birefringent patches | n = 1.487<br>$n_X$ = $n_E$ = 1.486<br>$n_Z$ = $n_O$ = 1.487<br>⊖Δ = 0.001 |
| 61a | Pollucite cub<br>$(\underline{Cs},Na)[AlSi_2O_6] \cdot (H_2O) < 1$ | {100} rare<br>xenomorphous, replacement-structure | # poor;<br>H ~ 6<br>D = (2.86) – 2.92 – (2.98) | isotropic | n = 1.507 – 1.526 |
| 61b | Wairakite mcl pseudocub<br>$Ca[AlSi_2O_6]_2 \cdot 2H_2O$ | crystals rare and poorly developed;<br>◈ (110), (100), lam.;<br>grating-structure;<br>grains, crusts | no good #;<br>H = 5½ – 6 D = 2.26 | X and Y diagonal to twin-grating,<br>X ~ ‖ b, Y ~ ‖ a,<br>Z ~ ‖ c | $n_X$ = 1.498<br>$n_Z$ = 1.502<br>⊕Δ = 0.004 |
| 61c | Faujasite cub<br>$Na_2Ca[AlSi_4O_{12}]_2 \cdot 16H_2O$ | {111};<br>◈ (111) | # {111} distinct;<br>H = 5 D = 1.92 | isotropic; after heating first weakly optically uniaxial positive, then uniaxial negative | n = 1.48 |
| 62 | Leucite tetr<br>$K[AlSi_2O_6]$ | paramorph. (cub):<br>{211};<br>◈ complex lam.<br>{110} tetr. | parting {110} cub, poor;<br>$K_1$=(110), $K_2$=(1$\bar{1}$0) cub<br>H = 5½ – 6 D ~ 2.47 | <600 – 620° complex pattern of tetr. lam. with abnormal optics (orh); small crystals metastably isotropic | $n_O$ = 1.508<br>$n_E$ = 1.509<br>⊕Δ = 0.001 |
| 63 | Sodalite group cub<br>$(Na,Ca)_8$<br>$[(Cl,SO_4)_{1-2}(AlSiO_4)_6]$<br>Sodalite: Na + Cl<br>Nosean: Na + $SO_4$<br>Hauyne: ($\underline{Na}$,Ca)+$SO_4$ | {110};<br>◈ {111};<br>corroded crystals, grains | # {110} distinct<br>H = 5 – 6 sod.: D = 2.30<br>nos.: D = 2.35<br>hau.: D = 2.45 | isotropic | sodalite:<br>n = 1.483 – 1.487<br>nosean:<br>n = 1.488 – 1.495<br>hauyne:<br>n = 1.496 – 1.507 |
| 64 | Volcanic Glasses amorphous<br>Rhyolite Glass<br>Pitchstone<br>Dacite Glass<br>Trachyte Glass<br>Andesite Glass<br>Leucitetephrite Glass<br>Basalt Glass | in interstices, groundmass | D =<br><br>(2.33) – 2.37 – (2.41)<br>(2.32) – 2.34 – (2.37)<br>(2.45) – 2.50 – (2.55)<br>(2.43) – 2.45 – (2.47)<br>(2.40) – 2.47 – (2.57)<br>(2.52) – 2.55 – (2.58)<br>(2.70) – 2.77 – (2.85) | isotropic;<br>sometimes weakly birefringent | n =<br><br>(1.48)–1.492–(1.51)<br>(1.492)–1.500–(1.506)<br>(1.504)–1.511–(1.592)<br>(1.488)–1.512–(1.527)<br>(1.489)–1.512–(1.529)<br>(1.525)–1.550–(1.580)<br>(1.506)–1.575–(1.612) |

**Optically Uniaxial,**

| No. | Name<br>Symmetry<br>Chem. Composition | Common Crystal Forms, Twinning<br>Crystal Habit<br>Aggregates | Cleavage<br>Translation Gliding<br>Deform. Twinning<br>Hardness, Density | Optic Orientation | Refractive Indices<br>Birefringence |
|---|---|---|---|---|---|
| 65 | Cassiterite tetr.<br>$SnO_2$ | {110}, {321}, {111};<br>◈ (101) lam.;<br>short prisms, grains | # (100) distinct<br># {110} poor<br>$K_1$=(101), $K_2$=($\bar{3}$01);<br>H = 6 – 7 D = 7.03± | Z ‖ c E = Z<br>l = (+) fig. p. 29 | $n_O$ = 1.997 – 2.001<br>$n_E$ = 2.093 – 2.098<br>⊕Δ = 0.096 – 0.097 |

| Angle and Dispersion of Optic Axes | Color in Thin Section Pleochroism | Chemical Reactions Special Remarks | Distinguishing Features of Similar Minerals | Occurrences | |
|---|---|---|---|---|---|
| | colorless, at best a pale yellow | soluble in acids; sometimes zonal structure | grossularite: gr. n | in contact-metam. marly limestones | **57** |
| | colorless, dull yellow or pink due to Fe-hydrates | soluble in HF and in hot KOH-solution; may contain cryptocrist. low-temp. cristobalite: X-ray | analcite, sodalite, glass: gr. n | hydroth. with tridymite and zeolites; alteration product; not in plutonites and metamorphic rocks | **58** |
| | colorless | soluble in HF | volcanic glasses: gr. n | at grain boundaries of quartz in fulgurites; with coesite, cristobalite in shock-metam. rocks near meteor craters | **58a** |
| | brown due to organic matter, zonal {100}; after heating black due to graphite | dull white crystals, mostly pseudomorph. altered to quartz | | authigen. in sulphur deposits | **59** |
| abnormal: $2V_X \sim 25^o$ | colorless (turbid) | soluble in boiling $Na_2CO_3$-solution; "ball"-structure; in opal as "low-temp. cristobalite": X-ray | tridymite: tabular, no ball-structure; chabasite: different paragenesis | pneumatol.-exhal. in cavities, devitrification product of acid glasses, not in plutonites and metam. rocks | **60** |
| ($2V_X$ = very small) | colorless | decomposed by HCl, gives $SiO_2$-powder; cannot be stained; turbid upon heating | sodalite group: other # and microstructure; glass and opal: no #; zeolites: Δ | late magmatic to hydrothermal in alkaline rocks; in vesicles; also in ore-bearing veins of zeolite facies | **61** |
| | colorless | decomposed by HCl; turbid upon heating | analcite: sm.n | in Na-Li-pegmatites with lepidolite, spodumene, tourmaline | **61a** |
| $2V_Z = 70^o - 105^o$ $r > v$ | colorless | gelatinizes with HCl-vapor, then stainable with aniline colors | analcite: isotrop. and sm. n; other zeolites: gr. Δ | hydrotherm. and very low grade reaction product of plagioclase in tuffaceous rocks, ignimbrites, spilites and sandstones | **61b** |
| | colorless, turbid | decomposed by HCl-vapor, then stainable | easily recognized by isotropic char. and sm. n | mainly in vacuoles of basaltic rocks, phonolites and metamorphites | **61c** |
| abnormal $2V_Z$ = very small | colorless | decomposed by HCl, gives powder; stained yellow by Na-hexanitrocobaltate III (like K-feldspar). Inclusions: "cinder-coronas"; in plutonites alteration to pseudoleucite | sodalite: sm. n, no birefringent lamellae | magmatic in potassic volcanics together with sodalite, nepheline; not with quartz, not in metamorphites | **62** |
| | usually colorless; also greenish, reddish, bluish or grey; often brownish black rims and patches | gelatinizes even with acetic acid, then stainable with aniline colors; reaction of $Cl^-$ or $SO_4^{2-}$; frequently opaque grating | fluorite: sm. n; analcite: poor # {001}, somewhat sm. n; leucite: "cinder-coronas" and gr. n, birefr. lamellae | magmatic in alkaline rocks; not together with quartz; not in metamorphites | **63** |
| | colorless, pale yellow, pale brown, the basic members up to light coffee-brown | acid glasses not easily, basic glasses easily corroded by cold HCl-vapor; often concentric contraction cracks | D=silica glass 2.203<br>K-feldsp. glass 2.351<br>leucite glass 2.410<br>Na-feldsp. glass 2.382<br>plag. $An_{50}$ glass 2.533<br>Ca-feldsp. glass 2.700<br>enstatite glass 2.743<br>hornbl.-glass 2.826<br>diopside glass 2.830<br>olivine glass 2.831<br>augite glass 2.835 | magmatic in the groundmass of volcanic rocks, also in veins, especially along the selvage; not in plutonic rocks or in medium or higher grade metamorphites | **64** |

## Tetragonal Minerals

| | | | | | |
|---|---|---|---|---|---|
| rarely abnormally biaxial: $2V_Z \sim 10^o$ ($38^o$), $v > r$ | colorless to yellowish brown, rarely reddish, greenish, zonal coloring; E ≈ O, somet.; E > O weak | soluble in KOH-melt; cannot be melted with blow pipe; resistant in acids; pleochr. haloes | allanite: sm. n and Δ; rutile: gr. n, gr. Δ see also Vol. 2 | with F-minerals, pneumatolytic and pegmatitic associated with granite intrusions | **65** |

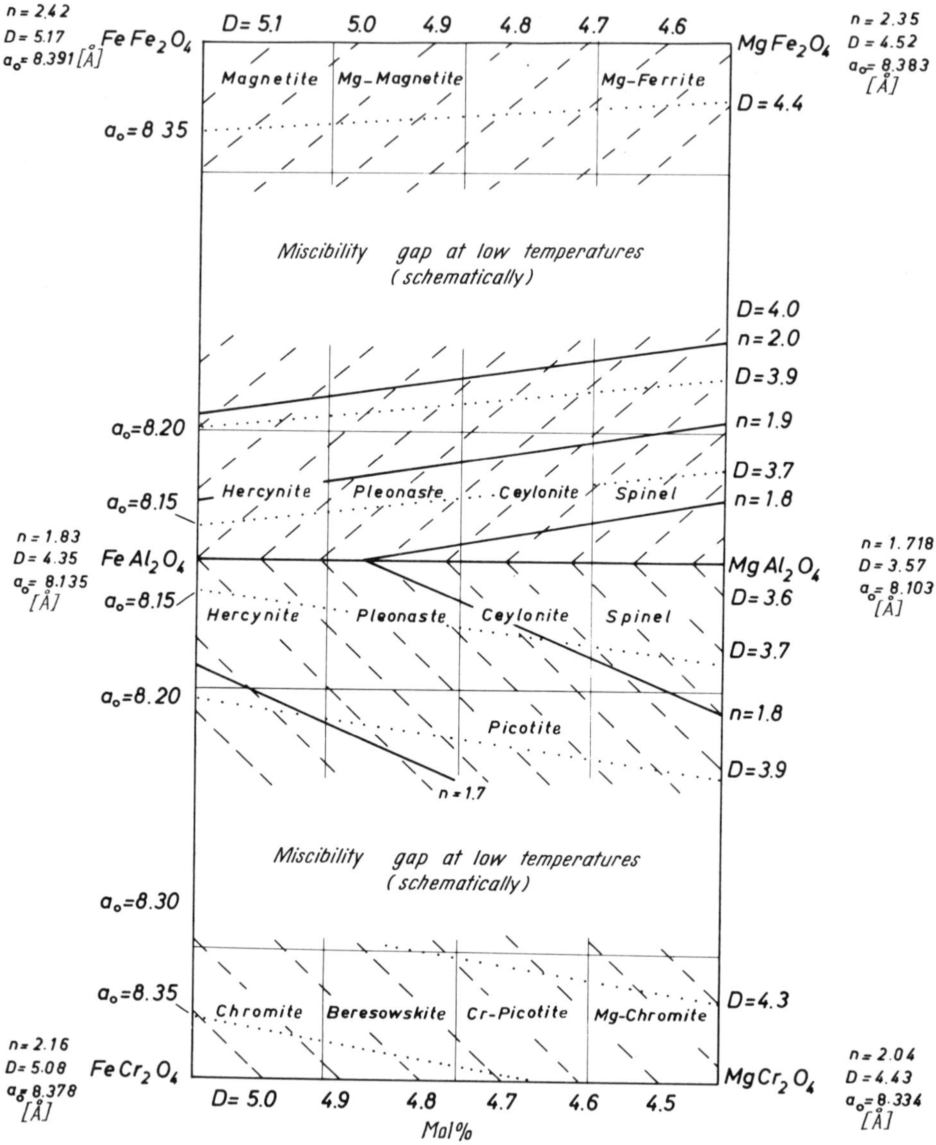


*50-1. Spinel group*

TRÖGER (1950); DEER, HOWIE, ZUSSMAN (1963)
(The linear relationship between chemical and physical properties is valid only as a first approximation, see 50-2.)

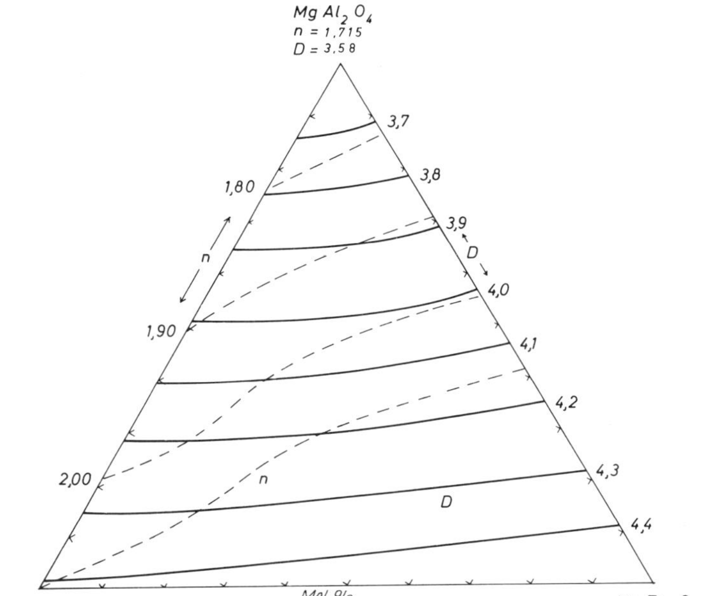


50-2. Spinel group (synthetic)
high temperature solid solutions
ALLEN (1966), Am. Min. 51, 239.

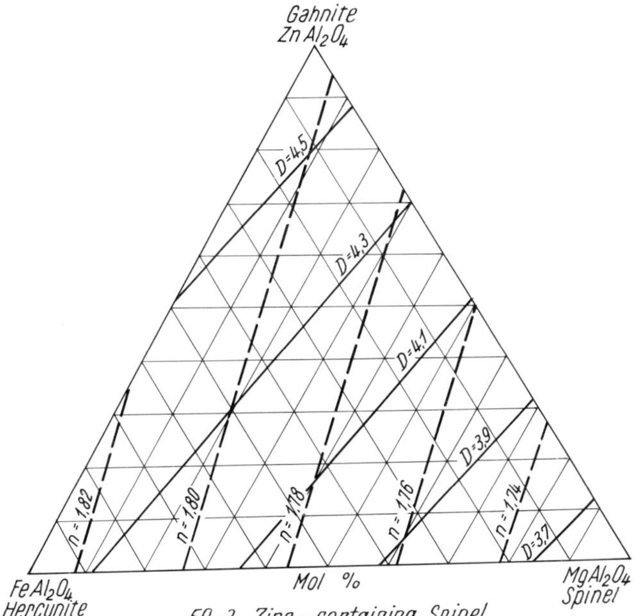


50-3. Zinc-containing Spinel
WINCHELL (1947)
(Copyrighted, 1951, by A. N. WINCHELL in Elements of Optical Mineralogy, part II; Reproduced by permission.)

(11)

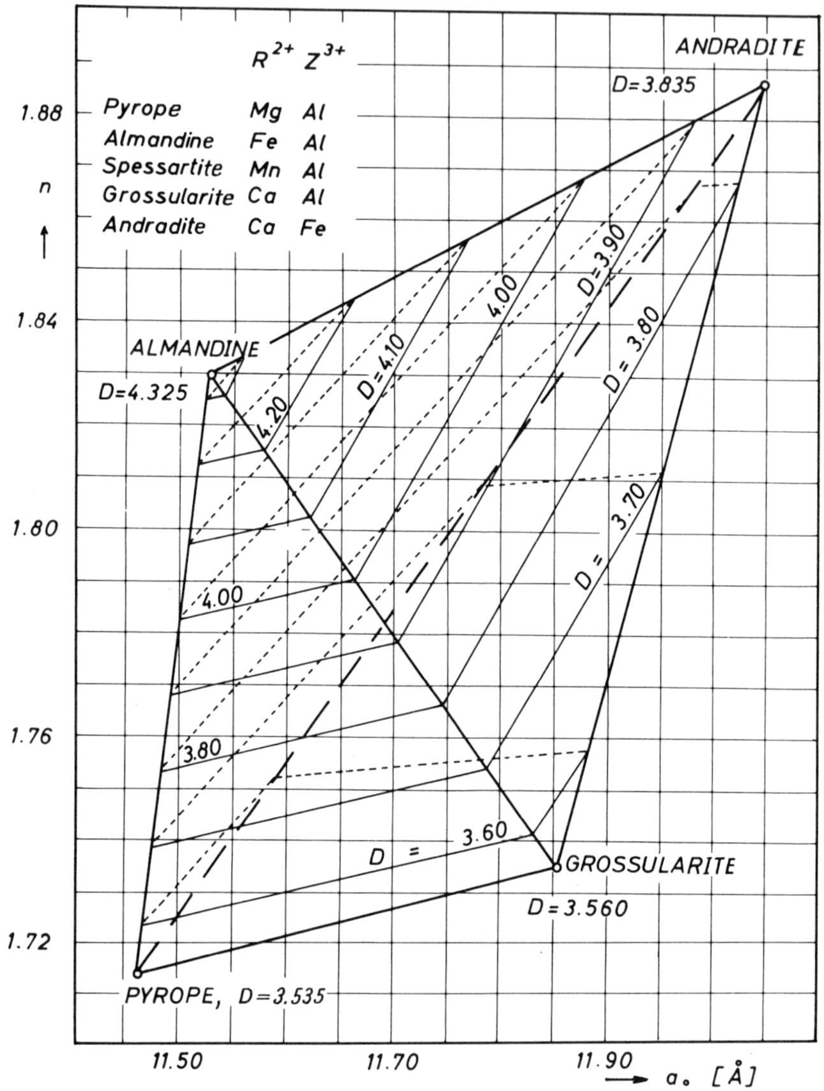


*56-1. Garnet group*
WINCHELL 1958 (Am. Min. 43, 595-600)

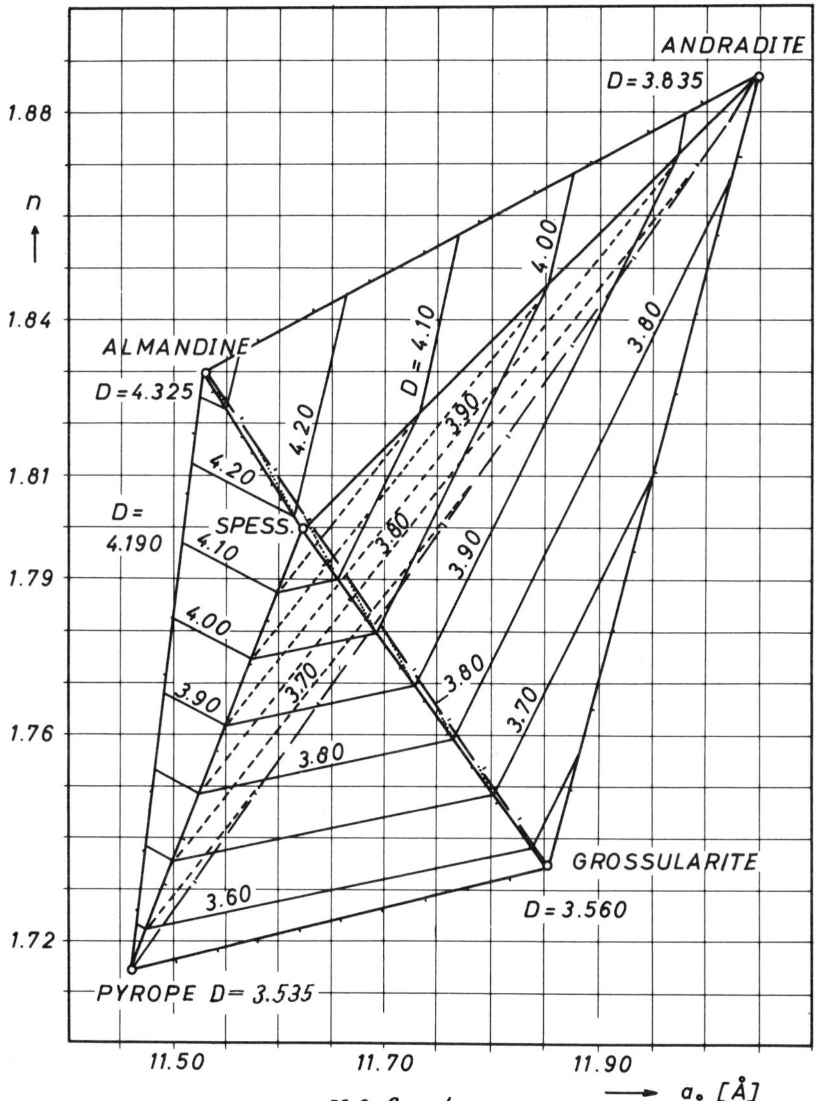


56-2. Garnet group
WINCHELL 1958 (Am. Min. 43, 595)

| No. | Name<br>Symmetry<br>Chem. Composition | Common Crystal Forms, Twinning<br>Crystal Habit<br>Aggregates | Cleavage<br>Translation Gliding<br>Deform. Twinning<br>Hardness, Density | Optic Orientation | Refractive Indices<br>Birefringence |
|---|---|---|---|---|---|
| 66 | Rutile tetr<br>$TiO_2$ | {110}, {100}, {hk0}<br>◈ (101) geniculated;<br>c Λ c = 65½° lam.<br>◈ (301) heart-shaped;<br>c Λ c = 54¾°<br>columnar ‖ c, granular;<br>"sagenite"-grating<br>= ◈ (101) + (301) | # {110} perf.<br># (100) good<br>$K_1$ = (101), $K_2$ = ($\bar{3}$01)<br>H = 6 – 6½ D = 4.23 | Z ‖ c E = Z<br>l = (+) fig. p. 29 | $n_O$ = 2.609 – 2.616<br>$n_E$ = 2.895 – 2.903 (Na-light)<br>⊕Δ = 0.286 – 0.287<br>disp. strong<br>r < v |
| 67 | Anatase tetr<br>$TiO_2$ | {111}, {hkl}, {001};<br>sharp pointed pyramids<br>or tabular ‖ (001) | # (001) perf.<br># {111} perf. } 68½°;<br>H = 5½ – 6 D = 3.90± | X ‖ c E = X<br>fig. p. 29 | $n_O$ = 2.561 – 2.562<br>$n_E$ = 2.488 – 2.489 (Na-light)<br>⊖Δ = 0.073 – 0.073<br>disp. Δ strong<br>r < v |
| 67a | Stishovite tetr<br>$SiO_2$ | extremely small grains,<br>may be submicroscopic,<br>acicular | D = 4.30 | abnormally:<br>Z Λ c = 0 – 5°, E = Z<br>l = (+) | 1. 2.<br>$n_O$ = 1.799 1.800<br>$n_E$ = 1.826 1.845<br>⊕Δ = 0.027 0.045 |
| 68 | Xenotime tetr<br>$Y[PO_4]$ | {111}, {110}, {100};<br>◈ (101)<br>short columns, grains | # {110} good;<br>H = 4 – 5 D = 4.55± | Z ‖ c E = Z<br>fig. p. 29 | $n_O$ = 1.721 – 1.720<br>$n_E$ = 1.816 – 1.827<br>⊕Δ = 0.095 – 0.107 |
| 69 | Zircon tetr<br>$Zr[SiO_4]$ | {110},{111},{100},<br>{131};<br>columns to rounded<br>grains, colorless<br>isomorph. zoning | # {110} distinct;<br>H = 7½<br>D = (3.91 – 4.68) – 4.72 | Z ‖ c E = Z<br>l = (+)<br>fig. p. 29 | $n_O$ = (1.848 – 1.911) – 1.926<br>$n_E$ = (1.855 – 1.943) – 1.985<br>⊕Δ = (0.007 – 0.032) – 0.059<br>metamict.: n → 1.826, Δ → 0.000 |
| 70 | Vesuvianite tetr<br>(Idocrase)<br>$Ca_{10}(\underline{Mg},Fe)_2Al_4$<br>$[(OH)_4/(SiO_4)_5/$<br>$(Si_2O_7)_2]$ | {110}, {100}, {111},<br>{001};<br>grains, acicular ‖ c | # (100) poor,<br># {110} poor;<br>H = 6½ D = 3.40± | X ‖ c E = X<br>optically inhomogen.<br>growth-zones, abnorm-<br>ally birefringent patches<br>fig. p. 29 | $n_O$ = 1.705 – 1.738<br>$n_E$ = 1.701 – 1.732<br>⊖Δ = 0.004 – 0.006<br>wiluite ⊕Δ – 0.005<br>diagr. p. 35 |
| 71 | Melilite group tetr<br>Gehlenite<br>$Ca_2Al[(Al,Si)_2O_7]$<br>Melilite ⇅<br>Åkermannite<br>$Ca_2Mg[Si_2O_7]$ | {001}, {100};<br>short columns, ↑ geh.<br>thick tabular<br>tabular<br>thin tabular ↓ åk. | # (001) distinct,<br># {110} distinct;<br>H = 5½<br>D = {3.05 ↑ ge.<br>2.94 ↓ åk} | X ‖ c E = X ↑ ge.<br>l = (+) (–)<br>Z ‖ c E = Z ↓ åk.<br>l = (–)<br>fig. p. 32 | ⊖ ge ⊕ åk<br>$n_O$ = 1.669 – 1.633<br>$n_E$ = 1.658 – 1.639<br>Δ = 0.011 – 0.006<br>mcl: partly Δ ~ 0.000<br>diagr. p. 34 |
| 72 | Meliophane tetr<br>(Meliphanite)<br>$(Ca,Na)_2(Be,Al)$<br>$[Si_2O_6F]$ | {001},<br>foliated, scaly | # (001) poor;<br>H = 5½ D = 3.01± | X ‖ c E = X<br>may be opt. abnormal<br>l = (+) fig. p. 32 | $n_O$ = 1.612 $n_X$ = 1.593<br>$n_E$ = 1.593 $n_Y$ = 1.612<br>⊖Δ = 0.019 $n_Z$ = 1.613<br>⊖Δ = 0.020 |
| 73 | Apophyllite tetr<br>$KCa_4[F(Si_4O_{10})_2]\cdot$<br>$8H_2O$ | {100}, {111}, {001};<br>tabular, granular, and<br>columnar, cubes | # (001) perf.,<br># {110} distinct;<br>H = 4½ – 5 D = 2.36± | Z ‖ c (rarely X ‖ c)<br>E = Z or X<br>sometimes opt.<br>abnormal patches,<br>l = (+) (–) fig. p. 32 | usually rarely<br>$n_O$ = 1.5345 – 1.5445<br>$n_E$ = 1.5365 – 1.5439<br>⊕Δ = 0.0020 – ⊖0.0006<br>also $n_O$ ~ $n_E$ = 1.542 |
| 74 | Scapolites tetr<br>$Na_8[(Cl_2,SO_4,CO_3)/$<br>$(AlSi_3O_8)_6]$ Mar.<br>$Ca_8[(Cl_2,SO_4,CO_3)_2/$<br>$(Al_2Si_2O_8)_6]$ Mei. | {110}, {100}, {111};<br>no ◈;<br>grains, short columns | # (100) good<br># {110} distinct;<br>brittle<br>H = 5½ – 6<br>D = 2.54 – 2.81 | X ‖ c E = X<br>l = (–)<br>fig. p. 32 | mar. mei.<br>$n_O$ = 1.539 – 1.596<br>$n_E$ = 1.537 – 1.557<br>⊖Δ = 0.002 – 0.039<br>diagr. p. 35 |

## Optically Uniaxial, Hexagonal

| No. | Name<br>Symmetry<br>Chem. Composition | Common Crystal Forms, Twinning<br>Crystal Habit<br>Aggregates | Cleavage<br>Translation Gliding<br>Deform. Twinning<br>Hardness, Density | Optic Orientation | Refractive Indices<br>Birefringence |
|---|---|---|---|---|---|
| 75 | Koenenite trig<br>$4NaCl \cdot 4(Mg,Ca)Cl_2 :$<br>$5Mg(OH)_2 \cdot 4Al(OH)_3$ | {0001};<br>barrel-shaped forms,<br>scaly | # (0001) perf.;<br>T = (0001): flexible;<br>H < 1 D = 1.98 | Z ‖ c E = Z<br>l = (–)<br>fig. p. 32 | $n_O$ = 1.540 – 1.555<br>$n_E$ = 1.565 – 1.585<br>⊕Δ = 0.025 – 0.030 |
| 76 | Corundum trig<br>$\alpha - Al_2O_3$ | {0001},{10$\bar{1}$1},{22$\bar{4}$1}<br>with l = 0 – 4;<br>◈ {10$\bar{1}$1} lam;<br>tabular, barrel-shaped,<br>rounded grains | parting {10$\bar{1}$1} 94°,<br>parting {0001};<br>T = (0001)<br>$K_1$ = (10$\bar{1}$1), $K_2$ = (01$\bar{1}$2),<br>$K_1'$ = (0001), $K_2'$ = (02$\bar{2}$1);<br>H = 9 D = 4.03 | X ‖ c E = X<br>fig. p. 32 | $n_O$ = 1.767 – 1.772<br>$n_E$ = 1.759 – 1.763<br>⊖Δ = 0.008 – 0.009 |

| Angle and Dispersion of Optic Axes | Color in Thin Section Pleochroism | Chemical Reactions Special Remarks | Distinguishing Features of Similar Minerals | Occurrences | |
|---|---|---|---|---|---|
| may be abnormal: small angle of opt. axes, O.A.P. ∥ (110) | O = yellow to red-brown E = brownish yellow to yellowish green; darker with increasing Fe-content, E > O weak | insoluble in HF; soluble in soda-, KF-, and $HKF_2$-melts; characteristic twinning; pseudomorphous as leucoxene; pleochr. haloes | cassiterite: sm. n and Δ; hematite, lepidocrocite, goethite, brookite: X-ray see Vol. 2 | with ilmenite in pegmatites; rarely magmatic (granites, foyaites, anorthosites); metam. in amphibolites; in sands, sandstones, and placers | 66 |
| may be abnormal: small angle of opt. axes, especially with deep colored crystals | O = light blue to bluish green, E = dark blue to greenish blue, O = green-grey to light brown, E = orange to brown; zonal coloring, E > O, rarely E < O | insoluble in acids; soluble in $HKSO_4$-melt; pseudomorphous as leucoxene | carborundum: small Δ and ⊕ see Vol. 2 | hydrothermal; by weathering of Ti-bearing min.; not magmatic; in sands and sandstones | 67 |
| | colorless | insoluble in HF | other $SiO_2$-modifications: sm. n, sm. Δ; X-ray | see Nr. 81 a coesite | 67 a |
| | colorless to brownish E ≈ O; also E > O, then: O = pink to pale yellow E = brownish yellow to greenish | decomposed by boiling $H_2SO_4$; absorpt. spectr. shows Er- and Y-lines; pleochroic haloes | zircon: gr. n; monazite: sm. n, biaxial; sphene: gr. n and Δ | magmatic in granites and syenites; in pegmatites and hydroth.; enriched in placers with zircon and monazite | 68 |
| often abnormal: $2V_Z \sim 10^o$ | colorless, rarely pale brownish, or greenish E ~ O | insoluble in HF; soluble in molten $K_2CO_3$; frequently pleochroic haloes in micas etc. | xenotime and monazite: sm. n | early magmat. precipitat. in granites, syenites, foyaites; resistant in sediments and metamorphites | 69 |
| often abnormal: $2V_X = 17^o - 33^o$ O.A.P. ⊥ (110) | colorless, greenish, brownish, yellowish O ≷ E | abnormal interference colors, sm. Δ; slightly decomposed by HCl | garnet (grossularite): usually isotropic; zoisite: # and sm. n and 2 V; melilite and apatite: sm. n | with garnet, epidote, augite contact-metam. and in fissures, rarely magmatic | 70 |
| | colorless, also yellowish or greyish, then O < E weak | gelatinizes with HCl, then stainable; melilites often zonal: core almost isotropic, rims ge-rich with abnormal deep blue interference colors | gehlenite: zoisite: gr. n; vesuvianite: gr. n; apatite: hex. idiom.; melilite: glass and opal sm. n | ge: contact-metam. in marbles; mel: magmatic only in Ca-rich ultrabasic rocks, usually together with perovskite; åk: only synthetically in slags | 71 |
| often abnormal: $2V_X < 22^o$ | O = light brownish yellow, E = light greenish yellow, E < O | insoluble in acids | similar minerals are biaxial see Vol. 2 | pegmatitic-pneumatolytic in foyaite-pegmatites | 72 |
| often abnormally biaxial | colorless | decomposed by HCl, gives $SiO_2$-powder; turbid after heating to red heat; poss. with abnormal interference colors | similar: uniaxial zeolites, these have no # (001) | hydrotherm. in vesicles of basaltic volcanites, rarely in granites or in lodes | 73 |
| | colorless, sometimes yellowish, O > E weak | only meionite decomposed by HCl | andalusite: gr. n; feldspars: oblique extinction, quartz: ⊕ | marialite: pneumatol. in tuffs; dipyre: contact-metam. schists; mizzonite: contact-metam., pegmatitic, post-magmat. in "saussurite"; meionite: contact-metam. limestones | 74 |

## and Trigonal Minerals

| | | | | | |
|---|---|---|---|---|---|
| | colorless, yellowish O = red-brown E = colorless | decomposes slowly in boiling $H_2O$; hematite-inclusions | brucite: r > v distinct, colorless; asharite: gr. n, fibrous; talc: X ⊥ #, opt. ⊖ | secondary in anhydrite and salt-cont. pelite deposits | 75 |
| abnormal $2V_X \sim 5^o - 7^o$ (up to > 30°) | colorless, also brownish, zonal and patchy distribution of color, O = blue, purple, E = light yellow, greenish yellow, O > E weak | insoluble in acids, very high melting point; decomposes in borax-melt, $KHSO_4$-melt | sapphirine: no ◇ lam., gr. $2V_X$; chrysoberyl: opt. ⊕; vesuvianite, apatite: sm. Δ, sm. n | in granite-pegmatites and syenite-pegmatites; contact-metam.; in schists, marbles, in placers | 76 |

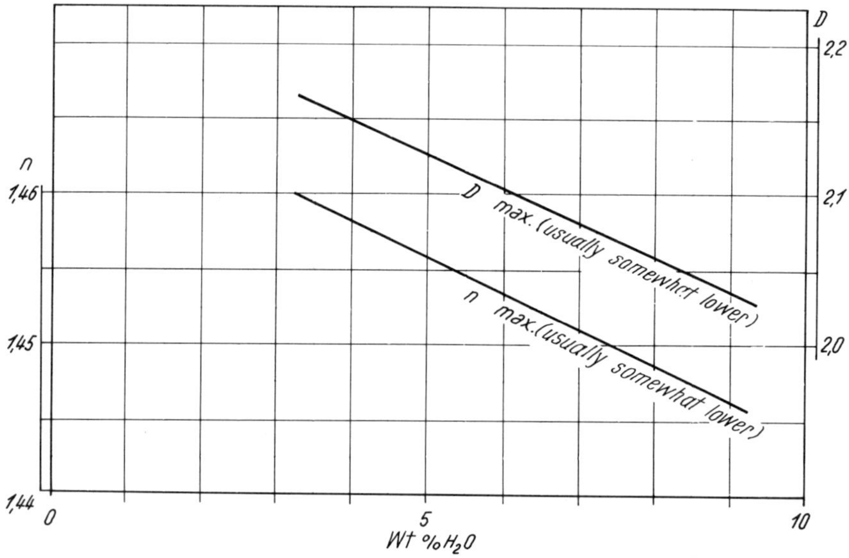


58. Opal
KOKTA 1930

65. Cassiterite

66-1. Rutile

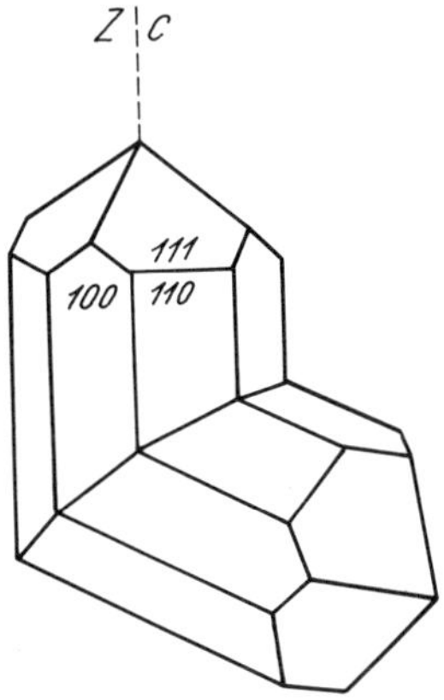


66-2. Rutile geniculated twin

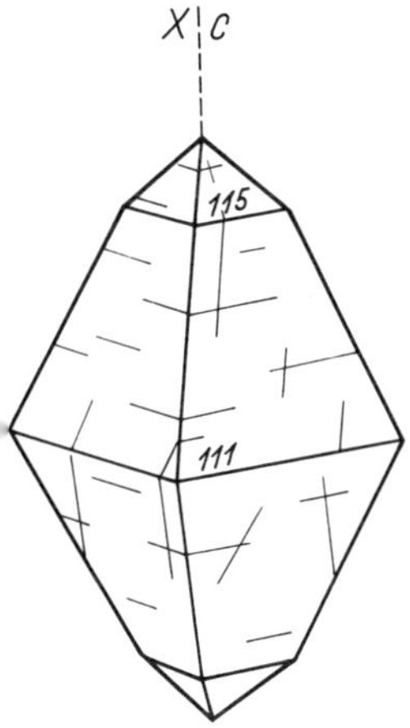


67-1. Anatase

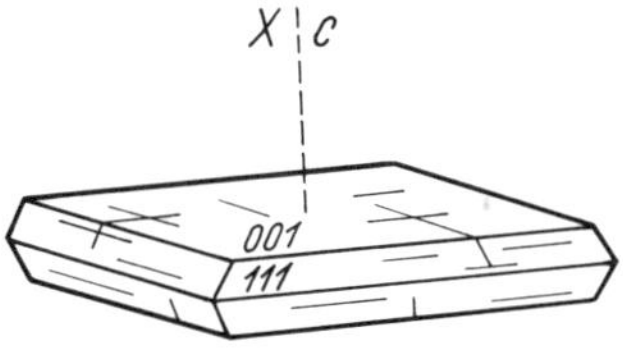


67-2. Anatase

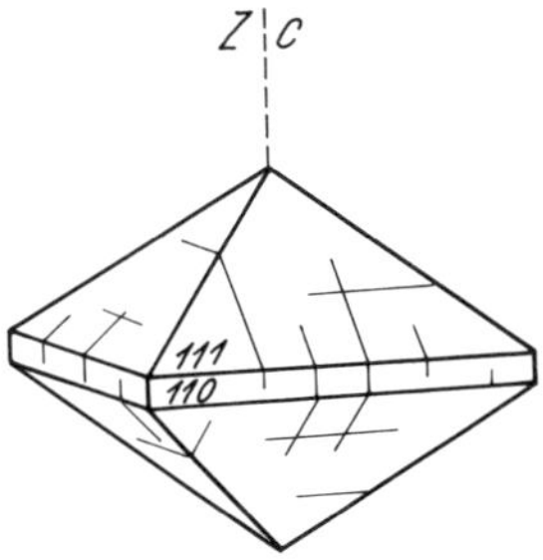


68. Xenotime

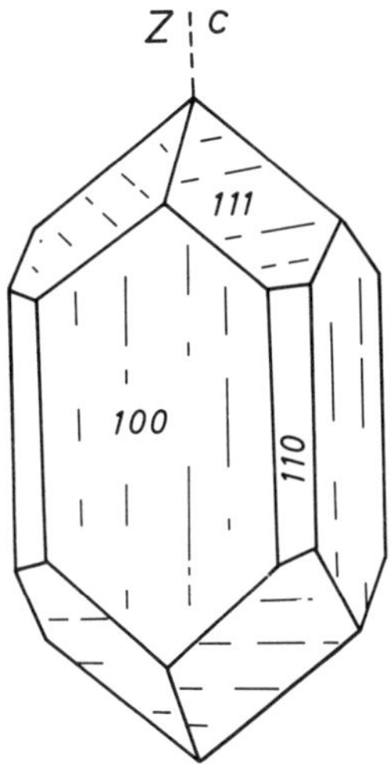


69-1. Zircon

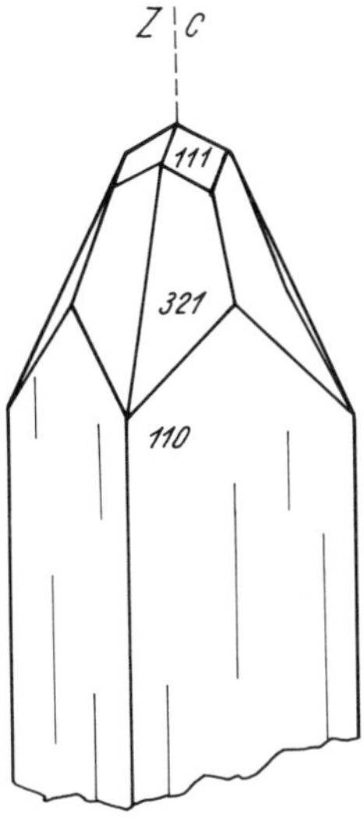


69-2. Zircon

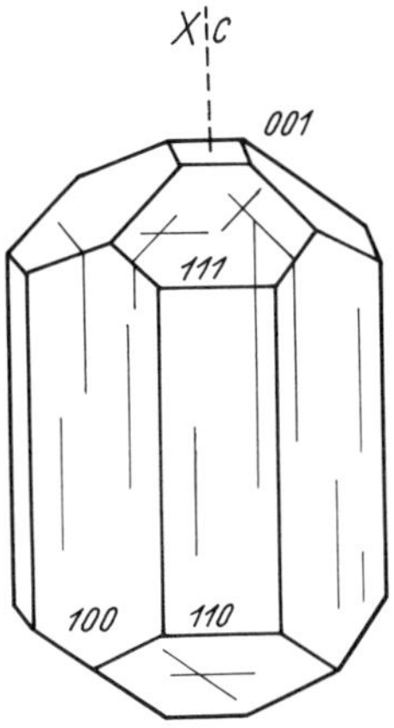


70-1. Vesuvianite

| No. | Name<br>Symmetry<br>Chem.Composition | Common Crystal Forms, Twinning<br>Crystal Habit<br>Aggregates | Cleavage<br>Translation Gliding<br>Deform. Twinning<br>Hardness, Density | Optic Orientation | Refractive Indices<br>Birefringence |
|---|---|---|---|---|---|
| 77 | Hematite trig<br>$a-Fe_2O_3$ | $\{0001\},\{10\bar{1}1\},\{11\bar{2}0\}$;<br>◈ $\{10\bar{1}1\}$ lam.<br>◈ $\{0001\}$;<br>tabular, foliated, scaly, granular, fibrous, dust particles, oolithic | parting $\{10\bar{1}1\}$ 94°,<br>parting $\{0001\}$;<br>T = (0001)<br>$K_1=(10\bar{1}1), K_2=(01\bar{1}2)$,<br>$K_1'=(0001), K_2'=(02\bar{2}1)$;<br>H ~ 6 D ~ 5.26 | X ∥ c E = X<br>usually l = (+)<br>fig. p. 32 | λ = 5890 Å λ = 7590 Å<br>$n_O$ = 3.22 – 2.90<br>$n_E$ = 2.94 – 2.69<br>⊖Δ = 0.28 – 0.21<br>disp. strong |
| 78 | Ilmenite trig<br>$FeTiO_3$ | $\{0001\},\{10\bar{1}1\},\{02\bar{2}1\}$;<br>◈ $\{10\bar{1}1\}$ lam.<br>◈ $\{0001\}$; flaky, granular, skeletal crystals | parting $\{10\bar{1}1\}$ 94½°<br>parting $\{0001\}$;<br>flexible<br>H = 5–6 D ~ 4.72 | X ∥ c E = X<br>fig. p. 32 | n ≫ 2.7<br>⊖Δ = very strong |
| 78a | Leucoxene<br>mixture of sphene, $SiO_2$-mod., rutile, anatase, brookite, hematite, goethite in different proport. | extremely fine grained, radiated | | | n high<br>Δ high |
| 79 | Pyrophanite trig<br>(<u>Mn</u>, Fe) $TiO_3$ | $\{0001\}$;<br>thin flakes, scales | # $\{02\bar{2}1\}$ perf.,<br># $\{10\bar{1}2\}$ good;<br>H = 5–6 D ~ 4.54 | X ∥ c E = X<br>l = (+)<br>fig. p. 32 | $n_O$ = 2.48<br>$n_E$ = 2.21<br>⊖Δ = 0.27 |
| 80 | Geikielite trig<br>(<u>Mg</u>, Fe) $TiO_3$ | $\{0001\}, \{10\bar{1}1\}$;<br>◈ $\{10\bar{1}1\}$ lam.;<br>granular | parting $\{10\bar{1}1\}$,<br>parting $\{0001\}$;<br>H = 5–6 D ~ 4.05 | X ∥ c E = X<br>fig. p. 32 | $n_O$ = 2.31<br>$n_E$ = 1.95<br>⊖Δ = 0.36 |
| 80a | Hoegbomite hex<br>(<u>Mg</u>, Fe)$(TiAl_2)O_{3-4}$ | $\{0001\}$;<br>◈ $\{0001\}$ lam.,<br>◈ $\{h0hl\}$ lam. rare;<br>thick or thin tabular, rounded grains | # (0001) perf.,<br>sometimes mica-like;<br>H = 6½ D ~ 3.8 | Z ∥ c E = Z<br>l = (–) | A B<br>$n_O$ = 1.83–1.84 1.85<br>$n_E$ = 1.81 1.80<br>⊖Δ = 0.02–0.03 0.05 |
| 81 | Quartz trig<br>>573° High Quartz hex<br>$SiO_2$ | $\{10\bar{1}1\},\{01\bar{1}1\},\{10\bar{1}0\}$;<br>◈ [0001],<br>◈ $\{11\bar{2}0\}$; granular columnar ∥ c, rarely pseudohex. paramorphs after high quartz | # $\{10\bar{1}1\},\{01\bar{1}1\}$ poor,<br>parting; brittle;<br>T = (0001), t = $[10\bar{1}0]$,<br>deformation-lam. with weakly diff. n;<br>H = 7 D = 2.65 | Z ∥ c E = Z<br>l = (+) fig. p. 33<br>often undulatory extinction; rotary polarization invisible in thin section | $n_O$ = 1.5442<br>$n_E$ = 1.5533<br>⊕Δ = 0.0091 |
| 81a | Coesite mcl<br>$SiO_2$ | pseudohexagonal,<br>β = 120°;<br>◈ {021}, (100);<br>tabular (010),<br>bladed ∥ (100),<br>elongated ∥ [001];<br>irregular, fine-grained | H ~ 7–8 D = 2.92 | X ∥ b<br>Y ∧ a = 24–26°,<br>O.A.P. ⊥ (010)<br>Z ∧ c = 4–6°, bi; l = (+)<br>sign of opt. extinction indeterminate because of β = 120° | $n_X$ = 1.594<br>$n_Y$ = 1.596<br>$n_Z$ = 1.599<br>⊕Δ = 0.005 |
| 81b | Stishovite tetr<br>$SiO_2$ | finest grains, sometimes submicroscopic, acicular | D = 4.30 | abnormal:<br>Z ∧ c = 0–5°, E = Z<br>l = (+) | 1. 2.<br>$n_O$ = 1.799 1.800<br>$n_E$ = 1.826 1.845<br>⊕Δ = 0.027 0.045 |
| 82 | Chalcedony<br>fibrous variety of Low Quartz | fibrous, radiated, massive, fibre-axis ⊥ c: chalcedony s.str.<br>fibre-axis ∥ c: quartzine | brittle;<br>H ~ 6 D = 2.58 – 2.63 | Z ∥ c E = Z<br>l = (–): chalcedony s.str.<br>l = (+): quartzine<br>fig. p. 33 | $n_O$ = 1.530 – 1.533<br>$n_E$ = 1.538 – 1.543<br>⊕Δ = 0.008 – 0.010 |
| 83 | Tridymite M<br>mcl < 100°C<br>>460° High Tridymite hex<br>$SiO_2$ | paramorph. (hex):<br>$\{0001\}, \{10\bar{1}0\}$;<br>◈ $\{30\bar{3}4\}$ twins and quadruplets,<br>◈ combination $\{30\bar{3}4\}$ and $\{10\bar{1}6\}$: "triplets" with (0001) ∧ (0001) = 35°; thin tabular | H = 6½ – 7 D = 2.27 | <100°C form. of up to 6 opt. orthorhombic sections;<br>(◈ $\{11\bar{2}0\}$);<br>O.A.P. ∥ $\{10\bar{1}0\}$;<br>l = (–)<br>fig. p. 33 | synth. nat.<br>$n_X$ = 1.469 1.477<br>$n_Y$ = 1.469 1.478<br>$n_Z$ = 1.473 1.481<br>⊕Δ = 0.004 0.004 |
| 84 | Brucite trig<br>$Mg(OH)_2$ | (0001) flakes, rarely fibrous ∥ $[10\bar{1}0]$ | # (0001) perf.;<br>T = (0001), flexible;<br>H = 2½ D = 2.39 | Z ∥ c E = Z<br>X ∥ fibre-axis;<br>l = (–)<br>fig. p. 33 | $n_O$ = 1.559 – 1.566<br>$n_E$ = 1.580 – 1.581<br>⊕Δ = 0.021 – 0.015<br>disp. r ≫ v |
| 85 | Calcite trig<br>$CaCO_3$ | $\{10\bar{1}0\},\{0001\},\{01\bar{1}2\}$<br>$\{02\bar{2}1\},\{hkl\}$;<br>◈(0001) and $(01\bar{1}2)$ lam.;<br>granular, fibrous, fills interstices, pseudomorphs | # $\{10\bar{1}1\}$ perf. 75°,<br>$\{01\bar{1}2\}$ and $\{0001\}$ parting;<br>$K_1=(01\bar{1}2), K_2=(0\bar{1}11)$;<br>plastic<br>H = 3 D = 2.715± | X ∥ c E = X<br>fig. p. 33 | $n_O$ = $1.658_4$<br>$n_E$ = $1.486_5$<br>⊖Δ = 0.1719<br>diagr. p. 39 |
| 86 | Magnesite trig<br>$MgCO_3$ | $\{10\bar{1}1\}, \{0001\}$;<br>◈ lamellae rare;<br>granular, fibrous, massive | # $\{10\bar{1}1\}$ perf., 72½°<br>T = (0001)<br>H = 4–4½ D = 2.96± | X ∥ c E = X<br>fig. p. 33 | pure Fe-cont.<br>$n_O$ = 1.700 – 1.719<br>$n_E$ = 1.509 – 1.527<br>⊖Δ = 0.191 – 0.199<br>diagr. p. 39 |

| Angle and Dispersion of Optic Axes | Color in Thin Section Pleochroism | Chemical Reactions Special Remarks | Distinguishing Features of Similar Minerals | Occurrences | |
|---|---|---|---|---|---|
| | O: brownish red, E: yellowish red, O > E; in extremely fine scales greyish yellow | slightly soluble in HCl; red streak | goethite, lepidocrocite: lighter colored, more brownish; ilmenite: black streak; ore microscope | with Fe-ores, magmatic, pneumatolytic and contact-metam., in low and medium grade schists, in sediments | 77 |
| | opaque, in very fine scales dark brown, E > O | slightly soluble in HCl; soluble in $KHSO_4$-melt; alteration to leucoxene | hematite: more red; pyrophanite, geikielite: X-ray, ore microscope | subordinate in nearly all magmatites, in sands and placers | 78 |
| | opaque, reflection brownish, yellowish, whitish | forms reaction rims around Ti-minerals | X-ray | pseudomorphous after ilmenite, sphene, perovskite, dysanalyte | 78a |
| | opaque, in thin scales deep red to yellowish red, O ~ E | slightly soluble in hot HCl | hematite, ilmenite, geikielite: X-ray, ore microscope | in Mn-skarn deposits | 79 |
| | opaque, in fine scales red-brown or purple, E > O weak | slightly soluble in hot HCl | see pyrophanite | in contact-metam. dolomit. marbles; with rutile in placers | 80 |
| | A / B<br>O: pale red / deep brown<br>E: pale orange / yellow-brown<br>O > E | decomposed by hot conc. acids to white isotr. subst.; replacement of spinel | biotite: sm. n; orthite, gadolinite: gr. 2V; cassiterite: gr. Δ | with spinel, corundum, titanomagnetite, mainly in hornfelses and emery rocks | 80a |
| sometimes abnormal: $2V_Z \sim 0^o - 10^o$ | colorless | soluble in HF without residue; no alteration products; liquid inclusions; phenocrysts often corroded | oligoclase, cordierite: residue if solved in HF, 2V; sanidine: sm. n, sm. Δ, 2V | common mineral under most p-T-conditions; not with foids; rarely primary with olivine | 81 |
| $2V_Z = 64^o$ horizontal disp. r < v weak | colorless | slightly soluble in HF | among other $SiO_2$-modifications only stishovite has gr. n; X-ray | with stishovite and lechatelierite in shock-metamorphosed quartzose rocks of meteor craters | 81a |
| | colorless | insoluble in HF | other $SiO_2$-mod.: sm. n, sm. Δ; X-ray | see Nr. 81a coesite | 81b |
| may be abnormal: $2V_Z \sim 0^o - 25^o$ | colorless, milky-blue, conspicuous banding | transitions to cristobalite-opal: X-ray | lussatine, lussatite, natrolite: sm. n; natrolite, gypsum: sm. H and different chem. reaction | hydrothermal in veins and vesicles: matter of silicification in magmatites and sediments; poss. with quartz and zeolites | 82 |
| $2V_Z = (35^o) - 70^o - (90^o)$ | colorless | soluble in HF, cold KOH-solution, hot $Na_2CO_3$-solution; crystals have often corroded rims | cristobalite, analcite, leucite: isometric outlines; zeolites: gr. n, gr. Δ; nepheline: gr. n, uniaxial $\ominus$; tridymites: X-ray | in place of quartz in acid volcanics; pneumat. to exhalat. in miarolitic cavities and vesicles; in sandstone xenoliths of basalts | 83 |
| may be abnormal: 2V small | colorless | soluble in HCl; after heating to red heat black color with $AgNO_3$ due to $Ag_2O$; subnormal interference colors | muscovite, talc: $2V_X$; alunite: diff. parag.; gypsum, sheridamite: $2V_Z$; gibbsite, hydromagnesite: sm. n, normal interference colors | hydrothermal alteration product of periclase in marbles; metam.-hydrothermal in joints of serpentine; rarely in low to medium-grade schists | 84 |
| $n_{E'(10\bar{1}1)} = 1.566$ abnormal: $2V_X$ up to $25^o (50^o)$ | colorless | soluble in acetic acid, effervesces in acids; slow reaction with $Co(NO_3)_2$; soluble in alum solution; ◊ lam. frequent | staining reactions see Vol. 2; other carbonates: not as easily soluble; aragonite: quicker staining with $Co(NO_3)_2$ | rarely magmatic; mainly biogenic, -sedimentary; in metam. together with CaMgFe-silicates; hydrothermal with quartz | 85 |
| $n_{E'(10\bar{1}1)} = 1.599$ | colorless rarely brownish (E < O) | almost insoluble in cold acids, readily soluble in hot HCl with effervescence | calcite, dolomite: sm. n, ◊ lam. more frequent; siderite: gr. n; staining methods: see Vol. 2 | metasomatic in limestones; metam. in talc-chlorite schists, in salt-rich pelites | 86 |

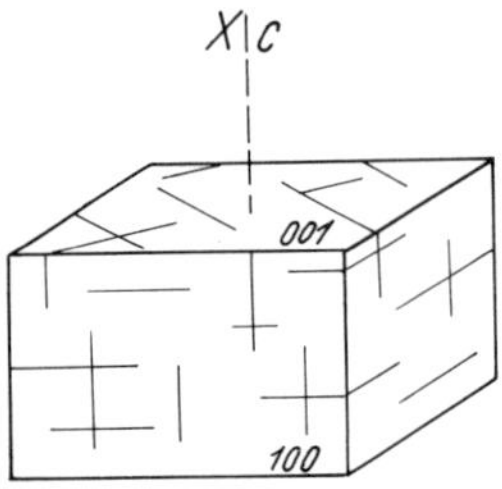

71-1. Gehlenite

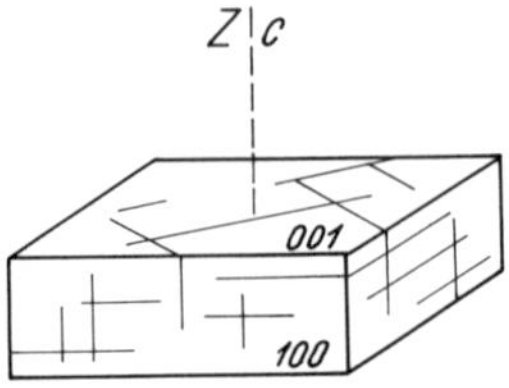

71-2. Åkermanite

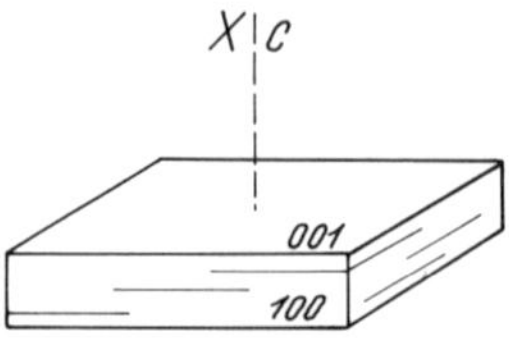

72. Melinophane

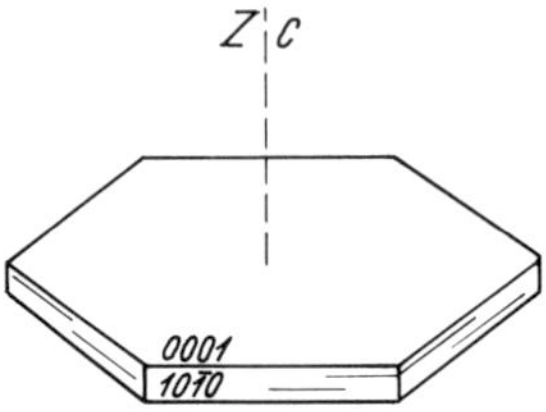

75. Koenenite

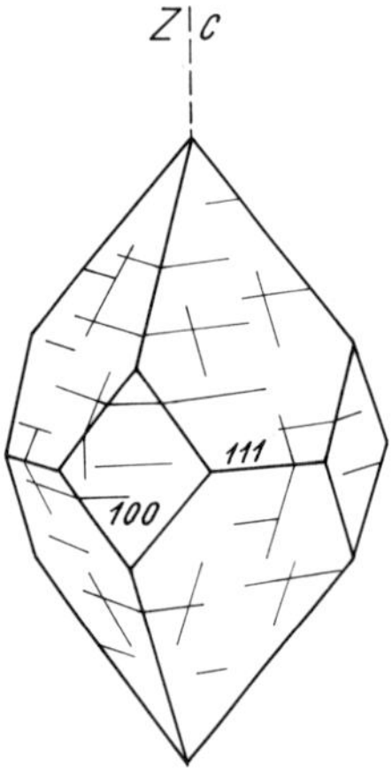

73. Apophyllite

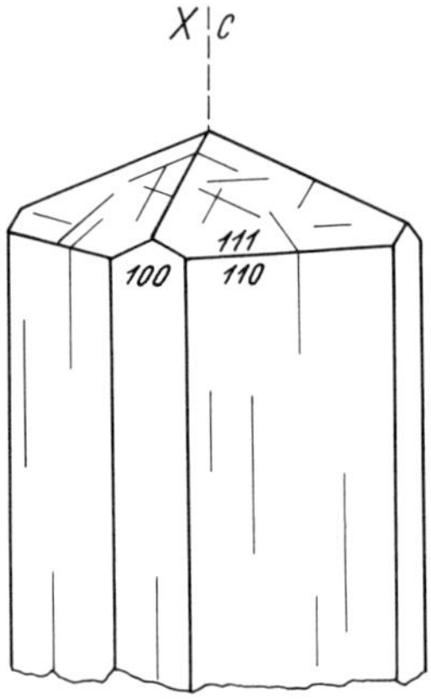

74-1. Scapolite

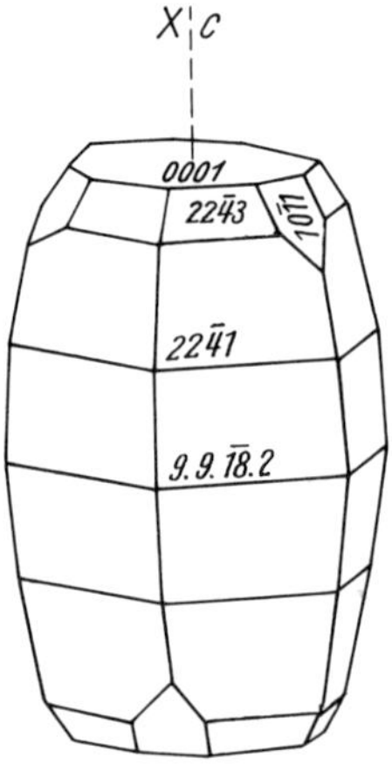

76. Corundum

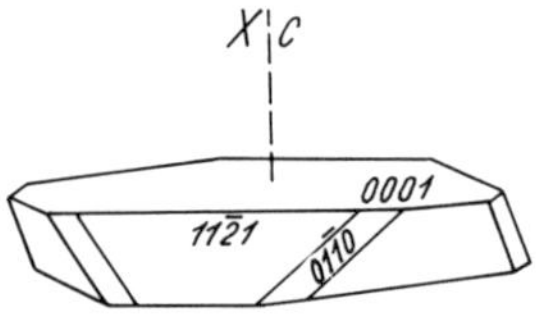

77. Hematite

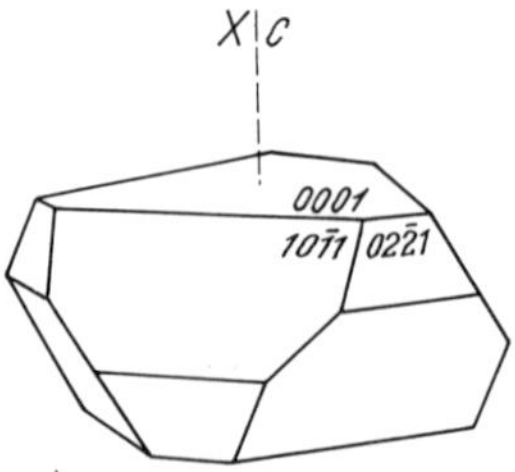

78. Ilmenite

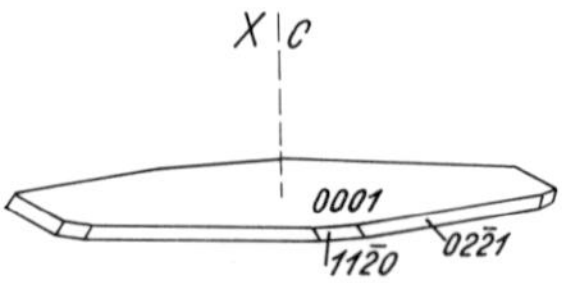

79. Pyrophanite

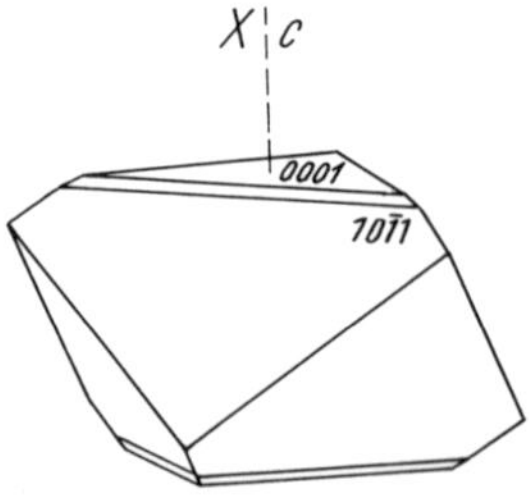

80. Geikielite

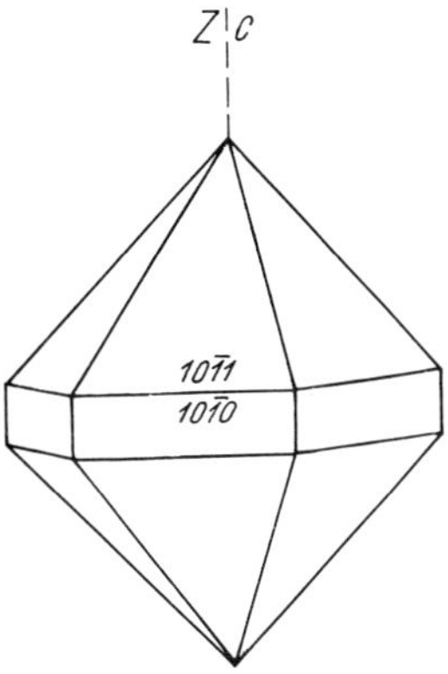


81-1. Morphology of High Quartz

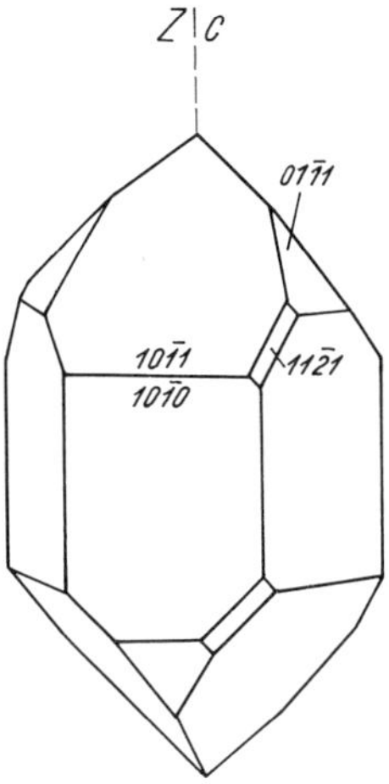


81-2. Low Quartz

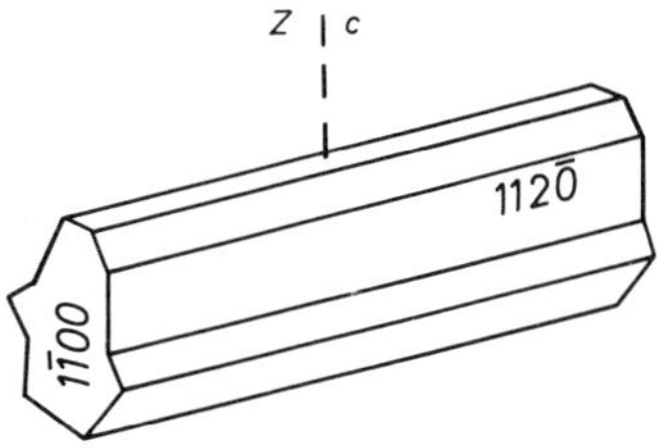


82–1. Chalcedony (sensu strictu)

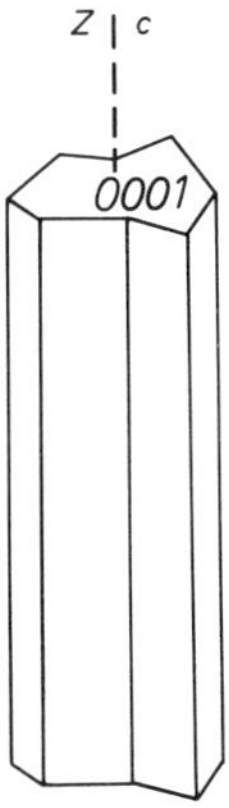


82-2. Quartzine

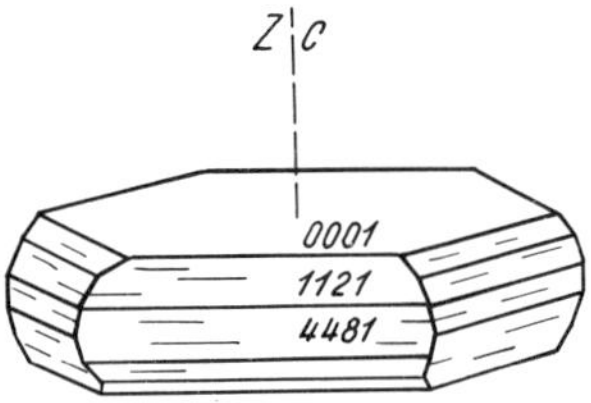


84. Brucite

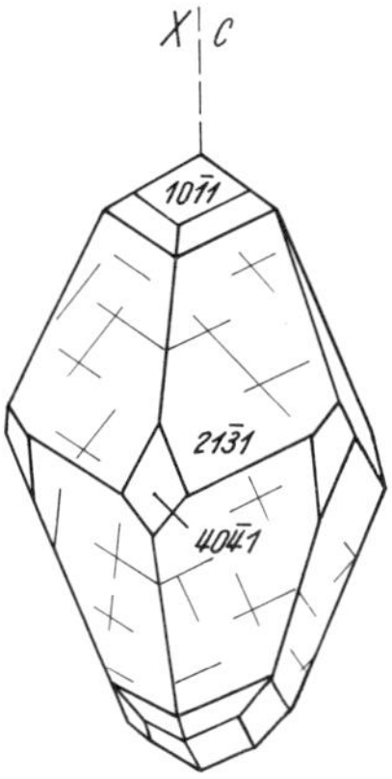


85-1. Calcite

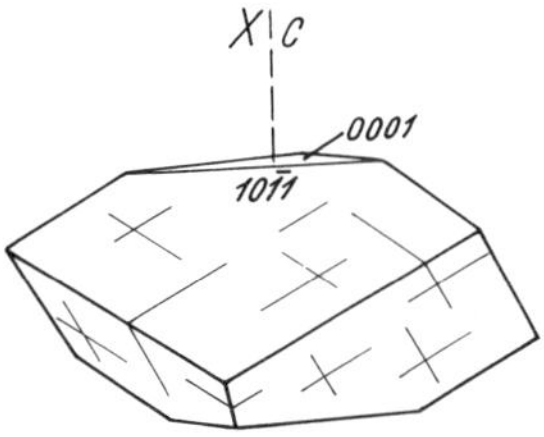


86-87. Magnesite, Siderite

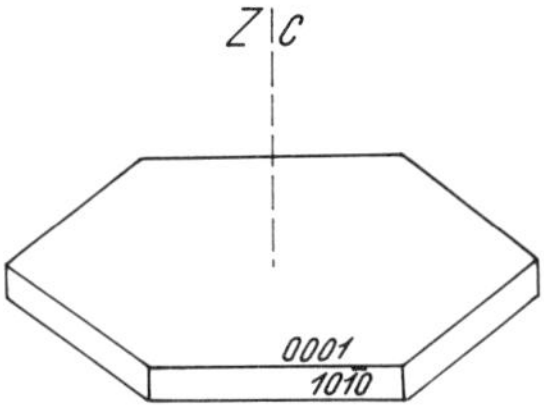


83. Morphology of High Tridymite

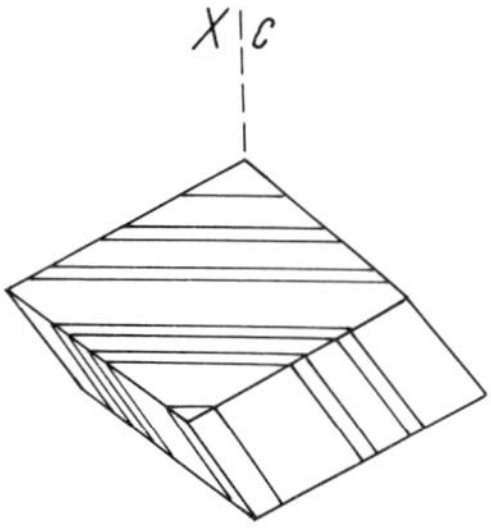


85-2. Calcite
#-rhombohedron $\{10\bar{1}1\}$; ◊ $(01\bar{1}2)$

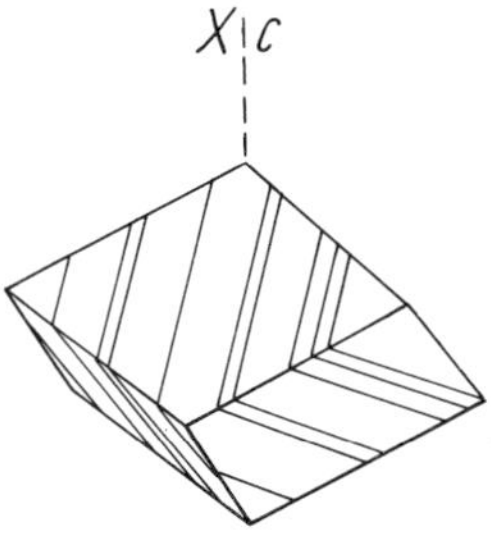


88. Dolomite
#-rhombohedron $\{10\bar{1}1\}$; ◊ $(02\bar{2}1)$

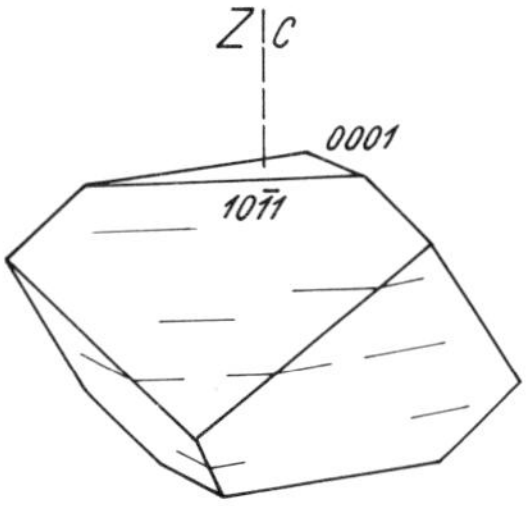


89. Alunite

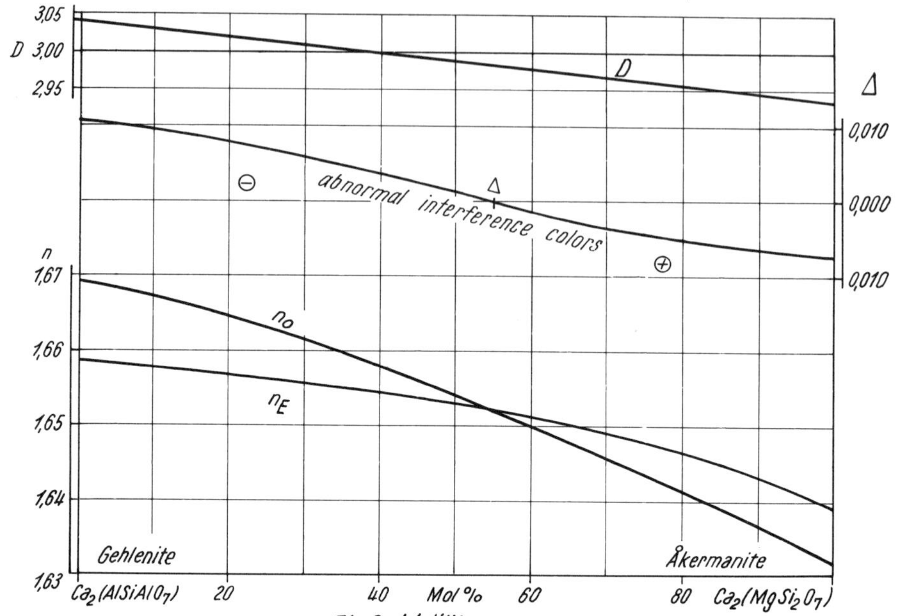


71-3. Melilite group

Na - and Fe - free

**WINCHELL 1927(Opt.Min.,267)**

(Copyrighted, 1951, by A.N. Winchell in Elements of Optical Mineralogy, pt. II; Reproduced by permission.)

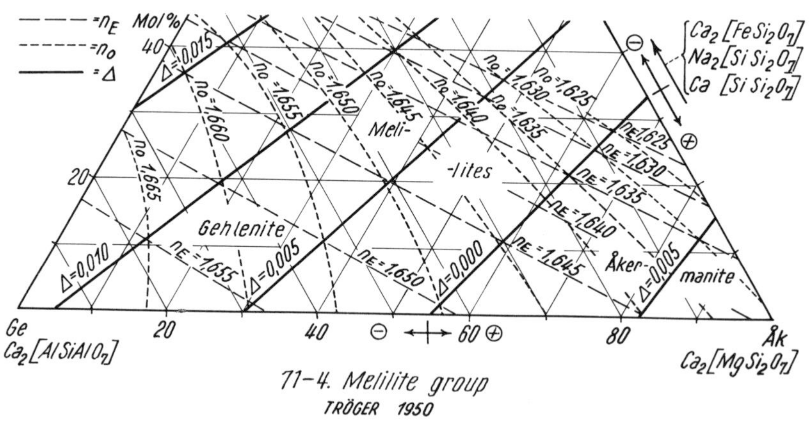


71-4. Melilite group

TRÖGER 1950

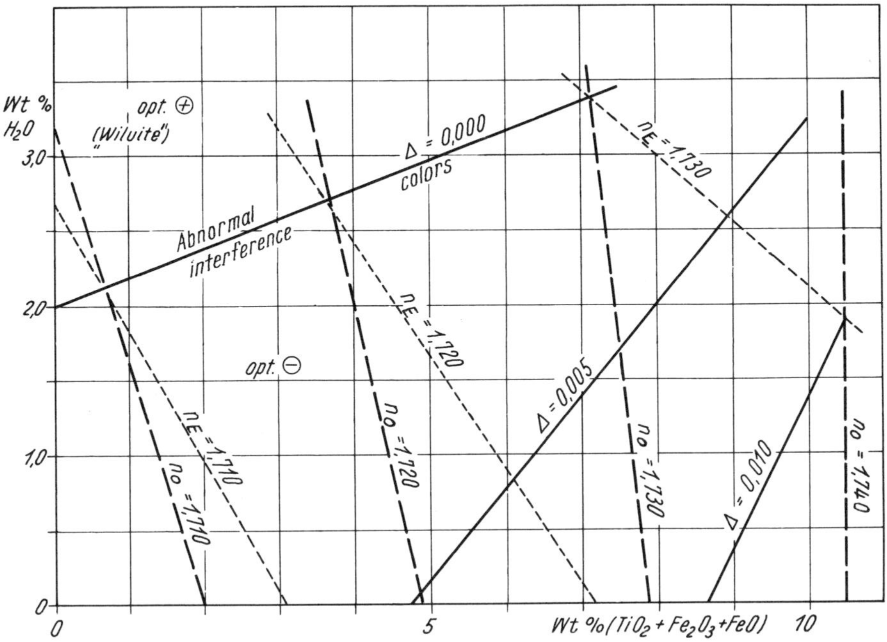


70-2. Vesuvianite
TRÖGER 1950

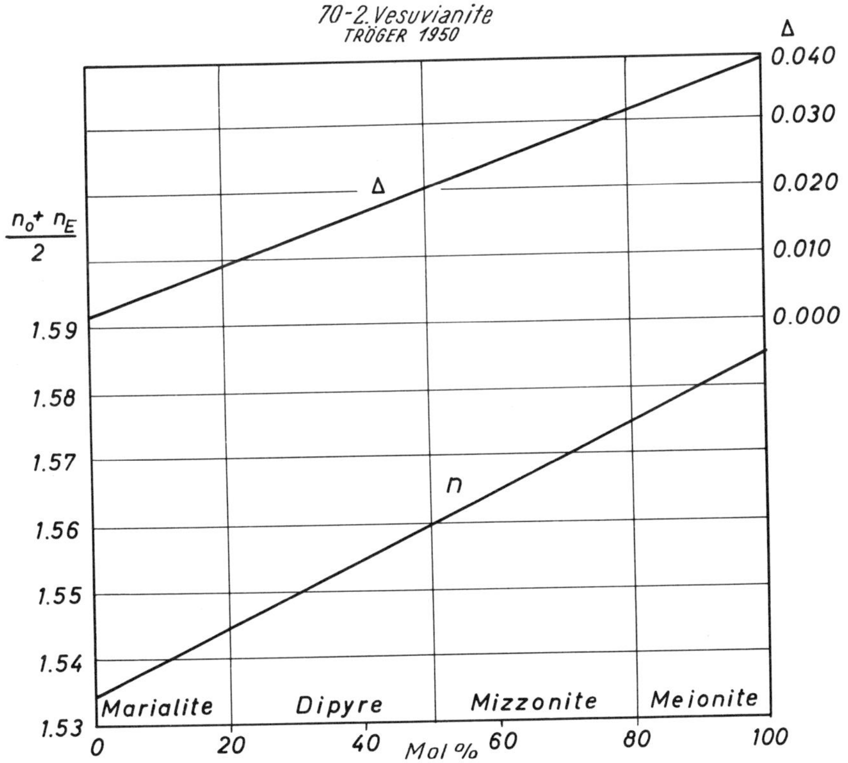


74 – 2. Scapolite group
SHAW 1960

(22)

| No. | Name<br>Symmetry<br>Chem. Composition | Common Crystal Forms, Twinning<br>Crystal Habit<br>Aggregates | Cleavage<br>Translation Gliding<br>Deform. Twinning<br>Hardness, Density | Optic<br>Orientation | Refractive Indices<br>Birefringence |
|---|---|---|---|---|---|
| 87 | Siderite trig<br>$(Fe,Mn)CO_3$ | $\{10\bar{1}1\}$, {0001};<br>◈ $(01\bar{1}2)$ lam;<br>grains, rhombohedra, spherulites, oolites | # $\{10\bar{1}1\}$ perf. 73°;<br>T = (0001);<br>H = 4–4½ D = 3.89± | X ∥ c E = X<br>fig. p. 33 | Mg-rich pure<br>$n_O$ = 1.851 – 1.875<br>$n_E$ = 1.612 – 1.633<br>⊖Δ = 0.239 – 0.242<br>diagr. p. 39 |
| 88 | Dolomite trig<br>$CaMg[CO_3]_2$ | $\{10\bar{1}1\}$, {0001};<br>◈ $(02\bar{2}1)$ lam.;<br>grains, rhombohedra | # $\{10\bar{1}1\}$ perf. 73½°<br>T = (0001);<br>H = 3½–4 D = 2.87± | X ∥ c E = X<br>fig. p. 33 | pure Fe-rich<br>$n_O$ = 1.679 – 1.698<br>$n_E$ = 1.502 – 1.513<br>⊖Δ = 0.177 – 0.185<br>diagr. p. 39 |
| 89 | Alunite trig<br>$KAl_3[(OH)_6(SO_4)_2]$ | $\{10\bar{1}1\}$, {0001};<br>granular, flaky, fibrous;<br>fig. p. 33 | # (0001) perf.;<br>H = 3½–4 D = 2.71± | Z ∥ c E = Z<br>l = (–)<br>fig. p. 33 | $n_O$ = 1.572<br>$n_E$ = 1.592<br>⊕Δ = 0.020 |
| 90 | Apatite hex<br>$Ca_5[(F,Cl,OH)(PO_4)_3]$ | $\{10\bar{1}0\}$, {0001}, $\{10\bar{1}1\}$;<br>needles, columns, grains;<br>"phosphorite": radiated, cryptocrystalline | # (0001) poor,<br># $\{10\bar{1}0\}$ poor;<br>brittle;<br>H = 5 D = 3.18± | X ∥ c E = X<br>l = (–) fig. p. 38<br>may have abnormally birefringent "panelling" | F OH Cl<br>$n_O$ = 1.634 – 1.651 1.667<br>$n_E$ = 1.631 – 1.646 1.665<br>⊖Δ = 0.003 – 0.005 0.002<br>diagr. p. 40 |
| 91 | Tourmaline trig<br>$Na(Mg,Fe,Mn,Li)_3Al_6$<br>$[(OH)_4 \mid (BO_3)_3$<br>$Si_6O_{18}]$ diagr. p. 39 | $\{11\bar{2}0\}$, $\{10\bar{1}0\}$, $\{10\bar{1}1\}$,<br>$\{01\bar{1}2\}$, $\{01\bar{1}0\}$, $\{02\bar{2}1\}$;<br>columnar, stellate, also short columns | brittle, no #;<br>parting ∥ (0001)<br>D = { 3.06± elbaite, 3.05± dravite, 3.16± schorlite,<br>H = 7 | X ∥ c E = X<br>l = (–)<br>fig. p. 38 | $n_O$ = 1.639 – 1.692<br>$n_E$ = 1.620 – 1.657<br>⊖Δ = 0.019 – 0.035<br>diagr. p. 39 |
| 91a | Phenakite trig<br>$Be_2[SiO_4]$ | $\{11\bar{2}1\}$;<br>◈ $(11\bar{2}0)$;<br>prismatic grains | # $\{11\bar{2}0\}$ good;<br>H = 7½–8 D = 2.98± | Z ∥ c E = Z<br>l = (+) | $n_O$ = 1.654 – 1.671<br>$n_E$ = 1.670 – 1.696<br>⊕Δ = 0.016 – 0.025 |
| 92 | Eudialyte trig<br>$(Na,Ca,Fe)_6 Zr$<br>$[(OH,Cl)(Si_3O_9)_2]$ | {0001}, $\{10\bar{1}1\}$, $\{11\bar{2}0\}$;<br>massive, granular, tabular | # (0001) distinct;<br>H = 5–6 D = 2.93± | Z ∥ c, E = Z "eudialyte"<br>Z ~ X "mesodialyte",<br>X ∥ c, E = X "eucolite";<br>zonal- and hour-glass structure<br>fig. p. 38 | eud. mes. euc.<br>$n_O$ = 1.593–1.607–1.643<br>$n_E$ = 1.597–1.607–1.634<br>Δ = 0.004–0.000–0.009<br>⊕ ⊖ |
| 93 | Catapleite mcl<br>pseudohex<br>$Na_2Zr[Si_3O_9] \cdot 2H_2O$ | {0001}, $\{h0\bar{h}l\}$;<br>◈ $(10\bar{1}0)$;<br>mimetic triplets, tabular (0001) | # $\{10\bar{1}0\}$ perf.,<br># $\{11\bar{2}0\}$ poor;<br>H = 6 D = 2.77± | between 20° – 200°C!<br>paramorphous: mcl domains<br>Z Λ c = 3½°, bi;<br>Y ∥ or ⊥ *a*, O.A.P. ∥ or ⊥ $(10\bar{1}0)$ fig. p. 38 | $n_X$ = 1.5905<br>$n_Y$ = 1.5921<br>$n_Z$ = 1.6269<br>⊕Δ = 0.0364 |
| 94 | Beryl hex<br>$Al_2Be_3[Si_6O_{18}]$ | $\{10\bar{1}0\}$, {0001}, $\{h0\bar{h}l\}$;<br>columnar, granular | # (0001) poor;<br>brittle;<br>H = 7½–8 D = 2.70± | X ∥ c E = X<br>l = (–)<br>fig. p. 38 | $n_O$ = 1.567 – 1.594<br>$n_E$ = 1.563 – 1.586<br>⊖Δ = 0.004 – 0.008<br>diagr. p. 41 |
| 95 | Nepheline hex<br>$(Na,K)[AlSiO_4]$ | $\{10\bar{1}0\}$, {0001};<br>~ isometric as phenocryst, in the groundmass in interstices | # $\{10\bar{1}0\}$ distinct,<br># (0001) distinct;<br>T = $(10\bar{1}2)$;<br>H = 5½–6 D = 2.62± | X ∥ c E = X<br>may have abnormally birefringent patches<br>fig. p. 38 | $n_O$ = 1.532 – 1.547<br>$n_E$ = 1.529 – 1.542<br>⊖Δ = 0.003 – 0.005<br>diagr. p. 40 |
| 96 | Kaliophilite hex<br>$K[AlSiO_4]$ | {0001}, $\{10\bar{1}0\}$;<br>short columns | # $\{10\bar{1}0\}$ distinct,<br># (0001) perf.;<br>H = 6 D = 2.60 | X ∥ c E = X<br>fig. p. 38 | pure Na-rich<br>$n_O$ = 1.532 – 1.537<br>$n_E$ = 1.527 – 1.533<br>⊖Δ = 0.005 – 0.004 |
| 97 | Cancrinite hex<br>$(Na_2Ca)_4[(CO_3)$<br>$(H_2O)_{0-3}(AlSiO_4)_6]$<br><$CO_3$ = Cancrinite<br><$SO_4$ = Wischnewite | $\{10\bar{1}0\}$, $\{10\bar{1}1\}$;<br>columnar, scaly | # $\{10\bar{1}0\}$ perf.,<br># {0001} perf.;<br>H = 5–6 D = 2.40± | X ∥ c E = X<br>l = (–)<br>fig. p. 38 | cancr. wischn.<br>$n_O$ = 1.524 – 1.4890<br>$n_E$ = 1.498 – 1.4885<br>⊖Δ = 0.026 – 0.0005<br>diagr. p. 41 |
| 98 | Davyne hex<br>$(K,Na)_6Ca_2$<br>$[(SO_4)_2(AlSiO_4)_6]$ | $\{10\bar{1}0\}$, $\{10\bar{1}1\}$, {0001}<br>$\{11\bar{2}0\}$;<br>acicular, scaly | # $\{10\bar{1}0\}$ perf.,<br># (0001) good;<br>H = 5½ D = 2.4± | Z ∥ c E = Z<br>(rarely: X ∥ c, E = X: potassium-rich);<br>l = (+) (–) fig. p. 38 | Na K<br>$n_O$ = 1.522 – 1.518<br>$n_E$ = 1.527 – 1.517<br>⊕Δ = 0.005 – ⊖Δ 0.001 |
| 99 | Hydronephelite<br>pseudohex<br>(?) uncertain alteration product | no crystal-faces;<br>fibrous, columnar, scaly, granular | | | n > 1.54<br>Δ ~ 0.013 |

| Angle and Dispersion of Optic Axes | Color in Thin Section Pleochroism | Chemical Reactions Special Remarks | Distinguishing Features of Similar Minerals | Occurrences | |
|---|---|---|---|---|---|
| $n_{E'(10\bar{1}1)}$ = 1.748 | colorless<br>ash-grey to brownish<br>E < O | soluble in hot HCl with effervescence; black and magnetic after heating >450°; insoluble in alum solution | cassiterite: sm. Δ, opt. ⊕; sphene: biaxial; other carbonates: < $n_{E'}$ | pegmatitic to hydrothermal (Mn-content); metasomatic in limestones, metamorphic in gneiss and schists | 87 |
| $n_{E'(10\bar{1}1)}$ = 1.587 | colorless<br>rarely grey, brownish<br>(E < O) | slightly soluble in cold acids with effervescence; not stainable with Lemberg's solution; more turbid than calcite; ◈ lam. | calcite: sm. n and different position of ◈ lam; siderite: gr. n; magnesite: no ◈ lam. | metasomatic-diagenetical from limestone, hydrothermal with quartz | 88 |
| | colorless | soluble in $H_2SO_4$ and KOH-solution, not sol. in HF or HCl; turbid after heating >520° | muscovite: l = (+); cancrinite: sm. n; brucite: different paragenesis | with diaspore and quartz hydrotherm. from feldspars and other K-minerals | 89 |
| may be abnormal:<br>weakly biaxial<br>$2V_X$ < 10°<br>v > r<br>Cl = mcl<br>$2V_Z$ = 20° – 25° | colorless, brownish, reddish; grey or black due to inclusions<br>E > O, rarely E < O | soluble in HCl, not in acetic acid; rarely pleochroic haloes | gehlenite: tetr. outline; zoisite: gr. n; quartz, nepheline, beryl: sm. n; phenakite: ⊕ and gr. Δ; pure Cl-apatite: $n_O$ = 1.666 | early magmatic precipitation in all types of rocks; hydrothermal; biogenic; in sediments | 90 |
| may be abnormal:<br>weakly biaxial | elbaite: alm. colorless<br>dravite: pale yellow, light brown; schorlite: olive-brown, bluish grey with zonal coloring, always O ≫ E | not attacked by HF; often pleochroic haloes; often zonal coloring | apatite: E > O weak; riebeckite: sm. Δ; aegirine: gr. n and gr. Δ; corundum: gr. n | pneumatolytic in granites and pegmatites; rarely in volcanic rocks; rarely magmatic: metamorphic; in sands and sandstones | 91 |
| | colorless | not attacked by acids | apatite: sm. Δ, ⊖; colorless tourmaline: l = (–) | pneumatol. to hydroth. in granites, gneisses and mica schists with other Be-minerals | 91a |
| often abnormal:<br>$2V_Z$ = 0° – 30°<br>in the same grain | pale bluish red to pale yellowish brown,<br>usually O > E weak | gelatinizes with HCl, then stainable; often abnormally birefringent domains | garnet: gr. n; melilite and vesuvianite: abnormal interference colors | magmatic in foyaites and tinguaites; late product in foyaite-pegmatites | 92 |
| $2V_Z$ = 5° – 60° diminishes upon heating (>140°) down to 2V = 0°,<br>r < v weak | light yellow to brownish (Ca-rich), colorless to brownish (Ca-poor) | easily soluble in HCl, gelatinizes; stainable; usually filled with inclusions | chondrodite: different paragenesis; pectolite: acicular, stellate | pegmatitic with other Zr-minerals in alkaline provinces; associated with syenites, foyaites | 93 |
| often abnormal:<br>$2V_X$ = 0° – 6° | usually colorless, rarely pale yellow or greenish, then E > O,<br>rarely E < O | not attacked by acids, even by HF | quartz: sm. n and gr. Δ, ⊕; apatite: gr. n; nepheline: sm. n; topaz: biaxial | with tourmaline and Li-micas in pegmatites of acid magmatic rocks and of metamorphic schists | 94 |
| may be abnormal:<br>$2V_X$ = 0° – 6° | colorless | gelatinizes with HCl, then stainable with aniline colors; dries to form NaCl-cubes; phenocrysts may have zonal structure | quartz: gr. Δ, ⊕; sanidine: oblique extinction; kaliophilite, kalsilite } K-staining | with other foids; magmatic to pegmatitic in sodic rocks; not together with quartz | 95 |
| | colorless | gelatinizes with HCl, then stainable with aniline colors | kalsilite (hex.):<br>D = 2.59 – 2.63<br>$n_O$ = 1.542 – 1.543<br>$n_E$ = 1.537 – 1.537<br>⊖Δ = 0.005 – 0.006<br>see Vol. 2 | pneumatolytic in potassic volcanics | 96 |
| may be abnormal:<br>2V small | colorless | effervesces and gelatinizes with HCl, then stainable with aniline colors; turbid upon heating to red heat; wischnewite: $SO_3$-reaction | davyne: usually ⊕; muscovite: gr. n, l = (+); nepheline: gr. n, sm. Δ<br>see Vol. 2 | magmatic to autopneumat. with nepheline in leucocratic alkaline rocks | 97 |
| | colorless | gelatinizes with HCl, then stainable with aniline colors | cancrinite: always ⊖<br>see Vol. 2 | autopneumatom. in sanidinites; in calcite rocks and in sarkolite rocks | 98 |
| $2V_X$ ~ 70° – 80° ? | colorless | gelatinizes with HCl, then stainable with aniline colors | natrolite: opt. biaxial | alteration product of nepheline | 99 |

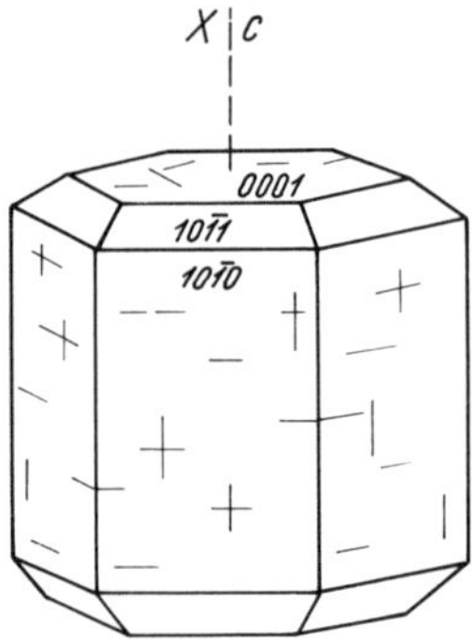


90-1. Apatite

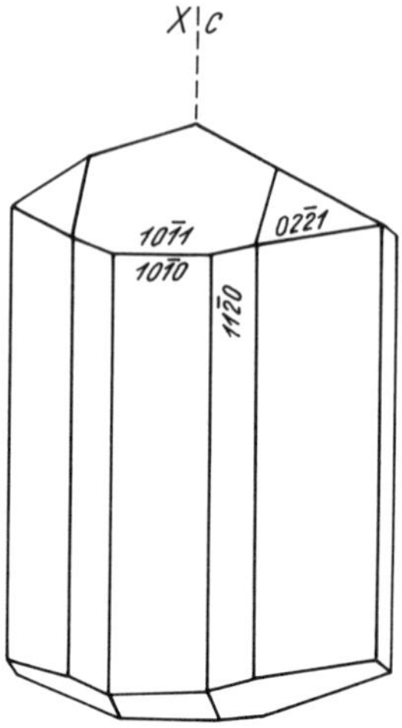


91-1. Tourmaline

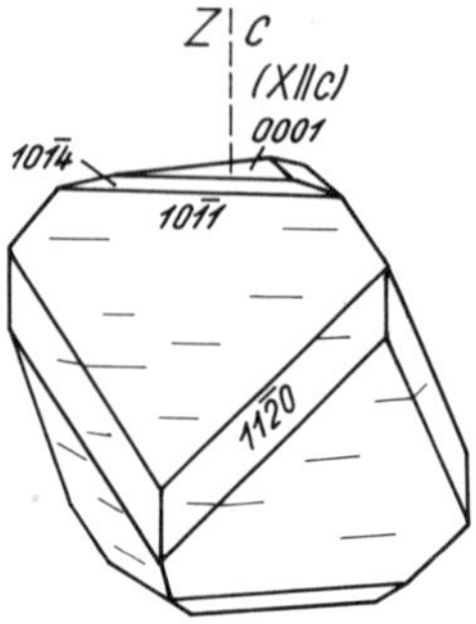


92. Eudialyte

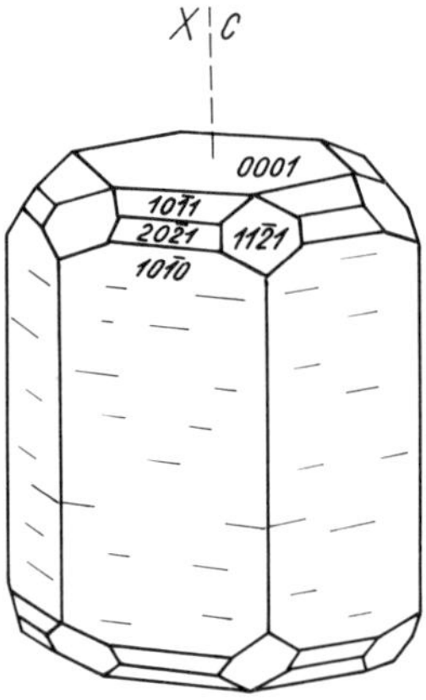


94-1. Beryl

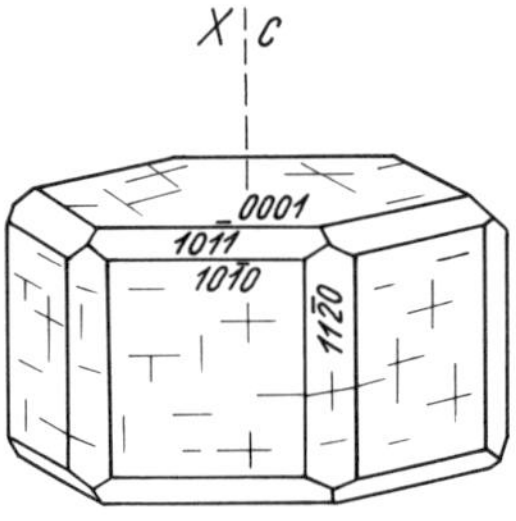


95-1.-96. Nephelite, Kaliophilite

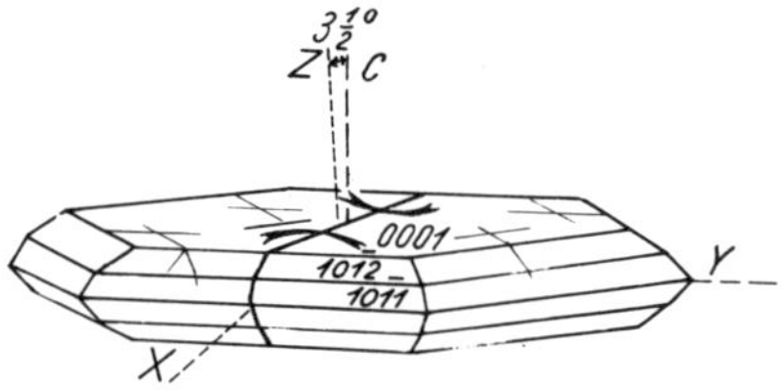


93. Katapleite

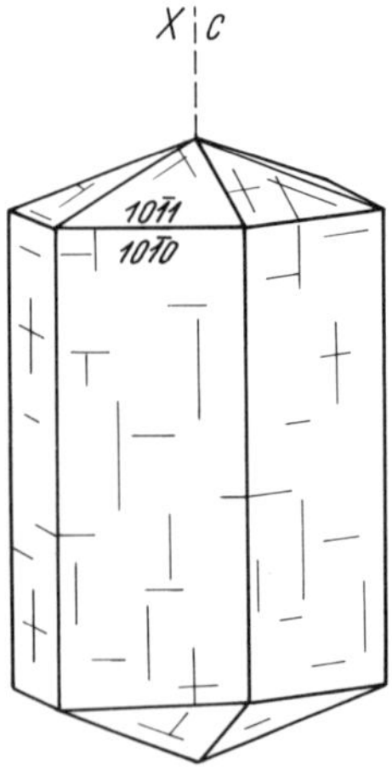


97-1. Cancrinite

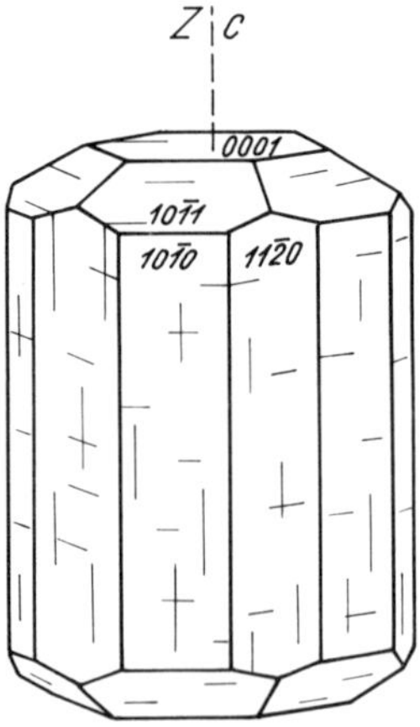


98-1. Davyne

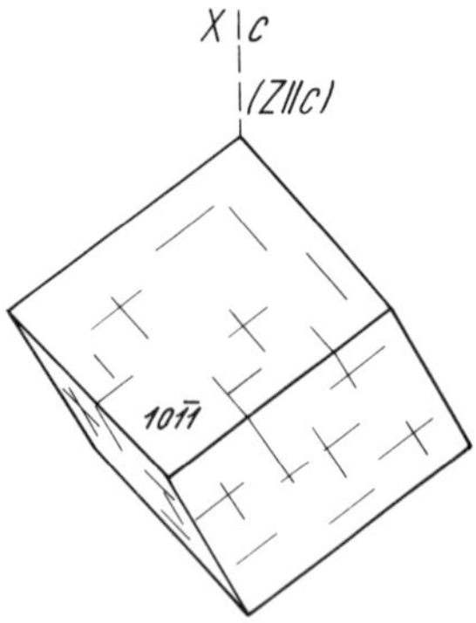


100. Chabasite

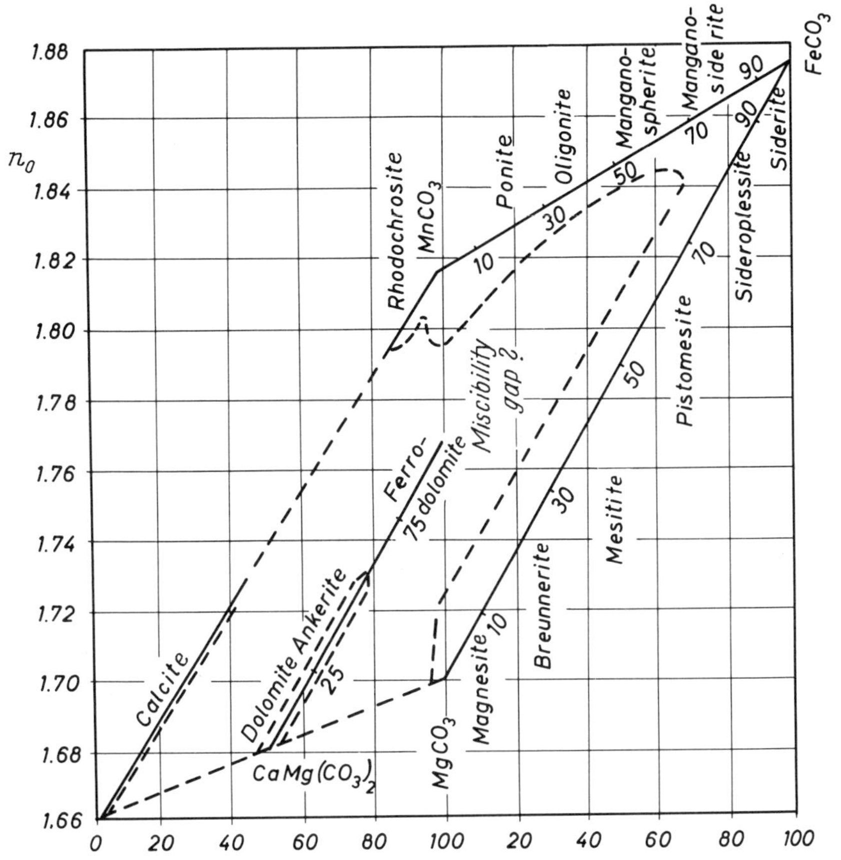


85-88. Calcite group

KENNEDY 1947, AMER. MINER. 32, 569. partly WINCHELL 1951, ELEMENT. OPT. MINER., supplemented

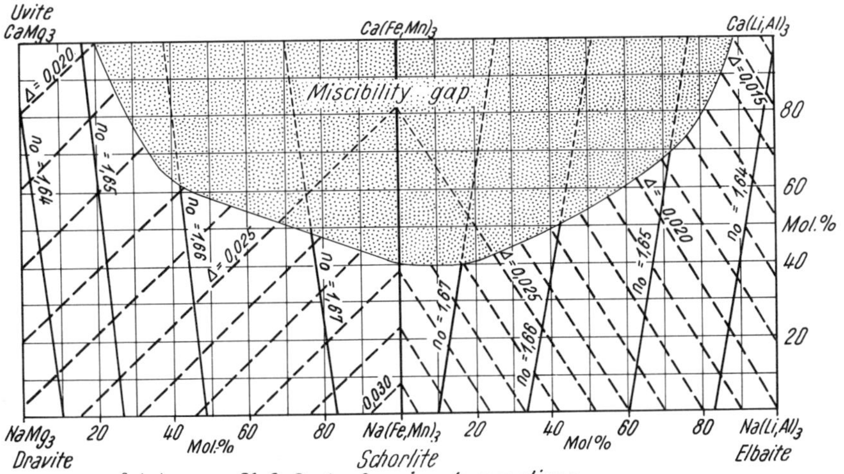


——— refr. index $n_o$
– – – $\Delta = n_E - n_o$

91-2. Rock-forming tourmalines

WINCHELL 1932 (Am. Min. 17, 475)

| | | | |
|---|---|---|---|
| 1 Wt %$H_2O$ | raises $n_o$ | by | 0.0051 |
| " Cl | " | | 0.0054 |
| " FeO | " | | 0.0040 |
| " MnO | " | | 0.0014 |
| " $(R.E.)_2O_3$ | " | | 0.0018 |
| " $SO_3$ | " | | 0.0003 |
| " SrO | " | | 0.0001 |

*after* YOUNG 1964

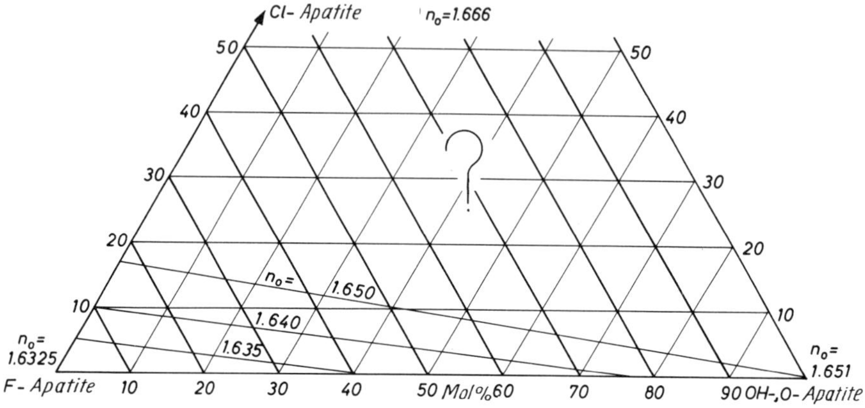


90-2. *Apatite*

TABORSZKY 1971

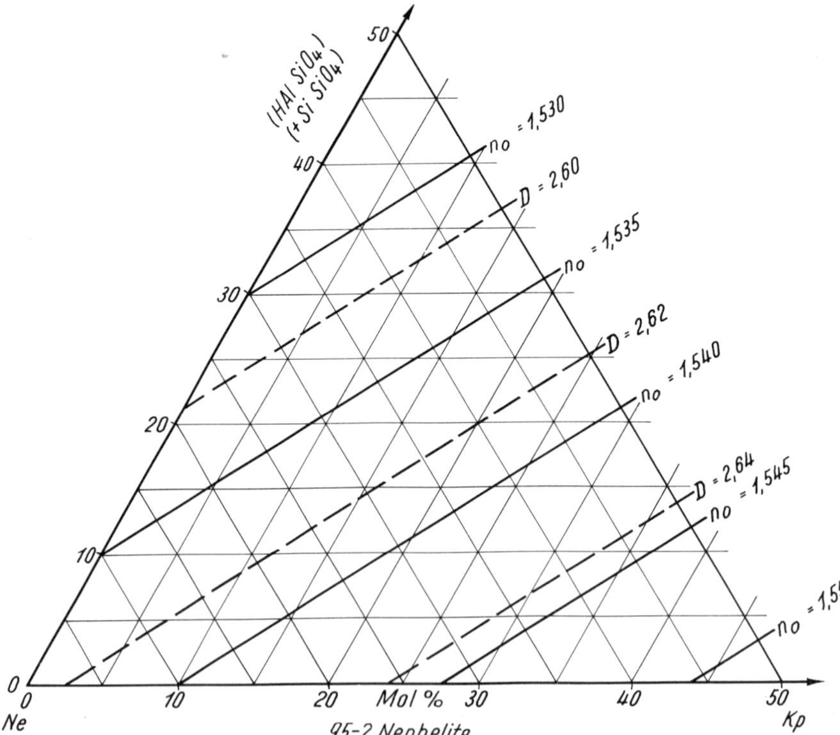


*95-2. Nephelite*

WINCHELL 1941(Am.Min.26,540)

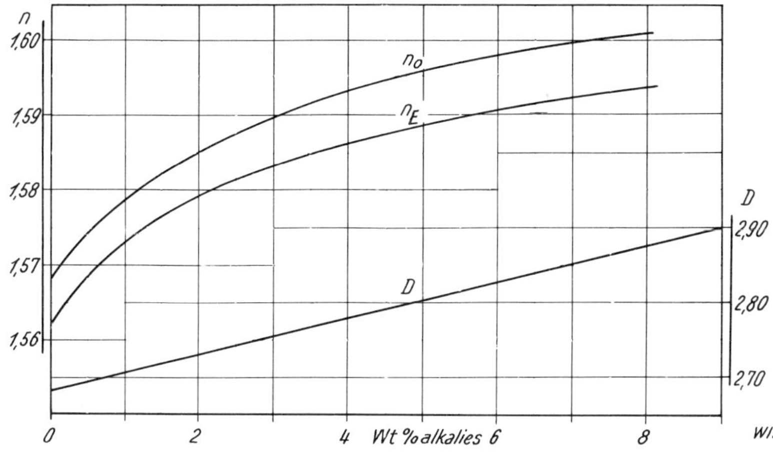


94-2. Beryl
WINCHELL 1927 (Opt.Min., 272)

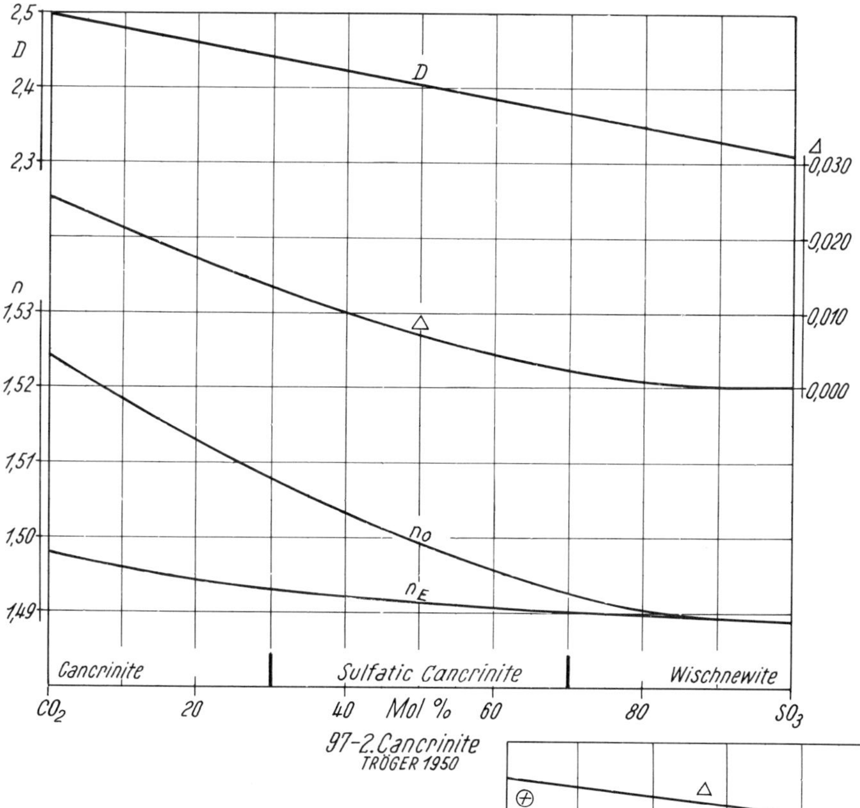


97-2. Cancrinite
TRÖGER 1950

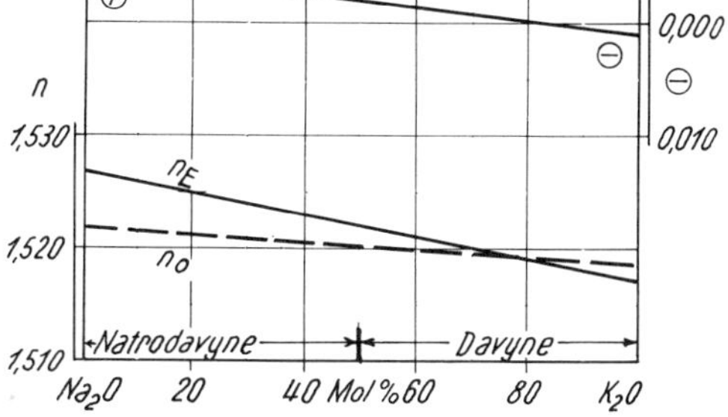


98-2. Davyne
TRÖGER 1950

| No. | Name<br>Symmetry<br>Chem. Composition | Common Crystal Forms, Twinning<br>Crystal Habit<br>Aggregates | Cleavage<br>Translation Gliding<br>Deform. Twinning<br>Hardness, Density | Optic<br>Orientation | Refractive Indices<br>Birefringence |
|---|---|---|---|---|---|
| 100 | Chabazite trig<br>(Ca,$Na_2$)[$Al_2Si_4O_{12}$]·$6H_2O$ | {10$\bar{1}$1}, {0001};<br>◇ (0001);<br>rhombohedra<br>(pseudocubes) | # {10$\bar{1}$1} good;<br>H = 4–5 D = 2.10± | X ∥ c E = X<br>(rarely Z ∥ c E = Z)<br>often abnormally birefringent "panelling" in 6 triclinic sectors;<br>fig. p. 38 | $n_O$ = 1.480 – 1.485<br>$n_E$ = 1.478 – 1.490<br>⊕Δ = 0.002 – 0.005 |
| 100b | Gmelinite hex<br>(Ca,$K_2$,$Na_2$)<br>[$Al_2Si_4O_{12}$]·$6H_2O$ | {11$\bar{2}$1};<br>◇ (0001), (10$\bar{1}$1);<br>prismatic, tabular | # {1010} good;<br>H = 4½ D = 2.10± | X ∥ c E = X | $n_O$ = 1.476 – 1.494<br>$n_E$ = 1.474 – 1.480<br>⊖Δ = 0.002 – 0.014 |
| 100c | Levynite trig<br>(Ca,$Na_2$,$K_2$)<br>[$Al_2Si_4O_{12}$]·$6H_2O$ | {0001}, {10$\bar{1}$1};<br>penetration twins | H = 4 D = 2.1 | X ∥ c E = X | $n_O$ = 1.499<br>$n_E$ = 1.505<br>⊖Δ = 0.006 |

## Optically Biaxial Minerals:

| No. | Name<br>Symmetry<br>Chem. Composition | Common Crystal Forms, Twinning<br>Crystal Habit<br>Aggregates | Cleavage<br>Translation Gliding<br>Deform. Twinning<br>Hardness, Density | Optic<br>Orientation | Refractive Indices<br>Birefringence |
|---|---|---|---|---|---|
| 101 | Chrysoberyl orh<br>$Al_2BeO_4$ | {100}, {010}, {011}, {120};<br>◇ and pseudohex. triplets = {031};<br>(100) striated ∥ c;<br>granular, tabular ∥ (100) | # {011} distinct<br># (100) and (010) poor<br>H = 8½ D = 3.72± | X ∥ a<br>Y ∥ b, O.A.P. (010)<br>Z ∥ c, bi;<br>fig. p. 46 | $n_X$ = 1.744 – 1.747<br>$n_Y$ = 1.747 – 1.749<br>$n_Z$ = 1.753 – 1.758<br>⊕Δ = 0.009 – 0.011 |
| 102 | Columbite group<br>(Niobite-Tantalite)<br>orh<br>(Fe,Mn) $(Nb,Ta)_2O_6$ | {100},{h01},{001}, {hk0};<br>◇ {021} and {023};<br>niobite: tabular ∥ (100), prismatic ∥ b; tantalite: columnar ∥ c, (100)<br>striated ∥ c; mass., stellate | # (100) distinct;<br>T = (001) ? ;<br>niobite:<br>H = 6–6½ D = 5.0<br>tantalite:<br>D = 7.9<br>diagr. p. 60 | X ∥ b<br>Y ∥ a, O.A.P. (100)<br>Z ∥ c, bi;<br>fig. p. 46 | nb ta<br>$n_X$ = 2.40 – 2.24<br>$n_Y$ = 2.45 – 2.30<br>$n_Z$ = 2.53 – 2.41<br>⊕Δ = 0.13 – 0.17 |
| 103 | Pseudobrookite orh<br>$Fe_2TiO_5$ | {100},{110},{103}, {010};<br>usually tabular on (100);<br>always idiomorphic | # (010) distinct;<br>H = 6 D = 4.39± | X ∥ b<br>Y ∥ c, O.A.P. (001)<br>Z ∥ a, bi;<br>fig. p. 46 | $n_X$ = 2.347 – 2.42 } Li<br>$n_Z$ = 2.375 – 2.47 }<br>⊕Δ = 0.028 – 0.05 |
| 104 | Brookite orh<br>$TiO_2$ | {100},{110},{122};<br>tabular ∥ (100) and usually elongated ∥ c;<br>[001] striated ∥ c;<br>{122} predominant<br>= arkansite | # {110} poor;<br>H = 5½–6 D = 4.14± | λ red: X ∥ b, Y ∥ c, O.A.P. (001), Z ∥ a, bi<br>λ blue: X ∥ c, Y ∥ b, O.A.P. (010), Z ∥ a, bi<br>rarely O.A.P. for all λ (001);<br>fig. p. 46 | $n_X$ = 2.583 – 2.583 }<br>$n_Y$ = 2.584 – 2.586 } Na<br>$n_Z$ = 2.700 – 2.741 }<br>⊕Δ = 0.117 – 0.158<br>disp. r > v, usually strong |
| 105 | Baddeleyite mcl<br>$ZrO_2$ | {100},{001},{h01};<br>{hk0}, {hkl}, β = 99½°;<br>◇ (100) and {110}, lam.;<br>tabular ∥ (100); usually elongated ∥ c; fibrous, divergent in "caldasite-favas" | # (001) good;<br># (010) poor,<br>{110} parting: 88½°;<br>H = 6½ D = 5.73± | X Λ c ≈ + 11° to + 13°, bi;<br>Y ∥ a ≈ –1° to –3°<br>fig. p. 46 | $n_X$ = 2.136<br>$n_Y$ = 2.236<br>$n_Z$ = 2.243<br>⊖Δ = 0.107 |
| 105a | Coesite mcl<br>$SiO_2$ | β = 120°;<br>◇ {021}, (100);<br>pseudohexagonal tabular (010), bladed ∥ (100), striated [001]; irregular, fine-grained | H = 7–8 D = 2.92 | X ∥ b<br>Y Λ a = 24° to 26°,<br>O.A.P. ⊥ (010)<br>Z Λ c = 4° to 6°, bi;<br>l = (+); character of extinction indeterminate due to β = 120° | $n_X$ = 1.594<br>$n_Y$ = 1.596<br>$n_Z$ = 1.599<br>⊕Δ = 0.005 |
| 106 | Goethite orh<br>α-FeOOH | {110},{hk0},{011};<br>acicular ∥ c, radiated or massive in colloform bodies; pseudomorphic | # (010) perf.,<br># (100) poor;<br>H = 5–5½ D = 4.28± | λ red: X ∥ b, bi, Y ∥ a, O.A.P. (100), Z ∥ c<br>λ green: X ∥ b, bi, Y ∥ c, O.A.P. (001), Z ∥ a<br>fig. p. 46 | pure $H_2O$-rich<br>$n_X$ = 2.275 – 2.15 }<br>$n_Y$ = 2.409 – 2.22 } Na<br>$n_Z$ = 2.415 – 2.23 }<br>⊖Δ = 0.140 – 0.08<br>changes with |
| 107 | Lepidocrocite orh<br>γ-FeOOH | {010};<br>tabular ∥ c, somewhat elongated; radiated, stellate | #(010) perf.,<br># (001) perf.,<br># (100) good;<br>H = 5 D = 4.09± | X ∥ b, bi;<br>Y ∥ c, O.A.P. (001)<br>Z ∥ a<br>fig. p. 46 | $n_X$ = 1.94 }<br>$n_Y$ = 2.20 } Na<br>$n_Z$ = 2.51 }<br>⊖Δ = 0.57 |
| 108 | Diaspore orh<br>α-AlOOH | {010},{hk0}, {hkl};<br>tabular, flaky ∥ (010);<br>rarely acicular [001] | # (010) perf.,<br># {210} good;<br>H = 6½–7 D = 3.36± | X ∥ c<br>Y ∥ b, O.A.P. (010)<br>Z ∥ a, bi;<br>l = (+)(–)<br>fig. p. 47 | $n_X$ = 1.702 – 1.704<br>$n_Y$ = 1.722 – 1.724<br>$n_Z$ = 1.750 – 1.752<br>⊕Δ = 0.048 – 0.048 |

| Angle and Dispersion of Optic Axes | Color in Thin Section Pleochroism | Chemical Reactions Special Remarks | Distinguishing Features of Similar Minerals | Occurrences | |
|---|---|---|---|---|---|
| often abnormal:<br>$2V_z = 0^o - 30^o$, then<br>$n_x = 1.485$<br>$n_y = 1.485$<br>$n_z = 1.488$ | colorless | gelatinizes with HCl, then stainable; smallest n of zeolite group; frequently abnormally birefringent domains | analcite: isotropic; sodalite, nosean, hauyne, davyne, apophyllite: gr. n | in cavities and joints of basaltic, but also in acid rocks due to hydrotherm. alteration of Ca-silicates | 100 |
| | colorless | decomposed by HCl | chabazite: different #; other zeolites: gr. Δ | in vesicles of basaltic rocks with chabazite and other zeolites | 100b |
| | colorless | | other zeolites: X-ray | frequently in vesicles of basaltic rocks, but always in small quantities | 100c |

## Oxides and Hydroxides

| | | | | | |
|---|---|---|---|---|---|
| $2V_z \approx 45^o$ ($0^o$ red, up to $90^o$ viol.); dispersion may be weak, usually however very strong<br>$r > v$ | X = colorless to pink<br>Y = colorless to light yellow<br>Z = colorless to light green | insoluble in acids; abnormal interference colors | apatite: sm. Δ and uniaxial ⊖;<br>corundum, vesuvianite: uniaxial ⊖;<br>merwinite: different paragenesis | in acid phosphate-poor aplites and pegmatites; contact-pneumatolytic in mica schists; rarely in sands and placers | 101 |
| $2V_z \approx 75^o$<br>$r < v$ | nb ta<br>X = red-brown to yellow<br>Y = dark brown to red-br.<br>Z = opaque to dark red-brown<br>X < Y < Z | soluble in $KHSO_4$-melt, insoluble in acids; nearly opaque in all sections if Fe-Nb-content high | hematite: uniaxial, ⊖;<br>rutile: uniaxial, gr. Δ; cassiterite: uniaxial, sm. Δ;<br>magnetite, uraninite, wolframite etc.: X-ray, ore microscopy | in granite- and syenite-pegmatites with cleavelandite, cassiterite, F-, Be-, B-, Li-, and W-minerals | 102 |
| $2V_z \approx 50^o$<br>$r < v$ | deep yellow to brownish red<br>Y > X = Z weak | insoluble in cold acids; soluble in hot $H_2SO_4$ + HF and in $KHSO_4$-melt | ilvaite: sm. n, soluble in HCl; brookite: lighter col., sm. 2V, strong disp.; rhoenite: sm. n, sm. Δ;<br>goethite: lighter col., ⊖Δ;<br>rutile: lighter col., uniax. | never primary magmatic; deuteric in Ti-rich plutonites; pneumatolytic in volcanics; pseudomorphous after ilmenite; in tuffs and river sands | 103 |
| disp. very strong (variable):<br>$2V_z$ ≈ $10^o$–$20^o$ (Na) in (001)<br>≈ $0^o$ for yellow-green<br>≈ $20^o$–$30^o$ for violet in (010) | ∥a: pale yellow, red-brown, clove-brown<br>∥b: colorless, olive, cinnamon-brown<br>∥c: light yellow, orange-brown, clove-brown<br>Z > Y > X weak | insoluble in acids (even HF); soluble in melts of KOH, $Na_2CO_3$, $KHSO_4$; extreme dispersion | rutile, anatase: uniaxial;<br>pseudobrookite: darker brown, constant 2V;<br>goethite: red-brown and ⊖Δ | hydrothermal or due to replacement during weathering of Ti-minerals; in alpine fissures; authigenic in sediments | 104 |
| $2V_x = 30^o \pm 1^o$<br>$r \gg v$ | colorless to yellow;<br>X = red-brown,<br>Y = olive-green,<br>Z = reddish brown,<br>X > Y > Z;<br>zonal coloring | slowly soluble in conc. $H_2SO_4$; soluble in $KHSO_4$-melt | goethite: reddish, straight extinction; lepidocrocite: much larger Δ and reddish; cassiterite, rutile: uniaxial ⊕ | in nepheline syenites and other $SiO_2$-poor alkaline rocks; in sanidinites, in carbonatites, in placers | 105 |
| $2V_z = 64^o$<br>horizontal dispersion<br>$r < v$ weak | colorless | slightly soluble in HF | among other $SiO_2$-modifications only stishovite has gr. n.;<br>X-ray | with stishovite and lechatelierite in shock-metam. quartz containing rocks near meteor craters | 105a |
| $2V_x$ = $23^o$ (for red) in (100)<br>$0^o$ (for λ = 605–620mμ)<br>$27^o$ (for green) in (001)<br>temperature | X = light yellow or brown<br>Y = yellow-brown,<br>Z = orange or olive,<br>Z > Y > X or<br>X > Z > Y | slightly soluble in HCl; easily soluble in $HNO_3$; DTA: peak at $400^oC$, due to hematite-formation; abnormal interference colors | lepidocrocite: gr. 2V and weaker dispersion;<br>hematite: gr. n and Δ, uniaxial;<br>similar are: brookite, pseudobrookite, ilvaite | weathering of Fe-minerals; in sedimentary Fe-ores; main component of limonite | 106 |
| $2V_x = 83^o$–$90^o$<br>dispersion weak | X = light yellow to yellow,<br>Y = red-orange to red,<br>Z = orange-red to brown-red,<br>X < Y < Z strong | like goethite (Nr. 106); DTA: peak at $350^o$; extremely high Δ | goethite: sm. 2V and strong disp.; hematite, pyrophanite, geikielite: uniaxial | like goethite, but rarer; weathering of sulfide ores | 107 |
| $2V_z = 84^o$–$85^o$<br>$r \approx v$ to $r < v$ | usually colorless; brownish due to (Fe,Mn), then<br>X < Y < Z | slightly soluble in HF; insoluble in other acids; soluble after heating to red heat in $H_2SO_4$ | astrophyllite: orange-yellow; låvenite: ⊖Δ; sillimanite: sm. n, sm. Δ | frequently in bauxites; in Al-rich metamorphites; usually together with pyrophyllite; contact-pneumatolytic in hornfelses | 108 |

(24, 30)

| No. | Name<br>Symmetry<br>Chem. Composition | Common Crystal Forms, Twinning<br>Crystal Habit<br>Aggregates | Cleavage<br>Translation Gliding<br>Deform. Twinning<br>Hardness, Density | Optic Orientation | Refractive Indices<br>Birefringence |
|---|---|---|---|---|---|
| 109 | Gibbsite mcl<br>(Hydrargillite)<br>$Al(OH)_3$ | {001},{100},{110};<br>β = 94½°;<br>◈ pseudohexagonal;<br>tabular ∥ (001);<br>chalcedony-like<br>aggregates | # (001) perf.,<br>mica-like<br>H = 2½ – 3 D = 2.38± | usually:<br>X ∥ b, O.A.P. ⊥ (010)<br>Y ∧ a = +25°<br>Z ∧ c = −21° bi;<br>l = (+) (−)<br>rarely:<br>X ∧ a = +30°<br>Y ∥ b, O.A.P. (010)<br>Z ∧ c = −25°, bi;<br>fig. p. 47 | $n_x$ = 1.566 – 1.568<br>$n_y$ = 1.566 – 1.568<br>$n_z$ = 1.587 – 1.589<br>⊕Δ = 0.021 – 0.021 |
| 109a | Boehmite orh<br>γ-AlOOH | {001};<br>tabular, flaky, rarely<br>well developed | # (010)?<br>H = 3½ – 4 D = 3.0± | uncertain | $n_x$ = 1.646<br>$n_y$ = 1.655<br>$n_z$ = 1.661<br>} decreas. with loss of $H_2O$<br>⊖Δ = 0.015 |
| 110 | Aragonite orh<br>$CaCO_3$ | {110},{010},{001},<br>{hkl};<br>◈ pseudohex. = {110}lam.;<br>columnar, fibrous ∥ c,<br>radiated, stellate | # (010) poor;<br>T = (010), t = [100],<br>$K_1$ = (110), $K_2$ = (1$\bar{3}$0);<br>H = 3½ – 4 D = 2.94± | X ∥ c, bi;<br>Y ∥ a, O.A.P. (100)<br>Z ∥ b;<br>l = (−)<br>fig. p. 47 | $n_x$ = 1.530 – 1.530<br>$n_y$ = 1.682 – 1.681<br>$n_z$ = 1.686 – 1.685<br>⊖Δ = 0.156 – 0.155 |
| 111 | Hydromagnesite mcl<br>$Mg_5[OH/(CO_3)_2] \cdot 4H_2O$ | {100},{110},{011};<br>β = 114°; ◈ (100) lam.;<br>tabular ∥ (100),<br>acicular ∥ c, radiated,<br>massive | # (010) perf.,<br>(100) parting;<br>H = 3½ D = 2.15± | X ∧ c = +47°<br>Y ∧ c = −43°,<br>O.A.P. ⊥ (010)<br>Z ∥ b, bi;<br>fig. p. 47 | $n_x$ = 1.515 – 1.523<br>$n_y$ = 1.530 – 1.527<br>$n_z$ = 1.544 – 1.545<br>⊖Δ = 0.029 – 0.022⊕ |
| 111a | Artinite mcl<br>$Mg_2[(OH)_2/CO_3] \cdot 3H_2O$ | {100},{001},{101};<br>β = 99½°;<br>acicular ∥ b,<br>radiated | # (100) perf. > 99°;<br># (001) dist.<br>H = 2 D = 2.03± | X ∧ a = −20°, bi;<br>Y ∥ b, O.A.P. (010)<br>Z ∧ c = +30° | $n_x$ = 1.488<br>$n_y$ = 1.534<br>$n_z$ = 1.556<br>⊖Δ = 0.068 |
| 112 | Ludwigite orh<br>$(Mg, Fe^{2+})_2Fe^{3+}[O_2/BO_3]$ | {110},{410},{310},<br>{210};<br>radiated ∥ c,<br>stellate | tough, resilient<br>H = 5 D = 3.6–4.3 | X ∥ a<br>Y ∥ b, O.A.P. (010)<br>Z ∥ c, bi;<br>l = (+) fig. p. 47 | Fe-rich Fe-poor<br>$n_x$ = 1.85 – 1.84<br>$n_y$ = 1.85 – 1.85<br>$n_z$ = 2.02 – 1.98<br>⊕Δ = 0.17 – 0.14 |
| 112a | Ascharite orh<br>$Mg_2[B_2O_5] \cdot H_2O$ | {010},{110};<br>acicular ∥ c,<br>bladed ∥ (010),<br>reticulated, bush-like | no obvious #;<br>H = 3½ D = 2.65± | X ∥ c, bi;<br>Y ∥ b, O.A.P. (010)<br>Z ∥ a;<br>l = (−) | $n_x$ = 1.58<br>$n_y$ = 1.64<br>$n_z$ = 1.65<br>⊖Δ = 0.07 |
| 113 | Anhydrite orh<br>$CaSO_4$ | {100},{010},{001};<br>isometric, granular;<br>spathic, massive;<br>{101},{010},{001}:<br>acicular, columnar ∥ c,<br>fibrous; ◈ {101} lam. | # (001) perf.,<br># (010) perf.,<br># (100) good;<br>$T_1$ = (001), $t_1$ = [010],<br>$T_2$ = (012),<br>$K_1$ = (101), $K_2$ = ($\bar{1}$01);<br>H = 3 – 3½ D = 2.96± | X ∥ c;<br>Y ∥ b, O.A.P. (010)<br>Z ∥ a, bi;<br>fig. p. 47 | $n_x$ = 1.5700<br>$n_y$ = 1.5757<br>$n_z$ = 1.6138<br>⊕Δ = 0.0438 |
| 114 | Celestite orh<br>$SrSO_4$ | {011},{001},{110},<br>{102};<br>prisms {011},<br>tabular ∥ (001)<br>granular, fibrous | # (001) perf.,<br># (010) distinct,<br># {110} good: 76°;<br>H = 3 – 3½ D = 3.95± | X ∥ c;<br>Y ∥ b, O.A.P. (010)<br>Z ∥ a, bi;<br>l = (+)<br>fig. p. 47 | $n_x$ = 1.622<br>$n_y$ = 1.624<br>$n_z$ = 1.631<br>⊕Δ = 0.009 |
| 115 | Barite orh<br>$BaSO_4$ | {001},{110},{011},<br>{102};<br>flaky to tabular;<br>{011}, {110},{010}:<br>columnar ∥ a, stellate,<br>also granular, massive;<br>deformation-◈ {110} | # (001) perf.,<br># (010) distinct,<br># {110} good: 78°;<br>$T_1$ = (001), $t_1$ = [100],<br>$T_2$ = (011), $t_2$ = [011];<br>H = 3 D = 4.47± | X ∥ c<br>Y ∥ b, O.A.P. (010)<br>Z ∥ a, bi;<br>l = (+)<br>fig. p. 47 | $n_x$ = 1.636<br>$n_y$ = 1.637<br>$n_z$ = 1.648<br>⊕Δ = 0.012 |
| 116 | Gypsum mcl<br>$CaSO_4 \cdot 2H_2O$ | {010},{120},{011};<br>β = 127½°; ◈ (100)<br>and {001}, not lam.;<br>tabular ∥ (010)<br>prismatic {120},<br>acicular ∥ c, fibrous<br>[010], granular, massive | # (010) perf.,<br># (100) good,<br># {111} perf. 41½°,<br>fibrous;<br>T = (010), t = [001],<br>flexible;<br>[001] : [101] = 66½°<br>H = 2 D = 2.32± | X ∧ c = −38°<br>X ∧ a = −15°<br>Y ∥ b, O.A.P. (010)<br>Z ∧ c = +52°, bi;<br>Z ∧ [101] = +14°;<br>l = (+) (−)<br>fig. p. 47 | $n_x$ = 1.5205<br>$n_y$ = 1.5226<br>$n_z$ = 1.5296<br>⊕Δ = 0.0091 |

| Angle and Dispersion of Optic Axes | Color in Thin Section Pleochroism | Chemical Reactions Special Remarks | Distinguishing Features of Similar Minerals | Occurrences | |
|---|---|---|---|---|---|
| $2V_Z = 0^o - 40^o$<br>(changing with $H_2O$-content, temperature-dependent)<br>r > v if O.A.P. ⊥ (010)<br>r < v if O.A.P. $\parallel$ (010) | colorless | easily soluble in cold HCl and in conc. $H_2SO_4$ | brucite } : uniaxial;<br>alunite }<br>wavellite: sm. n;<br>kaolinite: sm. Δ;<br>lepidolite: ⊖Δ | weathering product of other Al-minerals;<br>frequently in bauxites;<br>in clays;<br>diaphtoritic from corundum or diaspore | 109 |
| $2V_X = 80^o$ | colorless | in $H_2SO_4$ not as readily soluble as gibbsite; opt. diagnosis not reliable | DTA, X-ray | diagenetically from gibbsite; often the principal mineral of bauxites; rare in tropical soils and in pelites | 109a |
| $2V_X = 18^o$<br>r < v weak | colorless | soluble in acetic acid; lilac stain readily attained with $Co(NO_3)_2$-solution; DTA: inversion to calcite at >400°C | calcite: perf. #, slow reaction with $Co(NO_3)_2$-solution | hydrothermal; sedimentary, biogenic; metamorphic under high pressures | 110 |
| changing:<br>$2V_X = 70^o - 90^o$ to<br>$2V_Z \approx 50^o$! | colorless | $CO_2$-formation with acids; after heating to red heat reaction with $AgNO_3$ | wavellite: always⊕Δ;<br>artinite: gr. Δ; brucite: abnormal interference colors; gypsum: sm. Δ | hydrothermal; weathering product in serpentinite and dolomitic limestone | 111 |
| $2V_X = 60^o$ | colorless | soluble in acids;<br>one opt. axis<br>⊥ # (100) | hydromagnesite: sm. Δ,<br>see Nr. 111 above! | usually forming at the expense of hydromagnesite | 111a |
| $2V_Z$ very small,<br>r > v very strong | X = Y = dark green,<br>Z = dark red-brown,<br>X = Y < Z<br>Fe-rich: nearly opaque (see Nr. 40) | soluble in acids; becomes red after heating; strongly paramagnetic | ilvaite: sm. Δ;<br>columbite: gr. n;<br>tourmaline: uniaxial ⊖ | pneumatolytic; contact-metasomatic in limestones and skarns | 112 |
| $2V_X = 24^o$ | colorless | easily soluble in HCl;<br>3 dehydration reactions:<br>170°, 350°, 670° | stassfurtite: sm. Δ,<br>l = (+);<br>talc: flakes, l = (+);<br>koenenite, brucite: uniaxial ⊕ | hydrothermal; metasomatic in skarns;<br>diagenetic by react. with undersat. sol. in oceanic salt deposits | 112a |
| $2V_Z = 42^o$<br>r < v | colorless | soluble in HCl, slightly soluble in $H_2O$;<br>below 42°C with $H_2O$ reaction to gypsum; poikilitic intergrowth with halite | polyhalite: sm. n,<br>soluble in $H_2O$ | predominantly sedimentary; also hydrothermal and as condensates from volcanic vapors | 113 |
| $2V_Z = 51^o$<br>r < v | colorless | almost insoluble in $H_2O$, somewhat easier soluble in NaCl-solution; insoluble in HCl | barite: insoluble in $H_2O$,<br>sm. 2V | diagen. and segreg. in chem.-biogen. sediments;<br>hydrothermal;<br>rarely primary in salt deposits | 114 |
| $2V_Z = 36^o - 38^o$<br>r < v | colorless | insoluble in $H_2O$<br>and in cold conc. HF | celestite: gr. 2V,<br>slightly soluble in $H_2O$;<br>topaz: different paragenesis | hydroth. with sulfidic ores; in fissures and concretions, also as secondary cementation in sediments | 115 |
| $2V_Z = 58^o$ (at 19°C) dependent on temperature; oblique dispersion<br>r < yellow > v | colorless | soluble in HCl; almost insoluble in water; turbid upon heating; reaction to anhydrite above 42° | chalcedony: different soublity, H; anhydrite,<br>hydromagnesite: gr. Δ;<br>brucite: gr. n, Δ | primary in salt deposits, sec. from anhydrite by $H_2O$-addition, in pelites and marls, by weathering of sulfides | 116 |

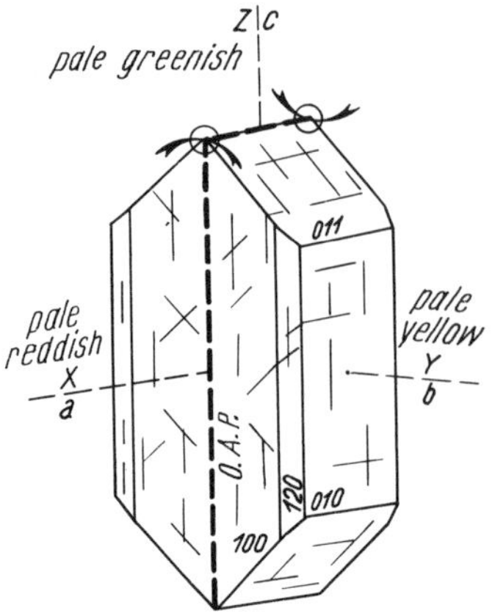


101. Chrysoberyl

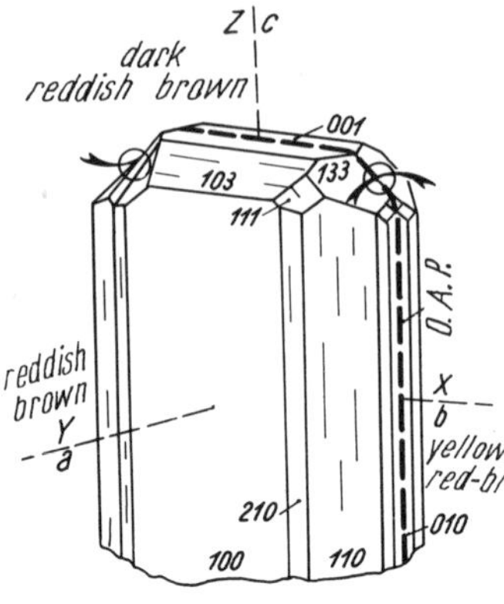


102-1. Columbite group (Niobite-Tantalite)

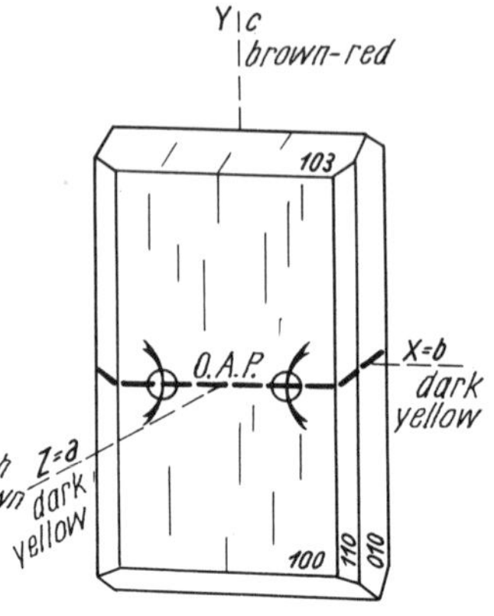


103. Pseudobrookite

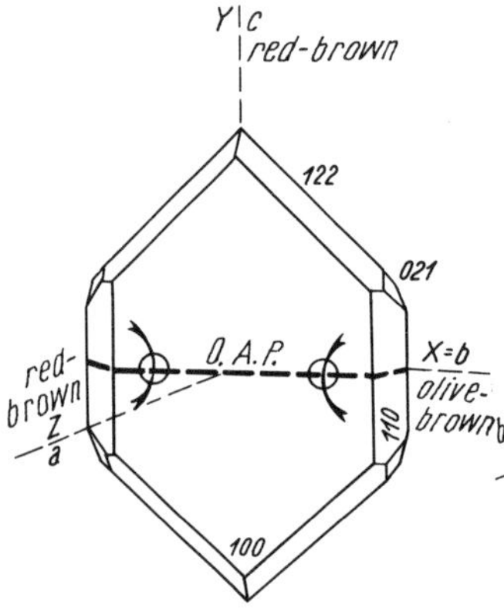


104-1. Brookite (λ = red)

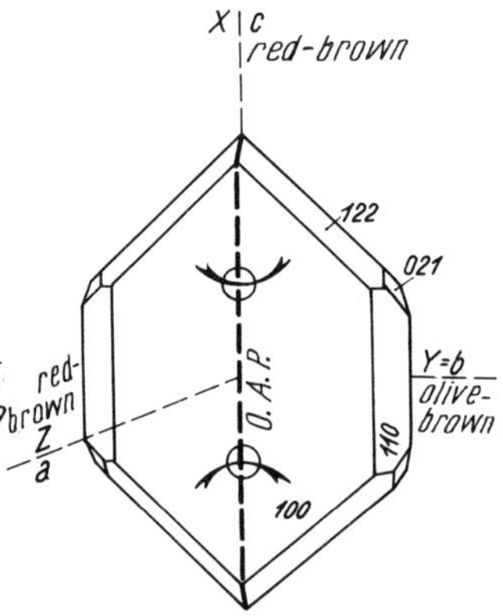


104-2. Brookite (λ = violet)

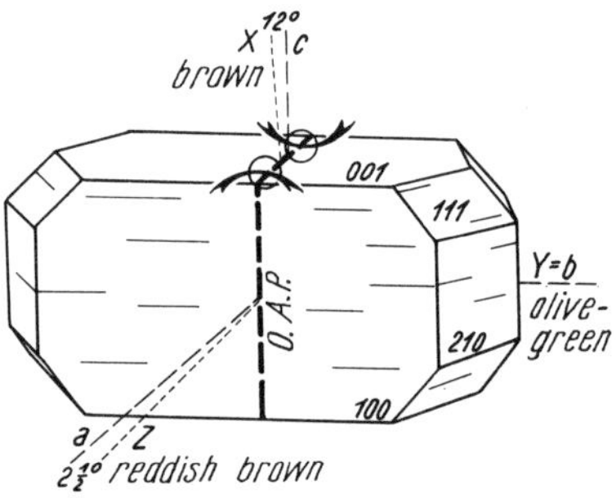


105. Baddeleyite

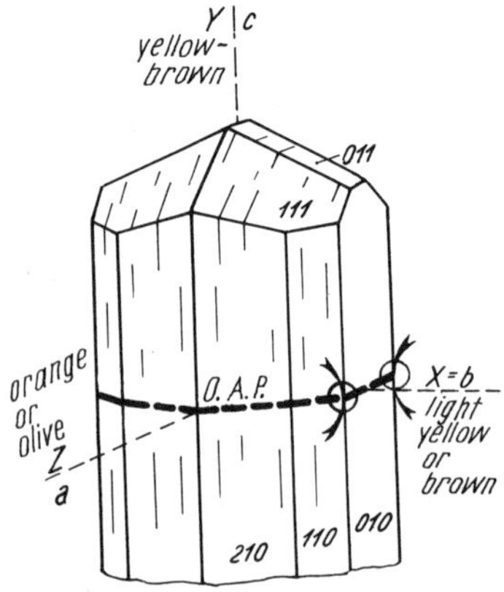


106-1. Goethite (λ = {yellow green blue)

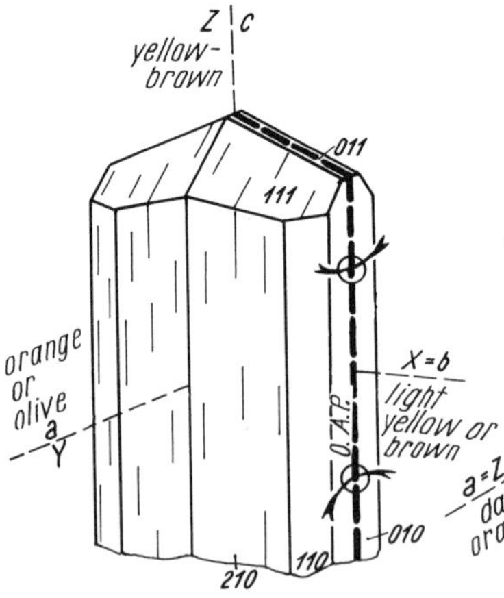


106-2. Goethite (λ = red)

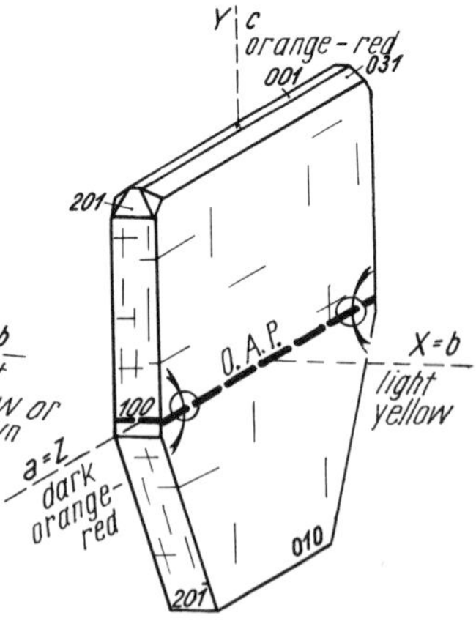


107. Lepidocrocite

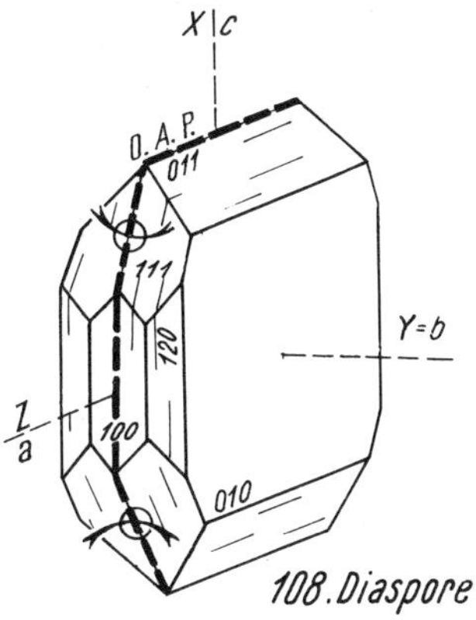


108. Diaspore

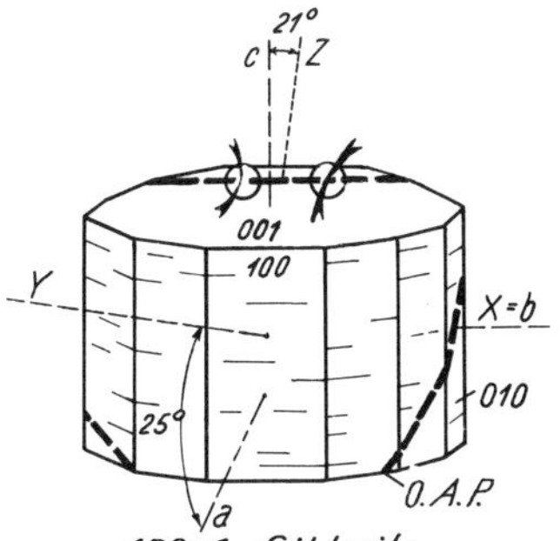


109-1. Gibbsite
O. A. P.⊥(010)

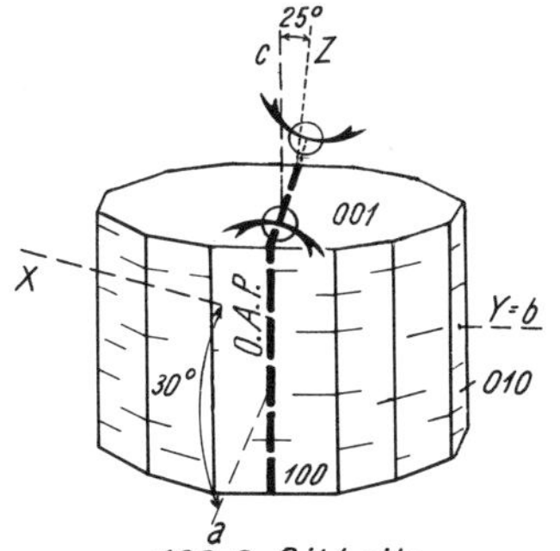


109-2. Gibbsite
O. A. P. ll (010)

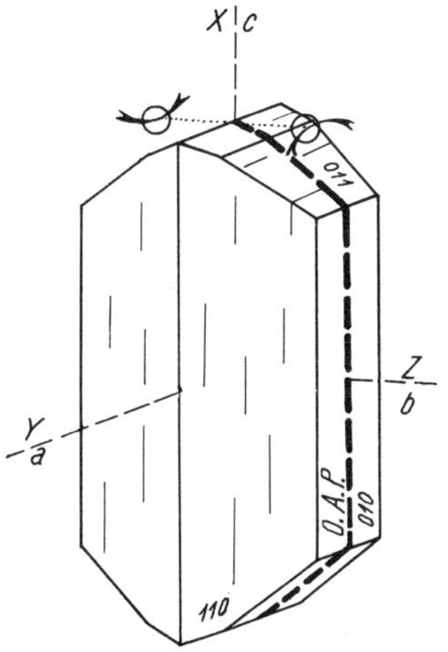


110. Aragonite

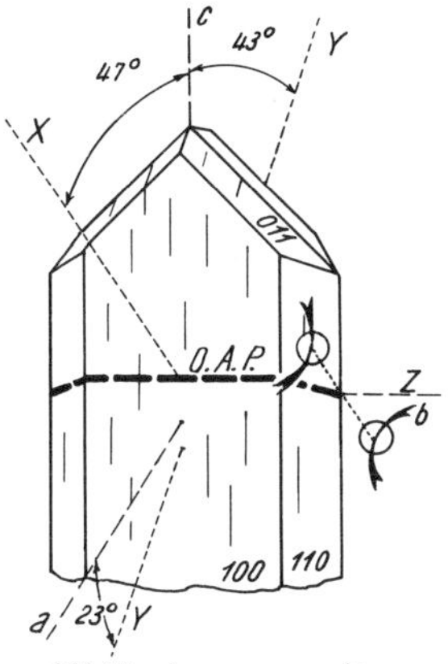


111. Hydromagnesite

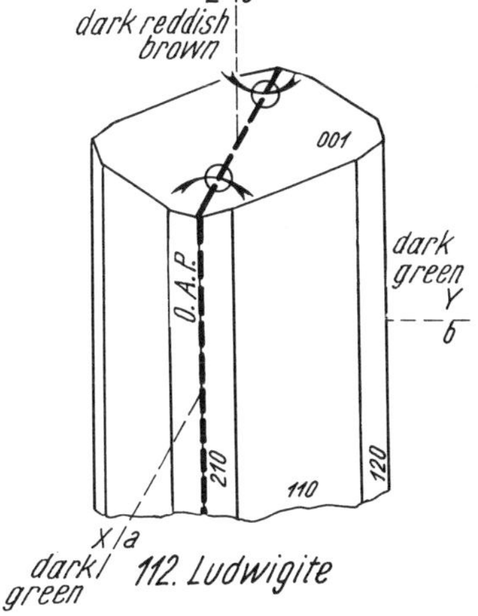


112. Ludwigite

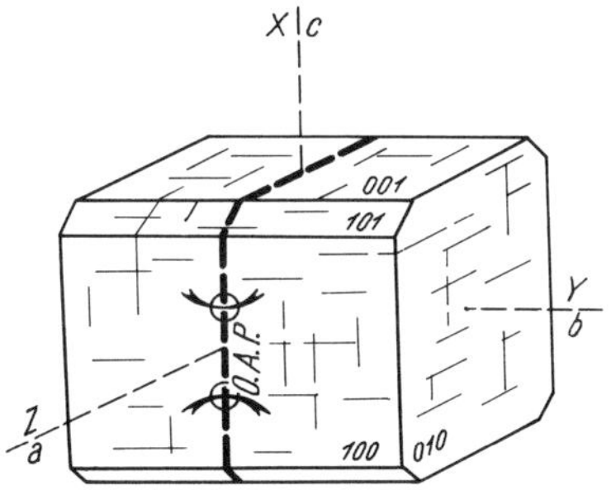


113. Anhydrite

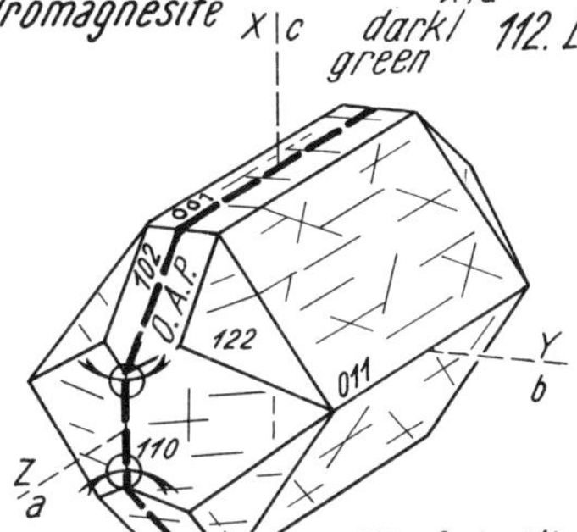


114. Celestite

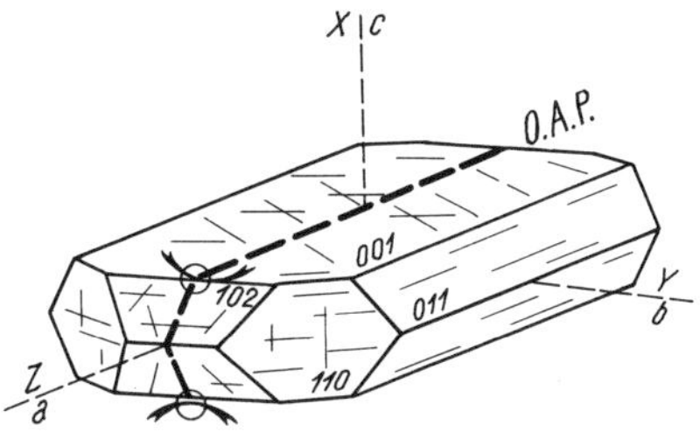


115. Barite

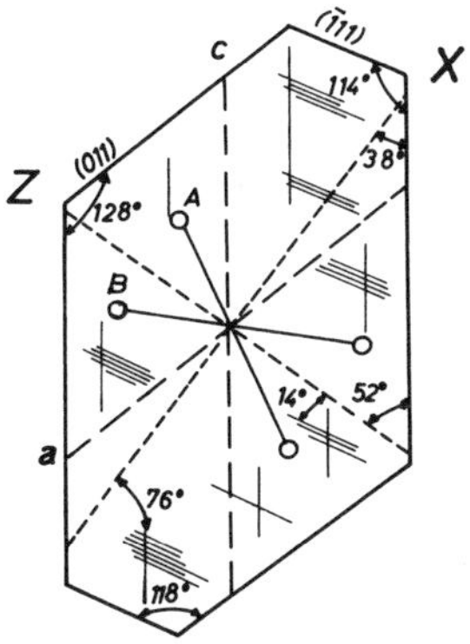


116-2. Gypsum, cleavage flake ll (010)

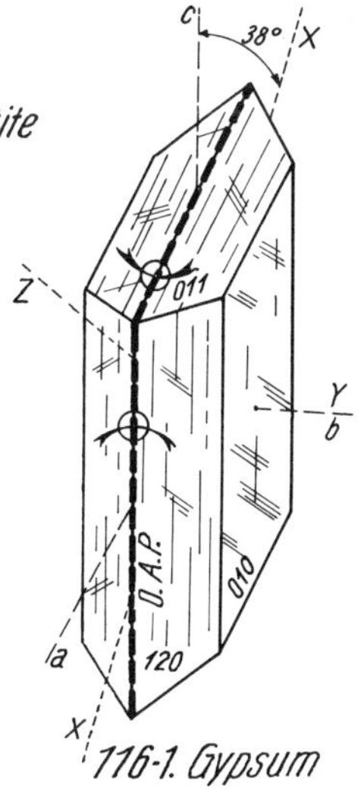


116-1. Gypsum

| | Name<br>Symmetry<br>Chem. Composition | Common Crystal Forms, Twinning<br>Crystal Habit<br>Aggregates | Cleavage<br>Translation Gliding<br>Deform. Twinning<br>Hardness, Density | Optic Orientation | Refractive Indices<br>Birefringence |
|---|---|---|---|---|---|
| 117 | Monazite mcl<br>$(\underline{Ce},La,Nd)PO_4$ | {100},{110},{101},<br>{011}, β = 104°;<br>◈ (100), ◈ (001), lam.;<br>granular, tabular ∥ (100),<br>columnar ∥ b | # (001) perf. } 76°;<br># (100) dist. }<br>H = 5 – 5½<br>D = 5.0 – 5.2 | X ∥ b, O.A.P. ⊥ (010)<br>Y Λ a = +8° – 11°<br>Z Λ c = +2° – 6°, bi;<br>fig. p. 52 | $n_x$ = 1.774 – 1.800<br>$n_y$ = 1.777 – 1.801<br>$n_z$ = 1.825 – 1.849<br>⊕Δ = 0.051 – 0.049 |
| 118 | Wagnerite mcl<br>$Mg_2[F/PO_4]$ | {100},{hk0},{$\bar{1}$01},<br>{001}; β = 108°;<br>granular, columnar ∥ c,<br>rarely tabular ∥ (001) | # (100) poor,<br># {210} poor;<br>H = 5 – 5½ D = 3.12 | X Λ a = +40°<br>Y ∥ b, O.A.P. (010)<br>Z Λ c = –21°, bi;<br>l' = (+) fig. p. 52 | $n_x$ = 1.559 – 1.568<br>$n_y$ = 1.561 – 1.572<br>$n_z$ = 1.571 – 1.582<br>⊕Δ = 0.012 – 0.014 |
| 118a | Triplite group mcl<br>$(Mn,Fe^{2+},Mg)[F/PO_4]$ | massive,<br>coarse-grained | # (001) good,<br># (010) distinct;<br>H ~ 5 D = 3.5 – 3.9 | X Λ a ~ +46° to ~ +58°<br>Y ∥ b, O.A.P. (010)<br>Z Λ c ~ –30° to ~ –46°<br>usually bi | Mn Mn,Mg Fe<br>$n_x$ = 1.662 – 1.643 – 1.696<br>$n_y$ = 1.673 – 1.647 – 1.704<br>$n_z$ = 1.684 – 1.668 – 1.713<br>Δ = ⊖0.022 ⊕0.025 ⊕0.017 |
| 119 | Lazulite group mcl<br>$(Mg,Fe)Al_2[OH/PO_4]_2$ | {111},{$\bar{1}$11},{101};<br>morph. β = 91°,<br>cryst. setting PRÜFER;<br>◈ (100), also lam.,<br>◈ {223} not frequent;<br>granular, sometimes<br>tabular ∥ (111) | # {110} distinct,<br># (101) poor;<br>H = 5 – 6<br>D = 3.06 – 3.22 | X Λ c ~ –9°, bi;<br>Y ∥ b, O.A.P. (010)<br>Z Λ c ~ +81°<br>fig. p. 52 | Mg Mg:Fe = 1:1<br>$n_x$ = 1.609 – 1.626<br>$n_y$ = 1.633 – 1.654<br>$n_z$ = 1.643 – 1.663<br>⊖Δ = 0.034 – 0.037<br>diagr. p. 54 |
| 120 | Amblygonite- trcl<br>Montebrasite<br>$LiAl[(F,OH)/PO_4]$ | {100},{001},{110},<br>{1$\bar{1}$0},{10$\bar{1}$},<br>cryst. setting DANA;<br>◈ (101) } lam. 90½°<br>◈ ($\bar{1}$01) }<br>massive, spathic | # (001) perf.,<br># (100) good,<br># (0$\bar{2}$1) poor;<br>somewhat flexible;<br>H = 6<br>D = { 3.11 ambl. / 2.99 mont. | Ambl. Mont.<br>X Λ b ~ 12° ~ 0°<br>fig. p. 52 | ambl. mont.<br>$n_x$ = 1.578 – 1.613<br>$n_y$ = 1.594 – 1.623<br>$n_z$ = 1.598 – 1.636<br>⊖Δ = 0.020 ⊕Δ = 0.023 |
| 120a | Triphylite orh<br>$Li(\underline{Fe},\underline{Mn},Mg)[PO_4]$ | {001},{110},{010};<br>cryst. setting DANA;<br>prismatic ∥ [001],<br>usually massive,<br>coarse-grained | # (001) perf.,<br># (010) poor;<br>H = 4 – 5<br>Mn Fe<br>D = 3.35 – 3.6 | a ∥ b ∥ c ∥ O.A.P.<br>X Z bi Y (001) ~ <$Fe_{60}$<br>Y Z bi X (100) ~ $Fe_{60-75}$<br>Y Z X bi (100) ~ $Fe_{75-82}$<br>Z Y X bi (010) ~ >$Fe_{82}$ | Mn-rich Fe-rich<br>$n_x$ = 1.663 – 1.696<br>$n_y$ = 1.666 – 1.700<br>$n_z$ = 1.673 – 1.702<br>⊕Δ = 0.010 ⊖Δ = 0.006 |
| 121 | Wavellite orh<br>$Al_3[(OH)_3/(PO_4)_2]\cdot 5H_2O$ | {110},{010},{101};<br>acicular-fibrous ∥ c,<br>spherulitic | # {110} good,<br># (010) distinct;<br>H = 3½ – 4 D = 2.35 | X ∥ b<br>Y ∥ a, O.A.P. (100)<br>Z ∥ c, bi;<br>l = (+) fig. p. 52 | $n_x$ = 1.520 – 1.535<br>$n_y$ = 1.526 – 1.543<br>$n_z$ = 1.545 – 1.561<br>⊕Δ = 0.025 – 0.026 |
| 121a | Variscite orh<br>$(\underline{Al},Fe^{3+})[PO_4]\cdot 2H_2O$ | {111};<br>usually fine-grained to<br>fibrous | # (010) good } ?<br># (001) poor }<br>H = 4½ D ~ 2.6 | X ∥ a, bi;<br>Y ∥ c, O.A.P. (001)<br>Z ∥ b | Al : Fe-rich<br>$n_x$ = 1.563 – 1.575<br>$n_y$ = 1.588 – 1.595<br>$n_z$ = 1.594 – 1.610<br>⊖Δ = 0.031 – 0.035 |
| 121b | Vivianite mcl<br>$Fe_3^{2+}[PO_4]_2\cdot 8H_2O$ | {100},{001},{$\bar{2}$21},<br>β = 104½°; columnar<br>to acicular ∥ c, rarely<br>tabular ∥ (100), stellate,<br>fibrous, earthy | # (010) perf.,<br># (100) poor;<br>fibrous fracture ~ ⊥ c<br>T = (010), t = [001];<br>H = 2 D = 2.6 – 2.7 | X ∥ b<br>Y Λ a = +33°,<br>O.A.P. ⊥ (010)<br>Z Λ c = –28½°, bi | $n_x$ = 1.579 – 1.582<br>$n_y$ = 1.602 – 1.604<br>$n_z$ = 1.629 – 1.635<br>⊕Δ = 0.050 – 0.053 |
| 122 | Olivine group orh<br>$(Mg,Fe)_2[SiO_4]$<br>Fo: $Mg_2[SiO_4]$<br>Fa: $Fe_2[SiO_4]$ | {100},{010},{110},<br>{021};<br>cryst. setting FLINK;<br>◈ {011} rare, c Λ c ~ 61°,<br>◈ {031},{012} even<br>rarer; granular | # (010) distinct,<br># (100) poor;<br>T = (010), t = [100];<br>H = 6½ – 7<br>D = { 3.16 fo / 4.33 fa | X ∥ b<br>Y ∥ c, O.A.P. (001)<br>Z ∥ a<br>fig. p. 53 | forsterite fayalite<br>$n_x$ = 1.636 – 1.827<br>$n_y$ = 1.651 – 1.869<br>$n_z$ = 1.669 – 1.879<br>⊕Δ = 0.033 ⊖Δ = 0.052 |
| 122a | Tephroite orh<br>$(\underline{Mn},Mg,Fe)_2[SiO_4]$ | {hk0},{010},{111},<br>{011};<br>cryst. setting FLINK;<br>usually granular | # (010) distinct<br># (100) poor;<br>H = 6 – 6½ D = 4.1 | X ∥ b, bi;<br>Y ∥ c, O.A.P. (001)<br>Z ∥ a | $n_x$ = 1.759 – 1.788<br>$n_y$ = 1.786 – 1.810<br>$n_z$ = 1.797 – 1.825<br>⊖Δ = 0.038 – 0.037<br>increasing with Fe<br>decreasing with Mg |
| 122b | Sinhalite orh<br>$MgAl[BO_4]$ | {hk0}, {010},{001},<br>{111},{021}; cryst.<br>setting CLARINGBULL-<br>HEY; short prisms ∥ c,<br>granular | H = 6 – 7 D = 3.48 | X ∥ b, bi;<br>Y ∥ c, O.A.P. (001)<br>Z ∥ a | $n_x$ = 1.667 – 1.676<br>$n_y$ = 1.696 – 1.704<br>$n_z$ = 1.705 – 1.712<br>⊖Δ = 0.038 – 0.036 |
| 123 | Monticellite orh<br>$Ca(\underline{Mg},Fe,Mn)[SiO_4]$ | {110},{120},{111},<br>{121};<br>cryst. setting: v. RATH;<br>short columns, grains,<br>tabular | # (010) poor;<br>H = 5 – 5½ D = 3.2 | X ∥ b, bi;<br>Y ∥ c, O.A.P. (001)<br>Z ∥ a<br>fig. p. 53 | CaMg 20%CaFe<br>$n_x$ = 1.641 – 1.654<br>$n_y$ = 1.646 – 1.664<br>$n_z$ = 1.652 – 1.674<br>⊖Δ = 0.011 – 0.020 |

| Angle and Dispersion of Optic Axes | Color in Thin Section Pleochroism | Chemical Reactions Special Remarks | Distinguishing Features of Similar Minerals | Occurrences | |
|---|---|---|---|---|---|
| $2V_Z = 6^{\circ}-19^{\circ}$<br>r < v weak,<br>rarely r > v weak;<br>horizontal dispersion | colorless;<br>rarely<br>X=pale yellow<br>Y=dark yellow<br>Z=greenish yellow<br>Y > X ~ Z weak | slightly soluble in HCl; decomposed by warm $H_2SO_4$; contains radio-active $ThSiO_4$; spectral absorption bands! | zircon: uniaxial ⊕, gr. n; orthite, epidote, fayalite: gr. $2V_X$; cassiterite, rutile, xenotime, sphene: see Vol. 2, p. 254 | accessory in plutonites, in pegmatites, aplites; hydroth. in fissures; in placers together with zircon, rutile, garnet | 117 |
| $2V_Z \sim 27^{\circ}-(46^{\circ})$<br>r > v distinct,<br>oblique dispersion | colorless | easily soluble in inorganic acids | cordierite: gr. $2V_X$;<br>apatite: gr. n, sm. Δ | hydrothermal in quartz segregations of phyllites; in rock salt deposits | 118 |
| $2V_Z \sim 25^{\circ}-92^{\circ}$<br>r > v<br>Mn + Fe: larger<br>Mg: smaller | variable pleochroism: yellowish and reddish-brownish | soluble in inorganic acids | triploidite and wolfeite: gr. n | in granite-pegmatites with triploidite, triphylite, apatite; in high temperature ore parageneses, e.g. with wolframite | 118a |
| $2V_X = 64^{\circ}-69^{\circ}$<br>r < v distinct<br>diagr. p. 54 | X = colorless<br>Y = azure blue<br>Z = azure blue<br>Z > Y ≫ X | insoluble in inorganic acids | dumortierite: pink-violet, gr. n; glaucophane, kyanite: good #, sm. Δ; serendibite, grandidierite: sm. $2V_X$ | in quartz veins; in quartzites and similar metamorphic rocks; in granite-pegmatites | 119 |
| $2V_X = 40^{\circ}-90^{\circ}$ ambl.<br>r > v ⇅<br>$2V_Z = 90^{\circ}-65^{\circ}$<br>r < v mont.<br>crossed and oblique dispersion strong<br>diagr. p. 55 | colorless | decomposed by HCl and $H_2SO_4$, $F^-$- and $PO_4^{3-}$-reactions; weakly abnormal interference colors; polysynth. ◊ | pargasite, tremolite: different #, diff. ◊;<br>melinophane: uniaxial ⊖;<br>topaz: sm. Δ;<br>bertrandite: X-ray | with Li-minerals pegmatitic to pneumatolytic | 120 |
| $2V_Z \sim 65^{\circ}-0^{\circ}$ r < v<br>$2V_Z \sim 0^{\circ}-90^{\circ}$ r > v<br>$2V_X \sim 90^{\circ}-0^{\circ}$ r < v<br>$2V_X \sim 0^{\circ}-90^{\circ}$ | pale greenish gray to brownish<br>X > Y ~ Z | easily soluble in HCl; sometimes abnormal interference colors | zoisite: different paragenesis:<br>apatite: uniaxial ⊖ | in Li-rich granite-pegmatites with amblygonite, spodumene, lepidolite; rarely in tin-ore lodes | 120a |
| $2V_Z = 71^{\circ}-84^{\circ}$<br>r > v | colorless | soluble in anorganic acids and in NaOH-solution | zeolites: sm. n, sm. Δ;<br>thomsonite: only sm. n!<br>prehnite: gr. n; gibbsite: oblique ext.; hydromagnesite: diff. paragenesis | hydrothermal in fissures, predominantly in sedimentary rocks | 121 |
| $2V_X \sim 50^{\circ}-60^{\circ}$<br>r < v<br>increases with increasing Fe-content | usually colorless | after heating, soluble in HCl | bertrandite: somewhat gr. Δ and n;<br>amblygonite: gr. D, oblique extinction | in cherts, pelitic schists and phyllites; filling veins together with quartz | 121a |
| $2V_Z = 81^{\circ}-84^{\circ}$<br>r < v weak | X=blue to indigo-blue<br>Y=pale yellow to green to bluish green<br>Z=pale-yellowish green | soluble in HCl; oxidizes easily; pleochroism becomes stronger with oxidation | see lazulite (Nr. 119) | weathering prod. under reducing conditions, especially in Fe-rich sediments | 121b |
| $2V_Z = 84^{\circ}$ fo<br>r < v ⇅<br>$2V_X = 50^{\circ}$ fa<br>r > v<br>diagr. p. 54 | almost colorless, may be stained due to oxidation;<br>see Vol. 2, p. 268 | gelatinizes slowly with HCl; alteration to phyllosilicates (serpentine), hematite, carbonates; "iddingsite" | monticellite: sm. Δ; pyroxenes: better #; humite: O.A.P. ∥ #; epidote: different color; tephroite, sinhalite:<br>see Nr. 122a, 122b | rarely primary with quartz; forsterite: contact-metam.; chrys. and hyal.: basalts, gabbros; fayalite: contact-metam., pegmatitic, in rhyolites | 122 |
| $2V_X = 60^{\circ}-71^{\circ}$<br>r > v<br>decreasing with Fe<br>increasing with Mg | almost colorless; in thick sections:<br>X = brownish red,<br>Y = reddish,<br>Z = green-blue<br>Y > Z > X | gelatinizes slowly with HCl | hortonolite: distinction difficult on optical grounds; chem. anal., X-ray | with rhodonite, bustamite in contact-metam. Mn-ore deposits | 122a |
| $2V_X = 51^{\circ}-56^{\circ}$ | almost colorless, in thick sections<br>X = brown,<br>Y = green,<br>Z = light brown,<br>X > Y > Z | soluble in HF and hot conc. $H_2SO_4$ | olivine: different relation n/$2V_X$, X-ray | in contact-metamorphic dolomitic marbles (with olivine, spinel) and calcite marble (with serendibite) | 122b |
| $2V_X = 70^{\circ}-82^{\circ}$<br>r > v weak<br>diagr. p. 55 | colorless | gelatinizes with HCl; alteration to "serpentine" | olivine: gr. Δ and better #; merwinite: gr. n and $2V_Z$; rankinite: $2V_Z$; chondrodite: gr. Δ | in contact-metamorphic marbles; in alnoites and polzenites | 123 |

| No. | Name<br>Symmetry<br>Chem. Composition | Common Crystal Forms, Twinning<br>Crystal Habit<br>Aggregates | Cleavage<br>Translation Gliding<br>Deform. Twinning<br>Hardness, Density | Optic<br>Orientation | Refractive Indices<br>Birefringence |
|---|---|---|---|---|---|
| 124 | Chondrodite mcl<br>$(\underline{Mg},Fe)_5[(OH,F)_2/(SiO_4)_2]$ | {221},{011},{021},<br>{010},{100}; β = 109°;<br>cryst. setting: GOLDSCHMIDT;<br>◊ (100) lam.;<br>lenticular, granular aggregates | # (100) distinct<br>H = 6 – 6½ D = 3.20 | X Λ c = +22° to +30°<br>Y Λ a = –3° to –10°,<br>O. A.P. ⊥ (010),<br>Z ∥ b, bi;<br>fig. p. 53 | $n_x$ = 1.593 – 1.619<br>$n_y$ = 1.603 – 1.632<br>$n_z$ = 1.623 – 1.653<br>⊕Δ = 0.030 – 0.034 |
| 125 | Humite orh<br>$(\underline{Mn},Fe)_7[OH,F)_2/(SiO_4)_3]$ | {001},{010},{hk0},<br>{hkl},{0kl};<br>cryst. setting: DANA;<br>◊ (107) lam. } pseudo-<br>◊ (307) lam. } hex.<br>granular, tabular ∥ (010) | # (001) good;<br>H = 6 – 6½ D ~ 3.25 | X ∥ a<br>Y ∥ c, O.A.P. (001)<br>Z ∥ b, bi;<br>fig. p. 53 | Fe-poor Fe-rich<br>$n_x$ = 1.607 – 1.643<br>$n_y$ = 1.619 – 1.653<br>$n_z$ = 1.639 – 1.675<br>⊕Δ = 0.032 – 0.032 |
| 126 | Clinohumite mcl<br>$(\underline{Mg},Fe)_9[(OH,F)_2/(SiO_4)_4]$ | {100},{001},{h0l},<br>{hkl}; β = 101°;<br>cryst. setting: GOLDSCHMIDT;<br>◊ (001) lam.;<br>granular | # (100) distinct;<br>H = 6 – 6½<br>D = 3.15 – 3.35 | XΛc = +7° to +15°<br>YΛa = ±4°, O.A.P.<br>⊥ (010)<br>Z ∥ b, bi;<br>fig. p. 53 | Fe-poor Fe-rich<br>$n_x$ = 1.623 – 1.664<br>$n_y$ = 1.636 – 1.673<br>$n_z$ = 1.651 – 1.698<br>⊕Δ = 0.028 – 0.034 |
| 126a | Titanclinohumite mcl<br>$(\underline{Mg},Ti)_9[(OH,F)_2/(SiO_4)_4]$ | {100},{hk0},{001},<br>{hkl}; β = 101°;<br>cryst. setting: GOLDSCHMIDT;<br>◊ ($\bar{1}$01) lam.;<br>columnar, lenticular, granular | # (100) distinct;<br>H = 6 D ~ 3.25 | XΛa = +11° (red) to<br>+27° (blue)<br>YΛc = ~0° (red) to<br>–16° (blue)<br>O.A.P. ⊥ (010)<br>Z ∥ b, bi; fig. p. 53 | Fe-poor Fe-rich<br>$n_x$ = 1.658 – 1.702<br>$n_y$ = 1.670 – 1.708<br>$n_z$ = 1.690 – 1.728<br>⊕Δ = 0.032 – 0.026 |
| 127 | Larnite mcl<br>$\beta-Ca_2[SiO_4]$ | ◊ (100) fine lam.,<br>◊ (010) lam. (rare);<br>granular, lenticular to tabular ∥ (010) | # (100) good,<br># (010) poor;<br>D = 3.29 | XΛc = 13 – 14°<br>Z ∥ b, O.A.P. ⊥ (010) | $n_x$ = 1.707<br>$n_y$ = 1.715<br>$n_z$ = 1.730<br>⊕Δ = 0.023 |
| 128 | Merwinite mcl<br>$Ca_3Mg[SiO_4]$ | {010},{110};<br>◊ {110} lam.: 42°<br>◊ (100) rare;<br>granular | # (010) perf.;<br>H = 6 D = 3.15 | XΛc = +36°, O.A.P.<br>⊥ (010)<br>Z ∥ b, bi;<br>fig. p. 53 | $n_x$ = 1.705 – 1.710<br>$n_y$ = 1.711 – 1.714<br>$n_z$ = 1.718 – 1.728<br>⊕Δ = 0.013 – 0.018 |
| 129 | Spurrite mcl<br>$Ca_5[CO_3/(SiO_4)_2]$ | ◊ (001) lam.,<br>◊ (h0l) lam.;<br>granular, columnar ∥ c | # (001) distinct,<br># (100) poor;<br>H = 5 D = 3.01 | X ∥ b, bi;<br>YΛa = +33°, O.A.P.<br>⊥ (010)<br>ZΛc = –22°<br>fig. p. 53 | $n_x$ = 1.637 – 1.641<br>$n_y$ = 1.672 – 1.676<br>$n_z$ = 1.676 – 1.681<br>Δ⊕ = 0.039 – 0.040 |
| 129a | Rankinite mcl<br>$Ca_3[Si_2O_7]$ | tabular, granular | H = 5½ D = 2.86 | XΛa = 15°<br>Y ∥ b, O.A.P. (010)<br>l = (–) | $n_x$ = 1.641<br>$n_y$ = 1.643<br>$n_z$ = 1.650<br>⊕Δ = 0.009 |
| 129b | Tilleyite mcl<br>$Ca_5[(CO_3)_2/Si_2O_7]$ | ◊ ($\bar{2}$01),<br>lamellar twinning rare | # (100) perf.;<br>D = 2.84 | Y ∥ b, O.A.P. (010)<br>ZΛc = 12 – 13° | $n_x$ = 1.612 – 1.617<br>$n_y$ = 1.632 – 1.635<br>$n_z$ = 1.653 – 1.652<br>⊕Δ = 0.041 – 0.035 |
| 130 | Andalusite orh<br>$Al^{[6]}Al^{[5]}[O/SiO_4]$ | {110},{001};<br>columnar ∥ c<br>(pseudotetragonal),<br>acicular, stellate,<br>granular | # {110} good: 89°;<br>brittle;<br>H = 7½ D = 3.14± | X ∥ c, bi;<br>Y ∥ b, O.A.P. (010)<br>Z ∥ a<br>l = (–) fig. p. 58 | pure Fe-cont.<br>$n_x$ = 1.629 – 1.640<br>$n_y$ = 1.633 – 1.644<br>$n_z$ = 1.638 – 1.651<br>⊖Δ = 0.009 – 0.011<br>diagr. p. 60 |
| 131 | Viridine orh<br>$(Al,Mn^{3+},Fe^{3+},Ti)_2[O/SiO_4]$ | short columns,<br>granular | # (001) good,<br># {110} good;<br>H = 6½ D = 3.21± | X ∥ a<br>Y ∥ b, O.A.P. (010)<br>Z ∥ c, bi;<br>l = (+) fig. p. 58 | $n_x$ = 1.649 – 1.662<br>$n_y$ = 1.654 – 1.671<br>$n_z$ = 1.662 – 1.691<br>⊕Δ = 0.013 – 0.029 |
| 132 | Kyanite trcl<br>(Disthene)<br>$Al^{[6]}Al^{[6]}[O/SiO_4]$ | {100},{010},{hk0},<br>{001};<br>◊ (100) and [001] lam.,<br>◊ [010] and (212);<br>bladed ∥ c,<br>flakes ∥ (100) | #(100) perf. > 74°<br>#(010) perf. > 87° > 79°<br>#(001) "fibrous"<br>T = (100), t = [001],<br>flexible;<br>H = 4 – 4½ ∥ c D = 3.59±<br>H = 6 – 7 ∥ b | in (100): Z'Λc = 27 – 32°<br>in (010): Z'Λc = 5 – 8°<br>in (001): X'Λc = 0 – 2½°<br>X ~ ⊥ (100), bi;<br>l' = (+)<br>fig. p. 58, 60 | $n_x$ = 1.710 – 1.713<br>$n_y$ = 1.720 – 1.722<br>$n_z$ = 1.727 – 1.729<br>⊖Δ = 0.017 – 0.016 |

| Angle and Dispersion of Optic Axes | Color in Thin Section Pleochroism | Chemical Reactions Special Remarks | Distinguishing Features of Similar Minerals | Occurrences | No |
|---|---|---|---|---|---|
| $2V_Z = 73^o - 89^o$<br>r > v weak | colorless to<br>X = golden to brown-yellow,<br>Y = light yellow,<br>Z = pale yellow,<br>X > Y > Z | alteration to serpentine, O.A.P. oblique to #; humite group: X-ray, ◈ lamellae | olivine: rarely ◈ lam., O.A.P. ⊥ to best #; clinohumite: sm. X∧c; spurrite: $2V_X$; tilleyite: O.A.P. ⊥ to # | with minerals of humite group, phlogopite, spinel, in metam. and contact-metam. dolomitic marbles, in alnoites | 124 |
| $2V_Z = 65^o - 84^o$<br>r > v weak | colorless to<br>X = golden yellow,<br>Y = pale yellow,<br>Z = pale yellow | humite group: stainable after HCl-treatment; alteration to "serpentine"; frequently ◈ lam., O.A.P. ∥ to # | see also chondrodite; olivine: O.A.P. ⊥ #; clinohumite: O.A.P. oblique to #; X-ray | rare; like chondrodite | 125 |
| $2V_Z = 62^o - 85^o$<br>r > v weak | colorless to<br>X = golden to brown-yellow,<br>Y = light yellow,<br>Z = pale yellow<br>X > Y > Z | gelatinizes with HCl; alteration to "serpentine"; O.A.P. oblique to #, ◈ lamellae | olivine: see chondrodite; humite group: X-ray | see chondrodite | 126 |
| $2V_Z = 52^o - 76^o$<br>r > v strong | X = red-orange to red,<br>Y = light yellow to orange,<br>Z = light yellow to orange,<br>X > Y ~ Z | gelatinizes with HCl; alteration to "serpentine"; very strong dispersion; ◈ lamellae | see also chondrodite; humite group: O.A.P. ~∥# or ∥ #, weaker dispersion, X-ray; staurolite: gr. n, sm. Δ; piemontite: gr. n, O.A.P. ⊥ #, ($2V_X$) | with olivine in serpentine and talc schists; rarely in marbles | 126a |
| $2V_Z = 70^o - 75^o$ | colorless | slowly decomposed by $H_2O$, quickly by acids; metastable, decomposes after heating to γ-modification (powdery, fibrous) | merwinite: sm. Δ, gr. X∧c, broader ◈ lam.; pigeonite: different paragenesis; salite: sm. n | high grade contact-metam. in siliceous dolomitic marbles; with spurrite or merwinite or gehlenite, rankinite, spinel | 127 |
| $2V_Z = 67^o - 76^o$ | colorless | gelatinizes with HCl; spurrite-rim; alteration to thaumasite and serpentine-like products; frequently ◈ lam. | hypersthene, monticellite: orh, $2V_X$; larnite, chrysolite: gr. Δ | with spurrite, gehlenite, wollastonite in contact-metam. calc-silicate rocks | 128 |
| $2V_X = 38^o - 41^o$<br>r > v distinct<br>crossed dispersion distinct | colorless | effervesces and gelatinizes with dilute HCl; weathering to carbonates; albite-like polysynthetic ◈ lam. | plagioclase: sm. n, sm. Δ, 2V; chrysolite: orh, no ◈ lam.; clinohumite: gr. $2V_Z$; chondrodite, tilleyite: sm. n, gr. $2V_Z$ | with gehlenite, wollastonite, merwinite in high grade contact-metam. marbles | 129 |
| $2V_Z = 64^o$ | colorless | gelatinizes with HCl; often poikilitic with inclusions of tabular gehlenite and drop-like larnite | monticellite: $2V_X$ | in high grade contact-metam. calc-silicate rocks with gehlenite, larnite, spinel, wollastonite | 129a |
| $2V_Z = 85^o - 90^o$<br>r < v | colorless | | monticellite: sm. Δ; merwinite: gr. n, sm. Δ | with gehlenite, also with spurrite and wollastonite in high grade metam. calc-silicate rocks and marbles | 129b |
| $2V_X = 83^o - 85^o$<br>r < v weak<br>(rarely r > v weak) | X = pink to colorless<br>Y = pale green to colorless<br>Z = pale green to colorless<br>patchy coloration | not attacked by HF; alteration to sericite; with symmetr. arrangement of carbonaceous inclusions = "chiastolite" | sillimanite: l = (+); kyanite: oblique extinction; prismatine: sm. 2V; enstatite: l = (+); scapolite: sm. n; thulite: gr. n | metam. with mica, cordierite, staurolite, sillimanite and others; also pegmatitic; in hornfelses; "chiastolite" in lower grade contact metam. pelites | 130 |
| $2V_Z = 52^o - 86^o$<br>r ≪ v strong | X = light yellow to yellowish-green<br>Y = green to emerald green<br>Z = dark yellow to golden yellow;<br>patchy coloration | vivid colors; strong pleochroism; abnormal interference colors | tawmawite: gr. n, sm. $2V_X$; aegirine: oblique extinction | metam. in quartzitic mica schists, kyanite and sillimanite quartzites; contact metam. in Mn-bearing hornfelses | 131 |
| $2V_X = 82\frac{1}{2}$<br>r > v very weak | colorless to greyish blue<br>X = colorless,<br>Y = pale violet-blue,<br>Z = pale cobalt-blue,<br>X < Y < Z | slightly soluble in HF; prominent intersecting cleavages; bending; alteration to sericite | hypersthene: sm. 2V; clinozoisite: diff. #; rhodonite: ⊕, and finer #; lawsonite: ⊕, diff. #; sapphirine: sm. Δ and sm. 2V | high pressure metam. (pelitic) rocks, often with staurolite, rutile; in mica schists and associated quartz-rich veins | 132 |

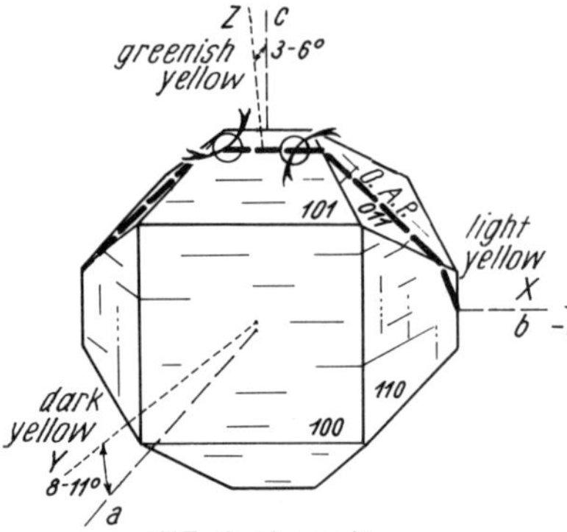

117-1. Monazite

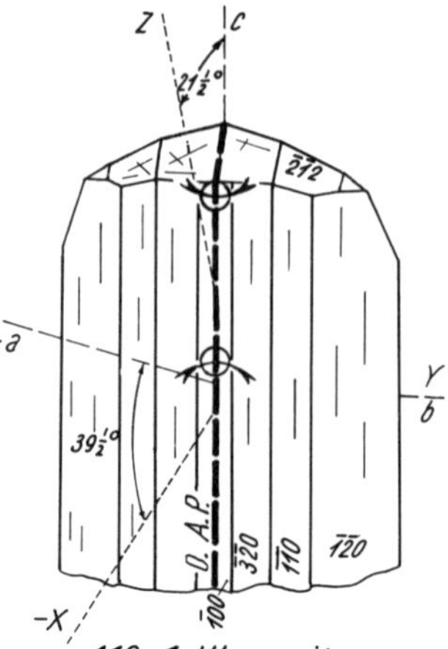

118-1. Wagnerite

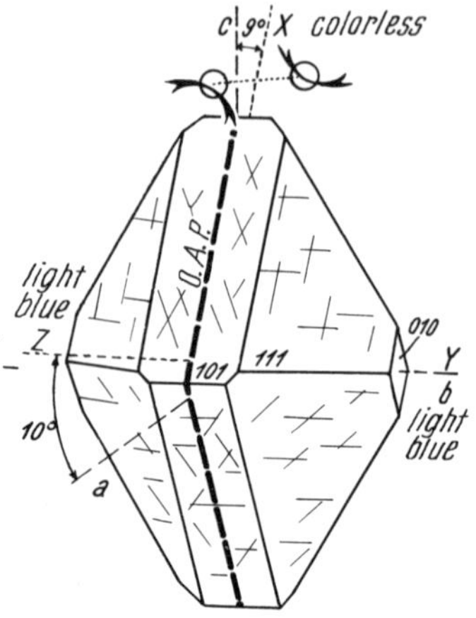

119-1. Lazulite (PRÜFER 1847)

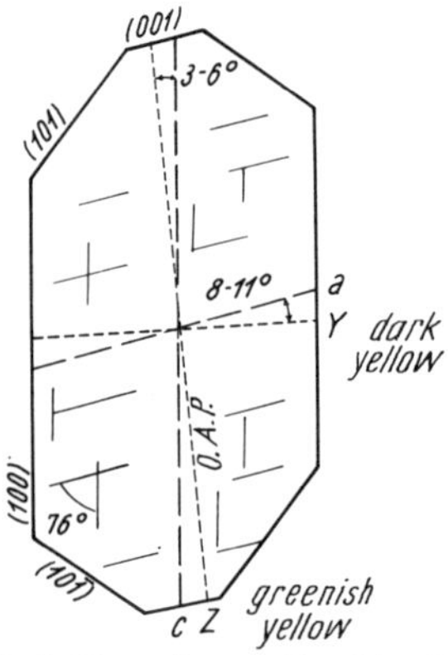

117-2. Monazite, section ‖ (010)

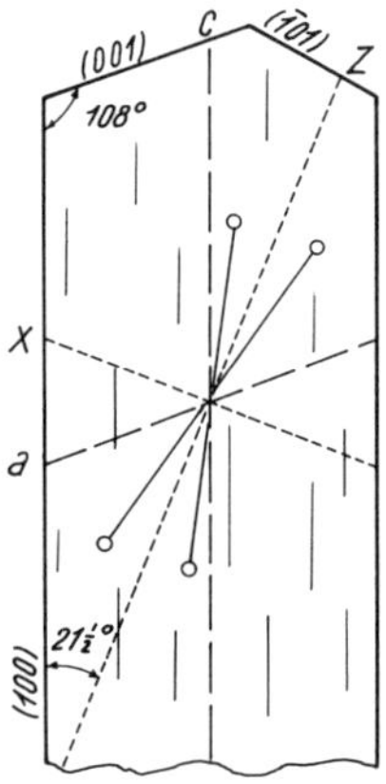

118-2. Wagnerite, section ‖ (010)

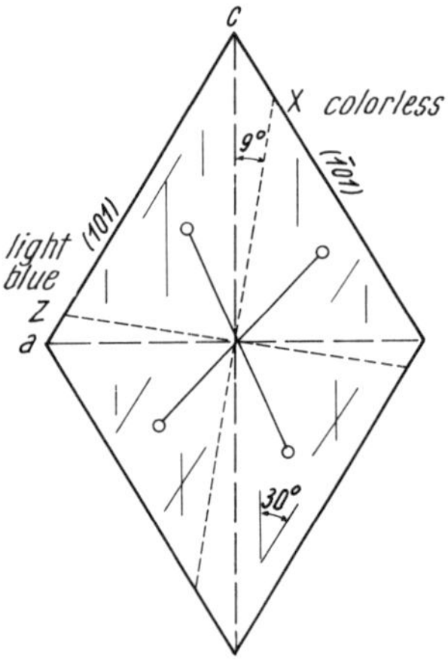

119-2. Lazulite, section ‖ (010)

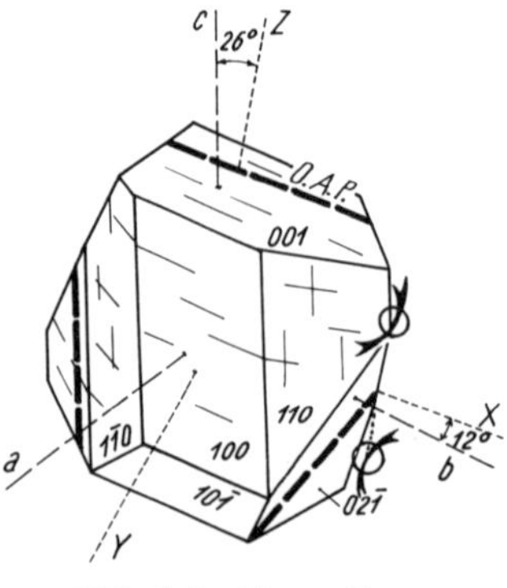

120-1. Amblygonite (DANA 1892)

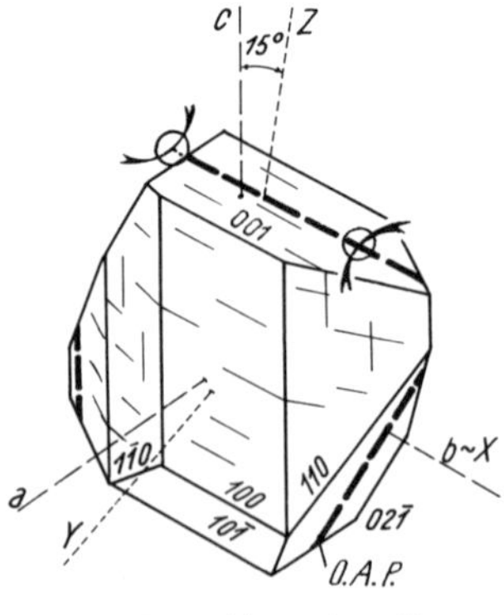

120-2. Montebrasite (DANA 1892)

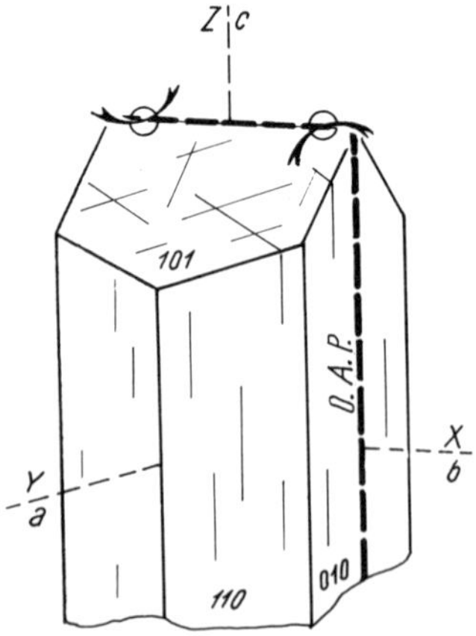

121. Wavellite

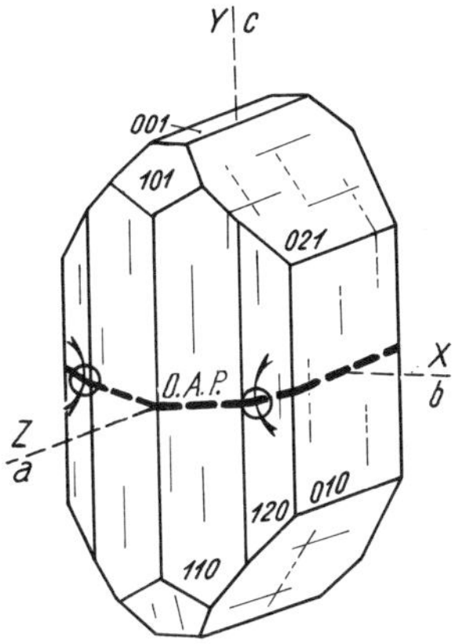


122-1. Forsterite
(FANTAPPIE 1905)

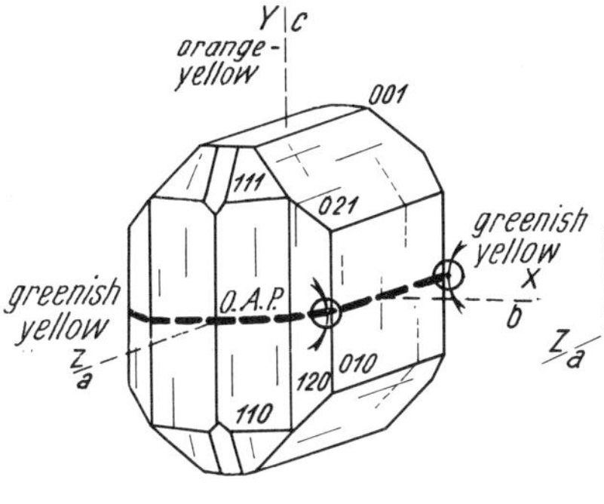


122-2. Fayalite
(SÖLLNER 1911)

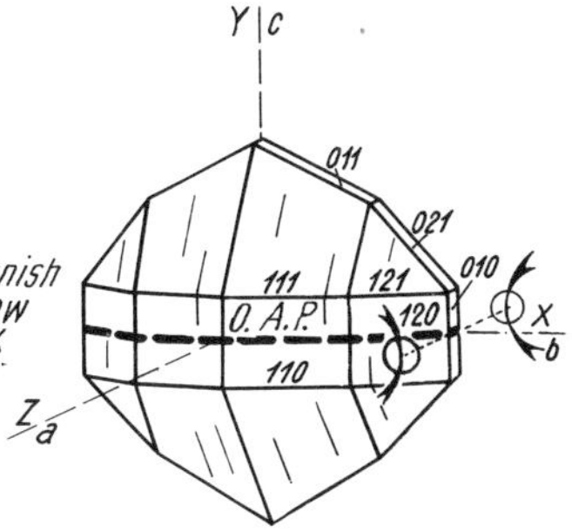


123. Monticellite
(v. RATH 1871)

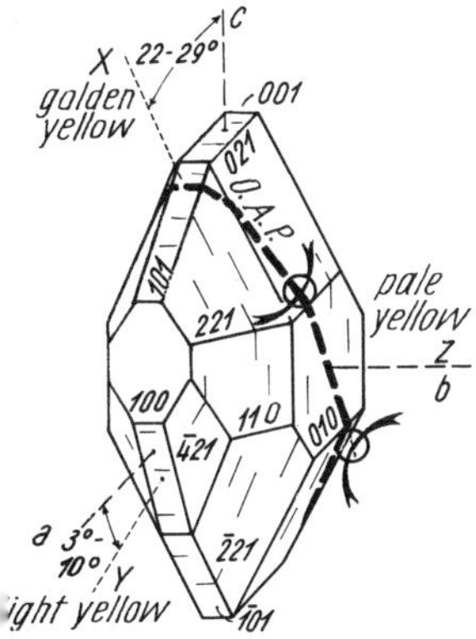


124-1. Chondrodite

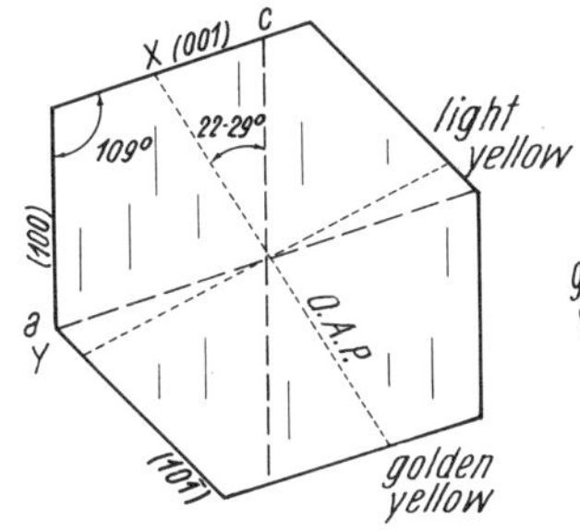


124-2. Chondrodite, section ǁ (010)

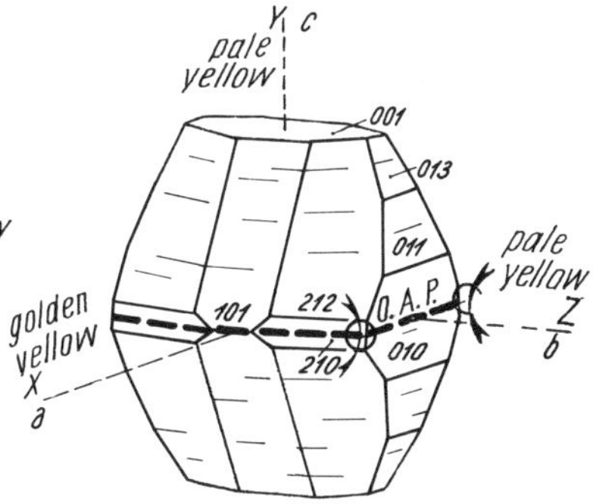


125. Humite
(SJÖGREN 1882)

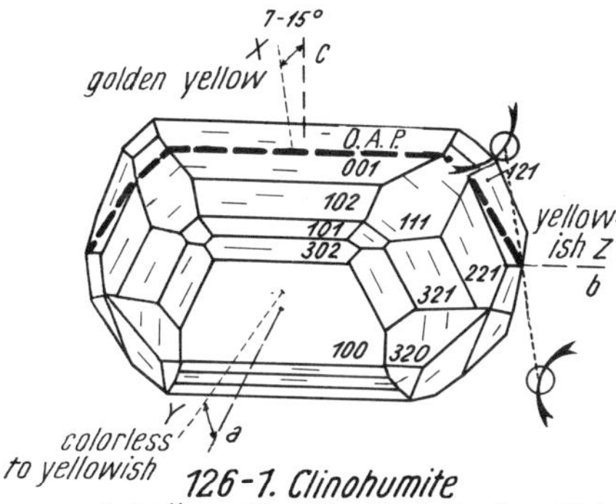


126-1. Clinohumite
(similar: No 127. Titanclinohumite)

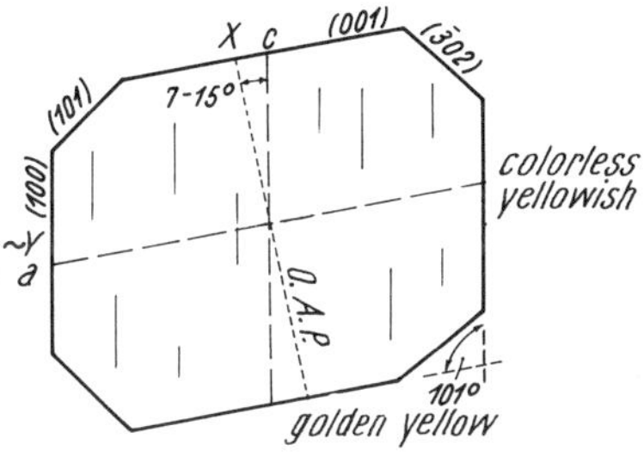


126-2. Clinohumite, section ǁ (010)

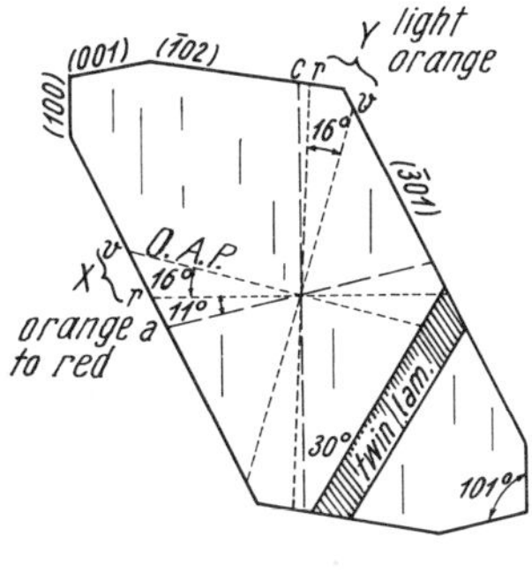


126a. Titanclinohumite,
section ǁ (010)

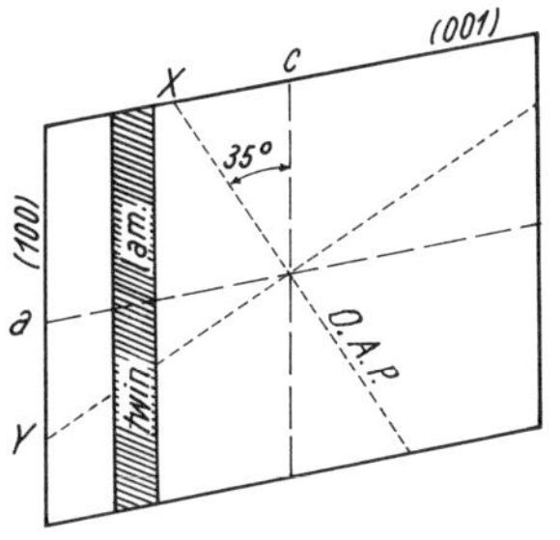


128. Merwinite, section ǁ (010)

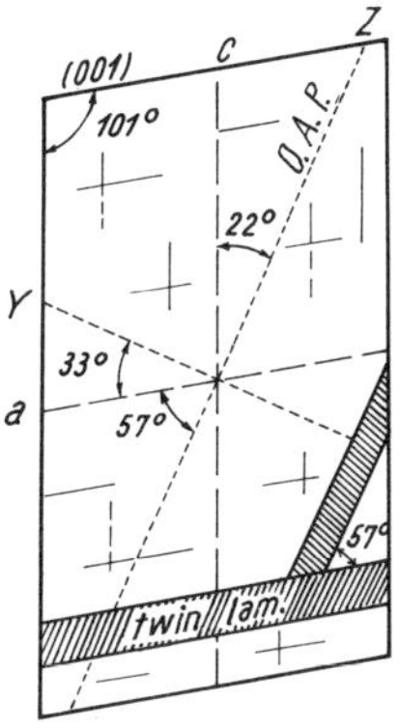


129. Spurrite, section ǁ (010)

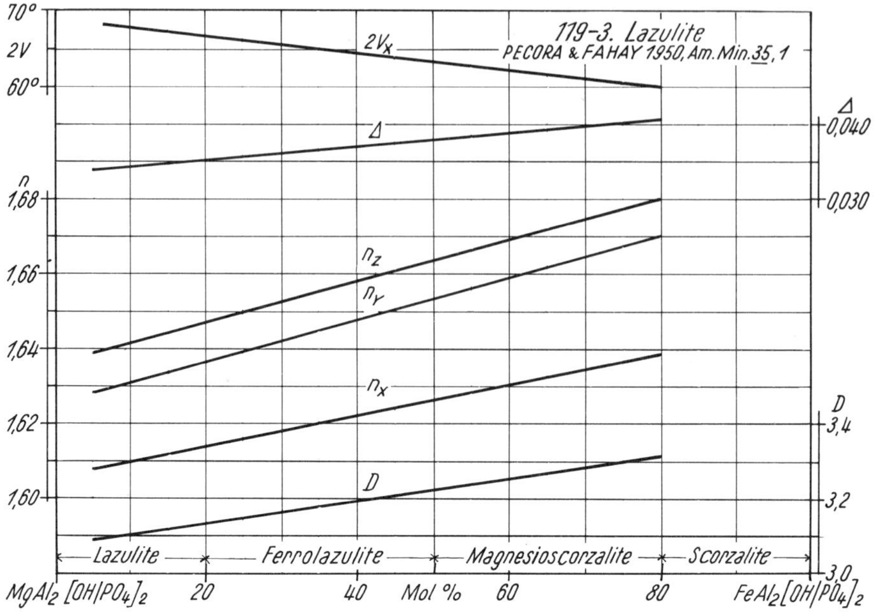
119-3. Lazulite
PECORA & FAHAY 1950, Am. Min. 35, 1
2V
70°
60°
$2V_X$
Δ
0,040
0,030
n
1,68
1,66
1,64
1,62
1,60
$n_Z$
$n_Y$
$n_X$
D
3,4
3,2
3,0
Lazulite
Ferrolazulite
Magnesioscorzalite
Scorzalite
$MgAl_2[OH|PO_4]_2$
20
40
Mol %
60
80
$FeAl_2[OH|PO_4]_2$

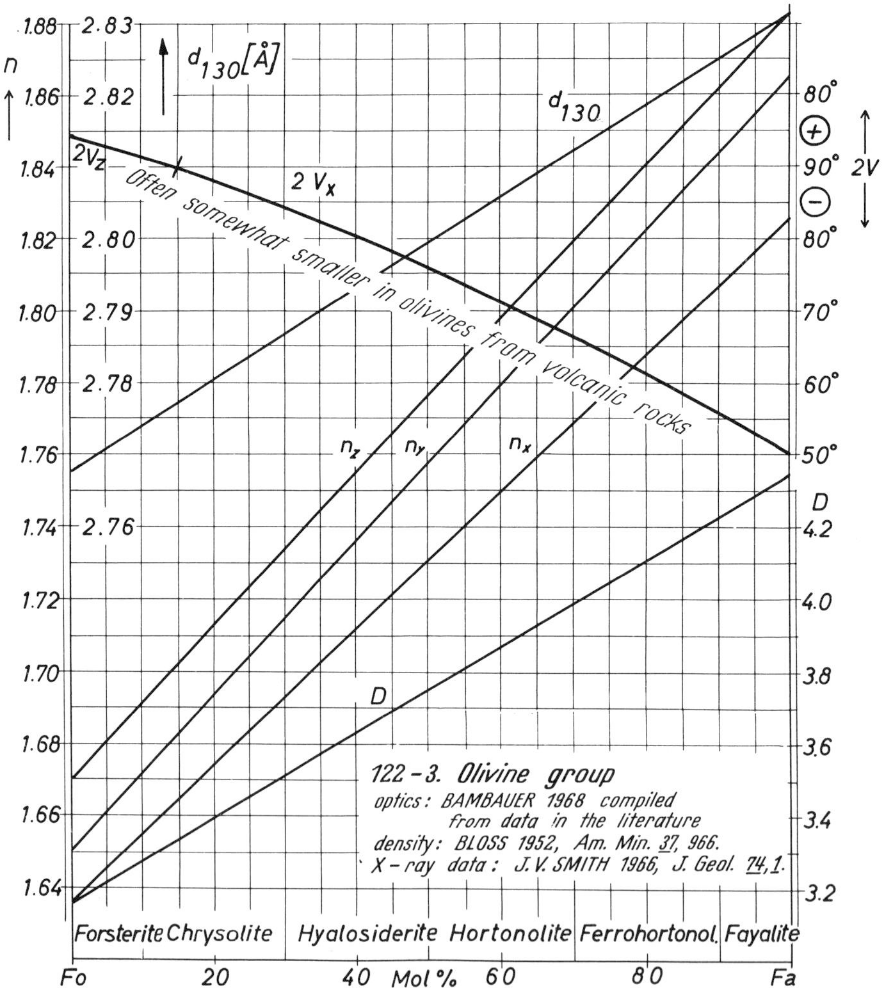
n
1.88
1.86
1.84
1.82
1.80
1.78
1.76
1.74
1.72
1.70
1.68
1.66
1.64
2.83
2.82
2.80
2.79
2.78
2.76
$d_{130}$[Å]
$d_{130}$
$2V_Z$
$2V_X$
Often somewhat smaller in olivines from volcanic rocks
80°
⊕
90°
2V
⊖
80°
70°
60°
50°
D
4.2
4.0
3.8
3.6
3.4
3.2
$n_Z$
$n_Y$
$n_X$
D
122-3. Olivine group
optics: BAMBAUER 1968 compiled from data in the literature
density: BLOSS 1952, Am. Min. 37, 966.
X-ray data: J.V. SMITH 1966, J. Geol. 74, 1.
Forsterite
Chrysolite
Hyalosiderite
Hortonolite
Ferrohortonol.
Fayalite
Fo
20
40
Mol %
60
80
Fa

123-2. Monticellite group
SCHALLER 1935 (Am. Min 20, 815)

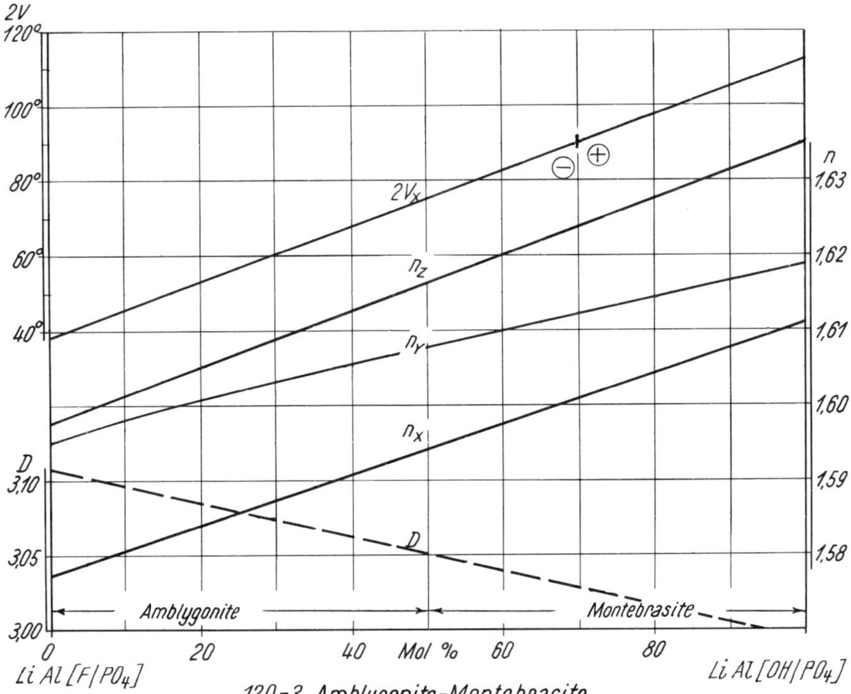


120-3. Amblygonite-Montebrasite
WINCHELL 1927, (Opt. Min. 149) corr. TRÖGER 1951.
(Copyrighted, 1951, by A. N. Winchell in Elements of Optical Mineralogy, part II; Reproduced by permission.)

| No. | Name<br>Symmetry<br>Chem. Composition | Common Crystal Forms, Twinning<br>Crystal Habit<br>Aggregates | Cleavage<br>Translation Gliding<br>Deform. Twinning<br>Hardness, Density | Optic Orientation | Refractive Indices<br>Birefringence |
|---|---|---|---|---|---|
| 133 | Topaz orh<br>$Al_2[(F,OH)_2/SiO_4]$ | {120},{hk0},{001},<br>{0kl}, {hhl};<br>short columns, grains,<br>acicular ∥ c: "pycnite" | # (001) perf.;<br>H = 8 D = { 3.56 (F) / 3.49 (F,OH)<br>diagr. p. 61 | X ∥ a<br>Y ∥ b, O.A.P. (010)<br>Z ∥ c, bi;<br>l = (−) fig. p. 58 | (F) (F,OH)<br>$n_x$ = 1.606 – 1.635<br>$n_y$ = 1.609 – 1.637<br>$n_z$ = 1.616 – 1.644<br>⊕Δ = 0.010 – 0.008 |
| 134 | Staurolite orh<br>$Al_4Fe^{2+}[O/OH/SiO_4]_2$ | {110},{010},{001};<br>⟐ {032} and {232},<br>no lamellae;<br>short prisms ∥ c | # (010) distinct,<br># {110} poor: 50½°<br>H = 7–7½ D = 3.76± | X ∥ b<br>Y ∥ a, O.A.P. (100)<br>Z ∥ c, bi;<br>l = (+) fig. p. 58 | $n_x$ = 1.739 – 1.747<br>$n_y$ = 1.745 – 1.753<br>$n_z$ = 1.752 – 1.762<br>⊕Δ = 0.013 – 0.015<br>diagr. p. 61 |
| 135 | Sapphirine mcl<br>$(Mg,Fe^{2+})_2Al_4$<br>$[O_6/SiO_4]$ | {010},{110},{100},<br>{001};<br>β = 110°;<br>tabular ∥ (010),<br>granular | # (010) distinct,<br># (100), (001) poor;<br>H = 7½ D = 3.49± | XΛa = +12° to +14°, bi;<br>Y ∥ b, O.A.P. (010)<br>ZΛc = +6° to +8°<br>l = (+) (−) fig. p. 58 | $n_x$ = 1.705 – 1.717<br>$n_y$ = 1.709 – 1.722<br>$n_z$ = 1.711 – 1.724<br>⊖Δ = 0.006 – 0.007 |
| 136 | Epidote group mcl<br>Clinozoisite: Fe-poor<br>Epidote s.str.: Fe-rich<br>$Ca_2(Al,Fe^{3+})Al_2$<br>$[O/OH/SiO_4/Si_2O_7]$ | {001},{100},{101},<br>{h0l}, {1̄11};<br>β = 115½°;<br>⟐ (100) lam.;<br>columnar ∥ b, granular | # (001) perf.,<br># (100) dist. > 115½°<br>H = 6½ D = 3.29± clinoz. / 3.45± epid. | XΛc = { +12°, 0° } clin. { −5° } epid.<br>Y ∥ b, O.A.P. (010)<br>ZΛa = { +15°, +25° } clin. { +30° } epid.<br>l = (+) (−) fig. p. 58, 62 | clin. epid.<br>$n_x$ = (1.697)–1.714–1.728<br>$n_y$ = (1.699)–1.722–1.758<br>$n_z$ = (1.702)–1.729–1.776<br>Δ = ⊕0.005–⊕/⊖0.015–⊖0.048<br>diagr. p. 62 |
| 136a | Tawmawite mcl<br>$Ca_2(Al,_{Cr,Fe})_3$<br>$[O/OH/SiO_4/Si_2O_7]$ | {100},{001},{101},<br>{h0l}, {1̄11};<br>β = 115½°;<br>⟐ (100);<br>columnar ∥ b | # (001) perf.,<br># (100) dist.;<br>H = 6½ D = 3.40± | XΛc = −24°, bi;<br>YΛa = +49½°,<br>O.A.P. ⊥ (010)<br>Z ∥ b fig. p. 58 | $n_x$ = 1.695 – 1.707<br>$n_y$ = 1.697 – 1.711<br>$n_z$ = 1.700 – 1.715<br>⊖Δ = 0.005 – 0.008 |
| 137 | Piemontite group mcl<br>$Ca_2(Mn^{3+},Fe^{3+})Al_2$<br>$[O/OH/SiO_4/Si_2O_7]$ | {100},{001},{101},<br>{h0l}, {1̄11};<br>β = 115½°;<br>⟐ (100) lam.;<br>columnar ∥ b, subparallel acicular, granular | # (001) good,<br># (100) and (010) dist.;<br>H = 6–6½ D = 3.45± | XΛc = −4° to −7½°<br>Y ∥ b, O.A.P. (010)<br>ZΛa = +29° to +33°, bi;<br>l = (+) (−)<br>fig. p. 58 and 62 | Mn-poor Mn-rich<br>$n_x$ = 1.730 – 1.756<br>$n_y$ = 1.747 – 1.789<br>$n_z$ = 1.765 – 1.829<br>⊕/⊖Δ = 0.035 – 0.073<br>diagr. p. 63 |
| 138 | Allanite mcl<br>(Orthite)<br>$(Ca,Ce,La)_2$<br>$(Al,Fe^{3+})Al_2$<br>$[O/OH/SiO_4/Si_2O_7]$ | {100},{001},{102},<br>{h0l};<br>β = 115°;<br>⟐ (100);<br>tabular ∥ (100),<br>granular | # (001) distinct,<br># (100) and {110} poor;<br>brittle;<br>H = 5½–6 D = 4.2–3.7<br>(isotropic up to D = 2.8)<br>diagr. p. 63 | XΛc = −(22°) to 35° to (47°), bi;<br>Y ∥ b, O.A.P. (010)<br>ZΛa = +(47°) to 60° to (72°);<br>partly strong dispersion;<br>fig. p. 59, 62 | fresh "brown" isotr.<br>$n_x$ = 1.788–1.727<br>$n_y$ = 1.810–1.739 } 1.54<br>$n_z$ = 1.820–1.751<br>⊖Δ = 0.032–0.024–0.000<br>diagr. p. 63 |
| 139 | Zoisite (α) orh<br>and Pseudozoisite (β)<br>$Ca_2Al_3$<br>$[O/OH/SiO_4/Si_2O_7]$ | {101},{210},{100},<br>{001};<br>cryst. setting:<br>BRÖGGER; no ⟐;<br>short columns ∥ b,<br>stellate, granular | # (100) perf.,<br># (001) poor;<br>parting ∥ (010);<br>H = 6–6½ D = 3.30± | α / β<br>X ∥ b / a<br>Y ∥ a / b<br>Z ∥ c, bi / c, bi<br>O.A.P. ∥ (100) / ∥ (010)<br>l = (+) (−) fig. p. 59 | α / β<br>$n_x$ = 1.701 / 1.695<br>$n_y$ = 1.702 / 1.695<br>$n_z$ = 1.707 / 1.702<br>⊕Δ = 0.006 / 0.007 |
| 140 | Ilvaite orh<br>(Lievrite)<br>$CaFe^{2+}_2Fe^{3+}$<br>$[OH/O/Si_2O_7]$ | {110},{120},{101},<br>{010};<br>acicular ∥ c,<br>fibrous, massive | # (010) distinct,<br># (001) and (100) poor;<br>H = 5½–6 D = 4.03± | X ∥ b<br>Y ∥ a, O.A.P. (100)<br>Z ∥ c, bi;<br>l = (+) fig. p. 59 | $n_x$ ~ 1.878<br>$n_y$ ~ 1.890<br>$n_z$ ~ 1.926<br>⊕Δ ~ 0.048 |
| 140a | Chevkinite mcl<br>$(Ce,La)_2Ti_2$<br>$[O_4/Si_2O_7]$ | {100},{001},{101};<br>β = 113°;<br>⟐ (001);<br>columnar ∥ b, granular | no #;<br>H = 5½ D = 4.3–4.6 | XΛc = −11 to −26°, bi;<br>YΛa = +34 to +49°<br>O.A.P. ⊥ (010)<br>Z ∥ b; | $n_x$ ~ 1.970<br>$n_y$ ~ 2.020<br>$n_z$ ~ 2.050<br>⊖Δ 0.080 |
| 141 | Lawsonite orh<br>$CaAl_2$<br>$[(OH)_2/Si_2O_7]\cdot H_2O$ | {010},{101},{001},<br>{0hl};<br>⟐ {101}: 67°;<br>lam.: undul. parquet-like; tabular ∥ (010),<br>elongated ∥ c, bladed acicular, stellate | # (010) perf.,<br># (100) good,<br># {101} distinct;<br>H = 6 D = 3.09± | X ∥ c<br>Y ∥ a, O.A.P. (100)<br>Z ∥ b, bi;<br>l = (−) fig. p. 59 | $n_x$ = 1.665 – 1.665<br>$n_y$ = 1.674 – 1.673<br>$n_z$ = 1.684 – 1.686<br>⊕Δ = 0.019 – 0.021 |

| Angle and Dispersion of Optic Axes | Color in Thin Section Pleochroism | Chemical Reactions Special Remarks | Distinguishing Features of Similar Minerals | Occurrences | |
|---|---|---|---|---|---|
| $2V_Z = 66^o$(F)$-44^o$ (F,OH)<br>r > v distinct<br>diagr. p. 61 | colorless | insoluble in cold conc. HF; slightly soluble in $H_2SO_4$ | phenakite: uniaxial, diff. #; quartz: sm. n, no #; andalusite: X ∥ c, ⊖, gr. 2V | pneumatolytic to pegmatitic with quartz, F- and B-minerals and cassiterite, especially in granites and their surroundings; heavy mineral in sands, placers | **133** |
| $2V_Z = 79^o-90^o$<br>r > v strong | X = pale yellow,<br>Y = light yellow,<br>Z = reddish yellow,<br>X ≈ Y < Z | insoluble in cold acids; usually poikiloblastically contaminated | dravite: uniaxial ⊖; epidote: lighter colored, more greenish; allanite: usually with isomorphic zoning; kaersutite: gr. Δ and # | with garnet, kyanite, rutile, micas in metamorphic schists and feldspar-poor paragneisses; heavy mineral in sands | **134** |
| $2V_X = 50^o-70^o$<br>r < v<br>oblique dispersion strong | X = colorless, pink, yellow,<br>Y = light blue, bluish green,<br>Z = deep blue, azure-blue,<br>X < Y < Z | insoluble in acids; decomposed by $KHSO_4$-melt; abnormal interference colors, especially in sections ≈ ⊥ X | corundum: uniaxial ⊖, gr. n; lazulite: sm. n, gr. Δ; kyanite: gr. Δ; serendibite: ⊕, ◊; chloritoid: ⊕, # (001) perf. | high grade metam. in Al-rich metamafitites; with anthophyllite, cordierite, spinel, corundum in plagioclase gneisses and biotite schists | **135** |
| $2V_Z = 65^o$ } clin.: r < v<br>$2V = 90^o$ }<br>$2V_X = 65^o$ } epid.: r ≫ v | clinoz.: colorless<br>epidote:<br>X = colorless to lemon yellow,<br>Y = greenish yellow,<br>Z = colorless to light yellowish green: X < Z < Y,<br>rarely: (X < Y < Z) | only epid. soluble in HCl; zonal structure; abnormal interference colors; no natural alteration products | zoisite: O.A.P. (001) and sm. 2V; fayalite: diff. color and #; pyroxene: ⊕ and # {110} perf.; vesuvianite: uniaxial ⊖; monazite: absorption spectrum! (see Vol. 2) | with augite and hornblende, quartz, albite in all basic metamorphic rocks; hydrothermal in fissures; clinozoisite also in saussurite | **136** |
| $2V_Z = 45^o-50^o$<br>r < v distinct | X = emerald-green to olive-green,<br>Y = yellow to brown,<br>Z = emerald-green | decomposed by HCl; conspicuous colors; abnormal interference colors; weathers easily in hydrous environments | viridine: sm. n, gr. $2V_Z$; epidote: O.A.P. (010), gr. 2V; aegirine: gr. 2V, different paragen. | low to medium grade metam. with jadeite; pneumatolytic with uvarovite in tremolite-quartzite; in metam. gabbros and diabases | **136a** |
| 2V ≈ $90^o \pm 20^o$;<br>$2V_X = 70^o-90^o$: $Fe^{3+}$-rich,<br>$2V_Z = 70^o-90^o$: $Mn^{3+}$-rich<br>r ≷ v diagr. p. 63 | X = yellow, orange,<br>Y = lilac, pink,<br>Z = carmine-red, purple,<br>Z > X > Y | after melting soluble in HCl; intergrown with epidote; strong pleochroism | thulite: sm. n, Δ; bustamite: oblique extinction; dumortierite: never yellow; glaucophane: sm. n, sm. Δ, #; kaemmererite: diff. #, sm. 2V | low to medium grade metam. with quartz, mica, glaucophane, spessartite etc.; autometam. in rhyolites; in metam. Mn-ore deposits | **137** |
| $2V_X = 70^o-90^o$<br>$-(120^o)$<br>r ≫ v very strong | fresh to metamict:<br>X = colorless to yellow-brown<br>Y = yellow-brown to red-brown<br>Z = light green to red-brown<br>X < Y ≦ Z | causes pleochroitic haloes; often metamict: sm. n and D, then soluble in HCl; isomorphous layering and intergrowth with epidote | melanite: gr. n; epidote: sm. X ∧ c; gadolinite: $2V_Z$, X-ray | pegmat.-pneumatolytic in granites, diorites; in metamorphic rocks (gneisses); heavy mineral in mech. conditioned sands | **138** |
| α: $2V_Z = 20-50^o$, r ≪ v<br>β: $0-30^o$, r > v | almost colorless;<br>thulite (β):<br>a = light greenish yellow,<br>b = lilac to pinc,<br>c = yellowish<br>fig. p. 59 | insoluble in acids; gelatinizes after red-hot heating; intergrowth with epidote; abnormal interference colors | vesuvianite, apatite: uniaxial ⊖; clinozoisite: abnorm. blue interference colors | with amphiboles in basic metam. rocks; in saussurite; in calc-silicate rocks; in quartz veins | **139** |
| $2V_Z \sim 40^o(-\sim 90^o)$<br>r ≪ v very strong | X = yellow-brown to brown<br>Y = dark brown to opaque<br>Z = dark green to opaque<br>X < Y < Z (see Nr. 41) | gelatinizes easily with HCl; hardly translucent; strong pleochroism | pseudobrookite: gr. n, Y = red; goethite: red-brown; ludwigite: gr. Δ and sm. 2V; kaersutite: weaker absorption; aenigmatite, rhoenite: oblique extinction | contact-metasomatic in skarns with hedenbergite, hornblende, epidote etc.; in diabase-hornfelses; deuteric in gabbros | **140** |
| $2V_X \sim 60^o \pm 20^o$<br>r > v distinct | pure / with $Fe^{3+}$<br>X = grey to yellow brown<br>Y = pale red-brown to opaque<br>Z = dark red-brown to opaque | unweathered cryst. not attacked by HCl; often metamict: sm. n. and soluble in HCl | allanite: paler col.; ilvaite: Z = green; columbite: X-ray; gadolinite: Y and Z = green; baddeleyite: uniaxial ⊕ | with gadolinite, sphene, monazite, allanite in granite-pegmatites; in rhyolite-tuffs | **140a** |
| $2V_Z = 79^o-85^o$<br>r ≫ v strong | colorless | slightly soluble even by HF, turbid after heating, intergrowth with pumpellyite: X of laws. ∥ Y of pu. | pumpellyite: sm. 2V; tremolite: oblique extinction; prehnite: gr. Δ; wollastonite: sm. $2V_X$ | with albite in metam. gabbros and diabases; in glaucophane-bearing rocks | **141** |

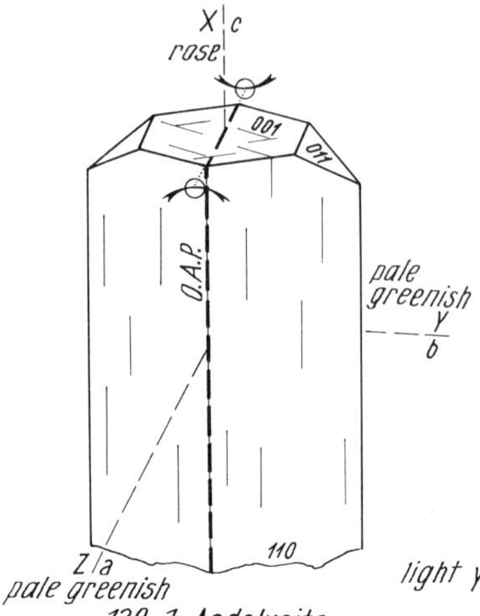


130-1. Andalusite

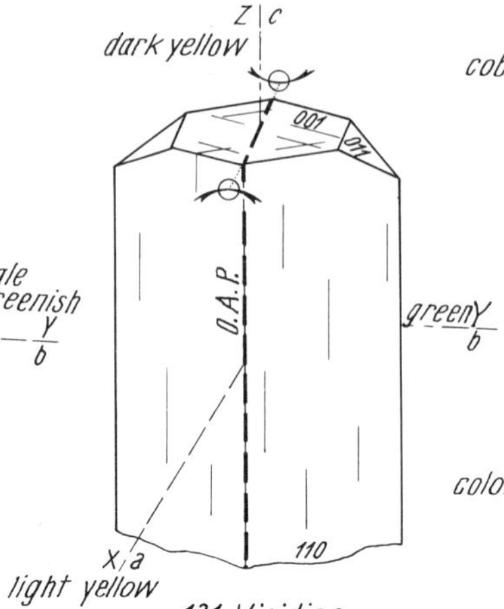


131. Viridine

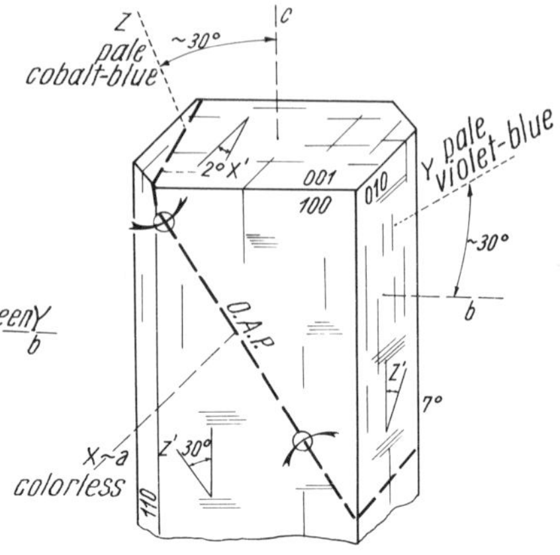


132-1. Kyanite

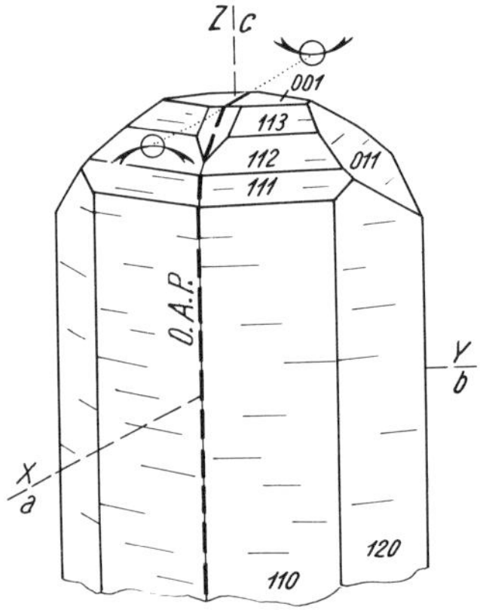


133-1. Topaz

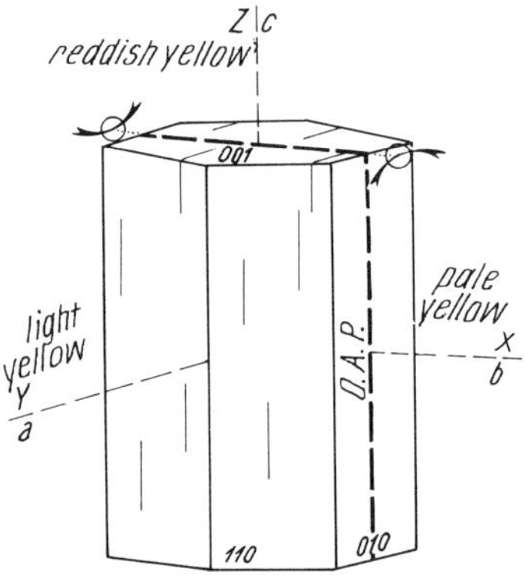


134-1. Staurolite

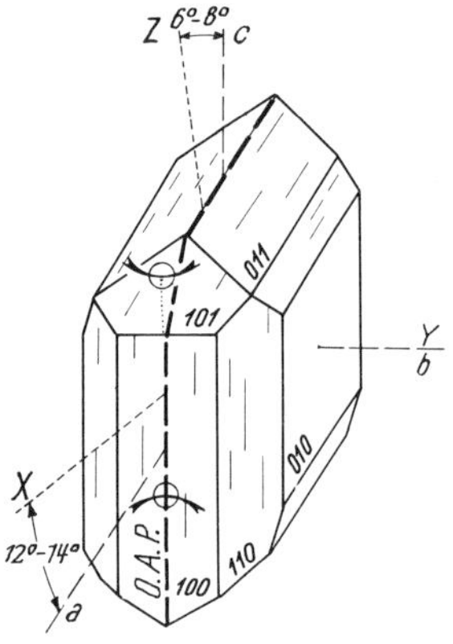


135-1. Sapphirine

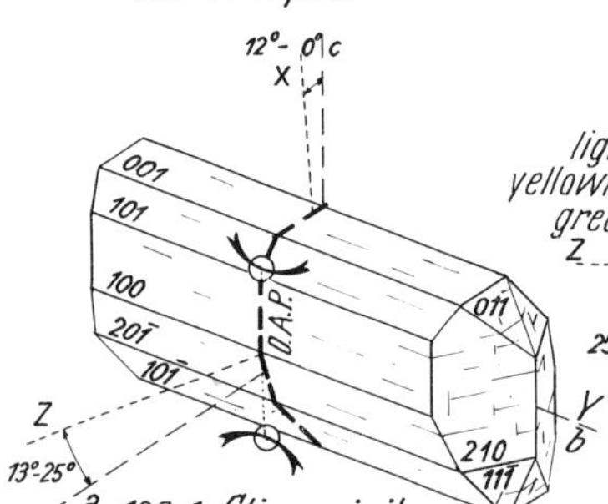


136-1. Clinozoisite

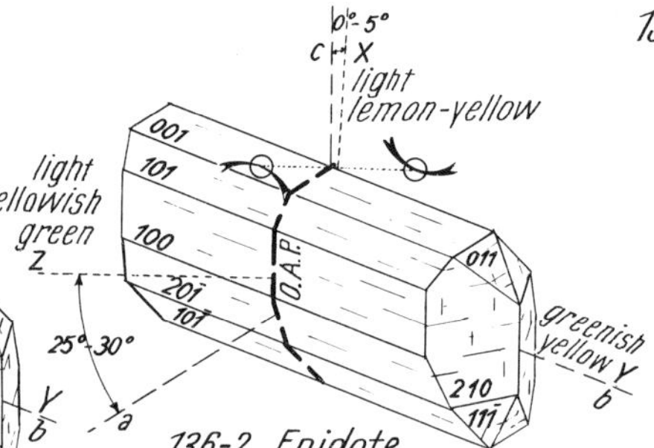


136-2. Epidote

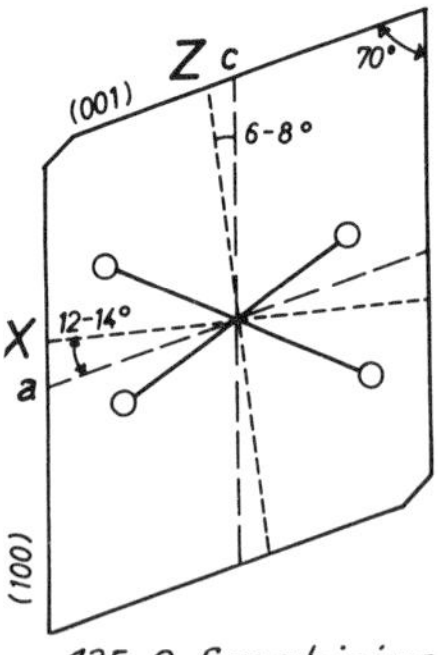


135-2. Sapphirine section ∥(010)

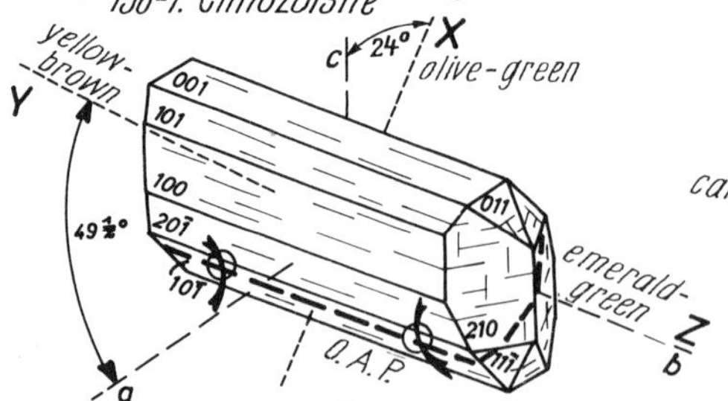


136a. Tawmawite

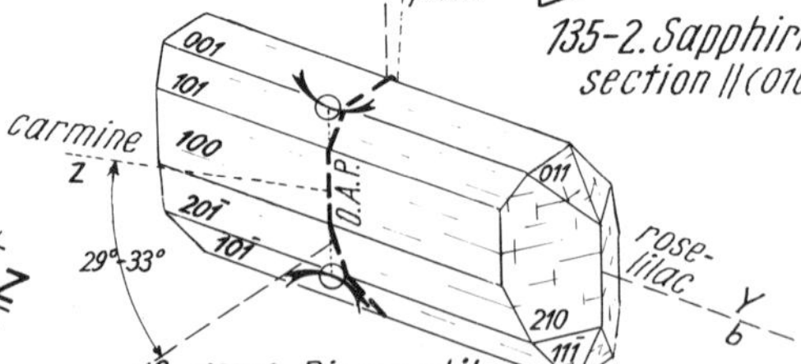


137-1. Piemontite

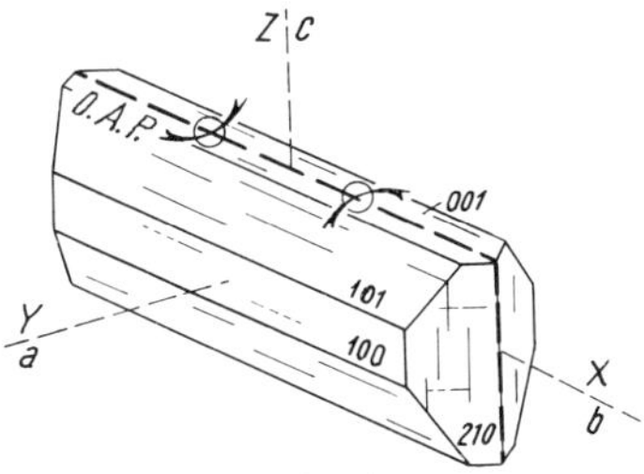


139-1. Zoisite (α)

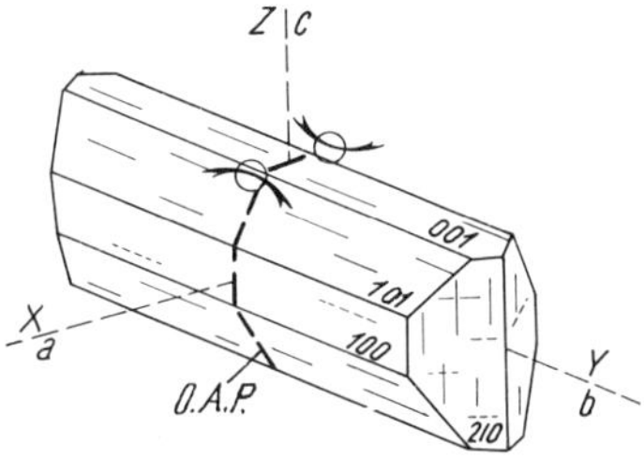


139-2. Pseudozoisite (β)

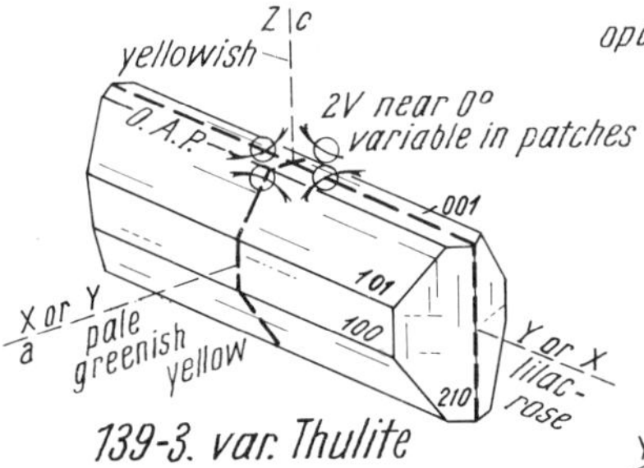


139-3. var. Thulite

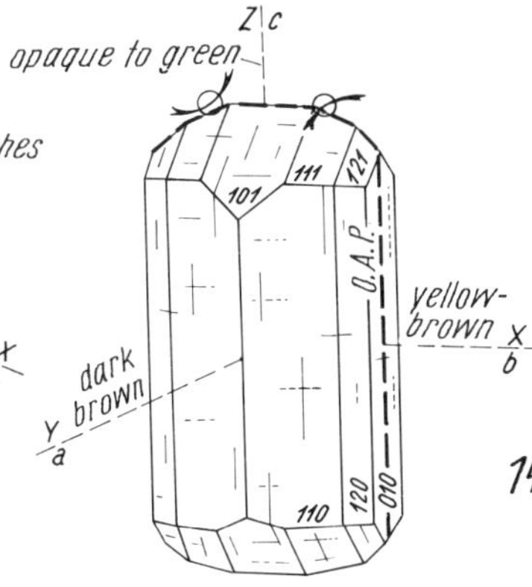


140. Ilvaite

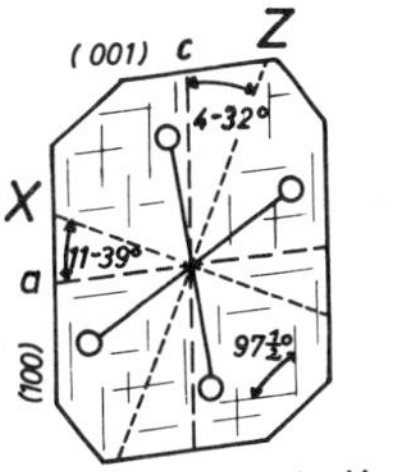


141a-2. Pumpellyite
section ‖ (010)

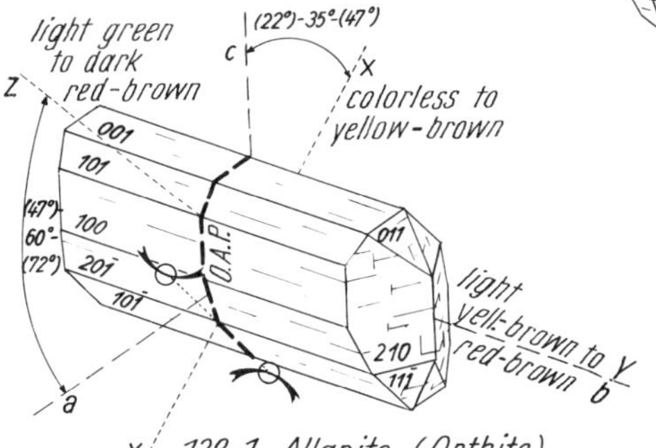


X 138-1. Allanite (Orthite)

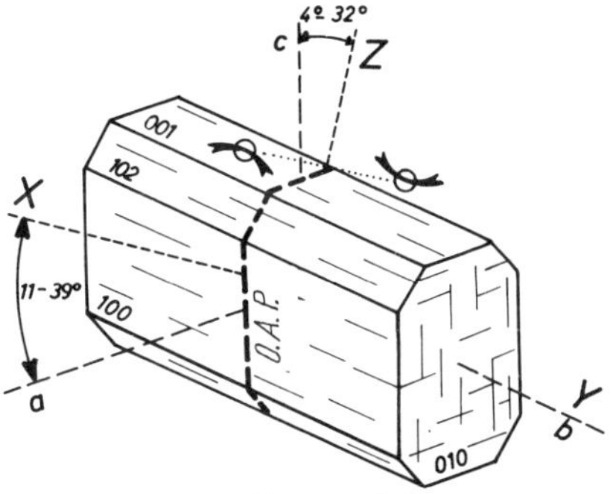


141a-1. Pumpellyite

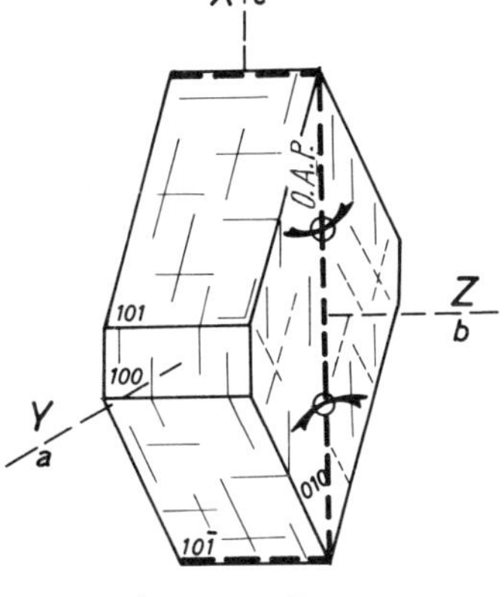


141. Lawsonite

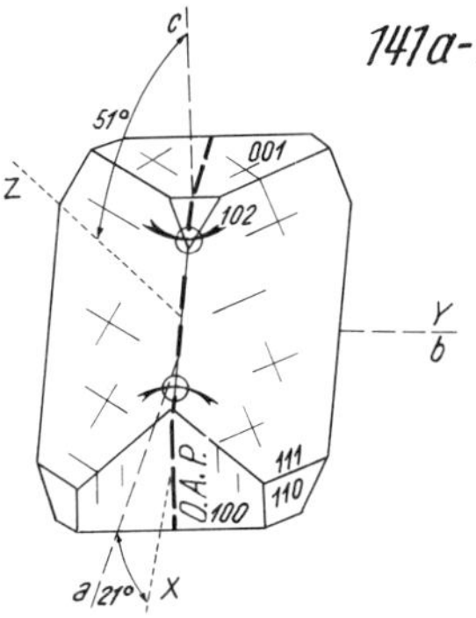


142-1. Titanite
(Sphene)

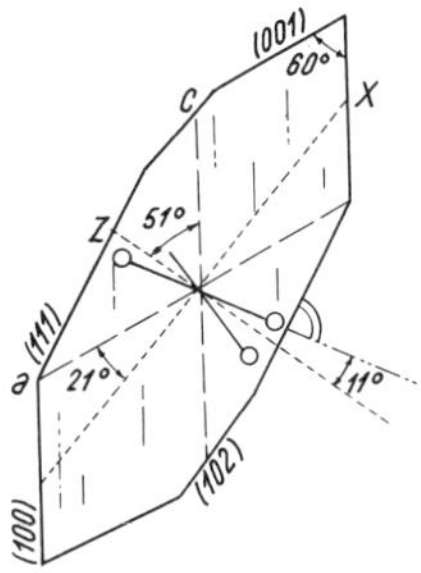


142-2. Titanite
section ‖ (010)

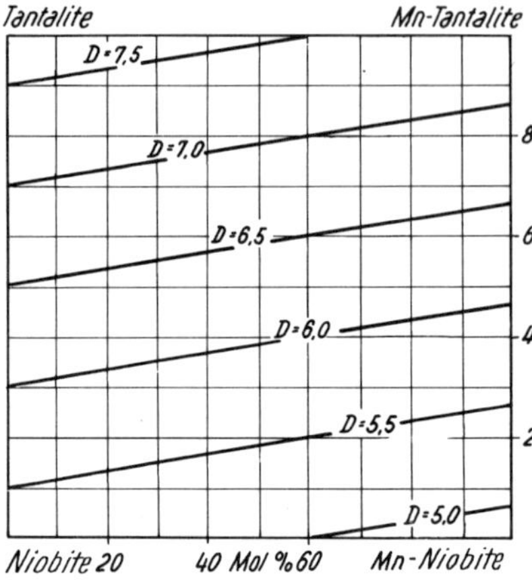


102-2. Columbite group
(Niobite-Tantalite)
WINCHELL 1927
(Copyrighted, 1951, by A. N. Winchell in Elements of Optical Mineralogy, part II; Reproduced by permission.)

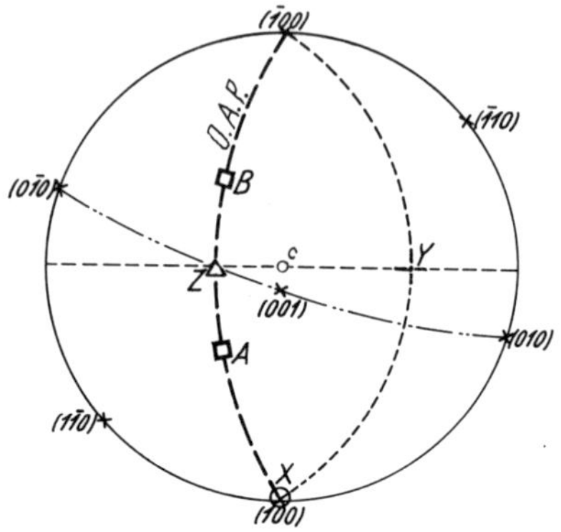


132-2. Kyanite

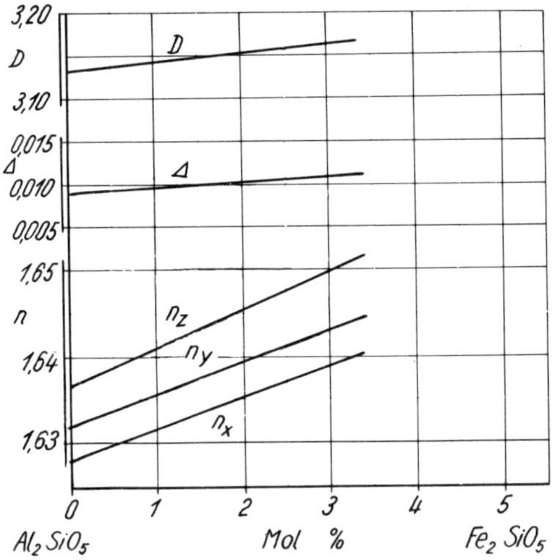


130-2. Andalusite
TRÖGER 1950

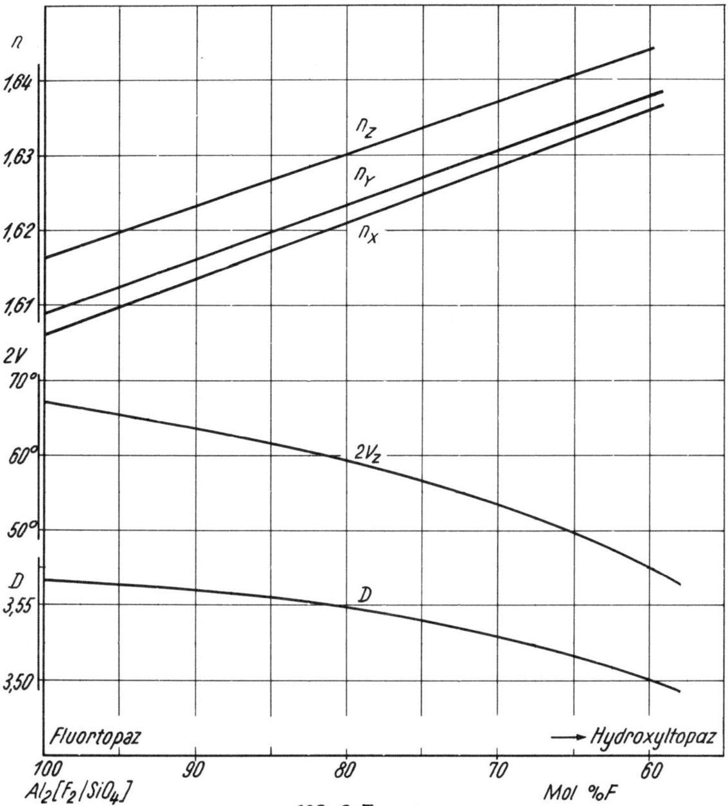


133-2. Topaz

TRÖGER 1950

See also RIBBE & ROSENBERG 1971 (Am. Min. 56, 1812 – 1821)

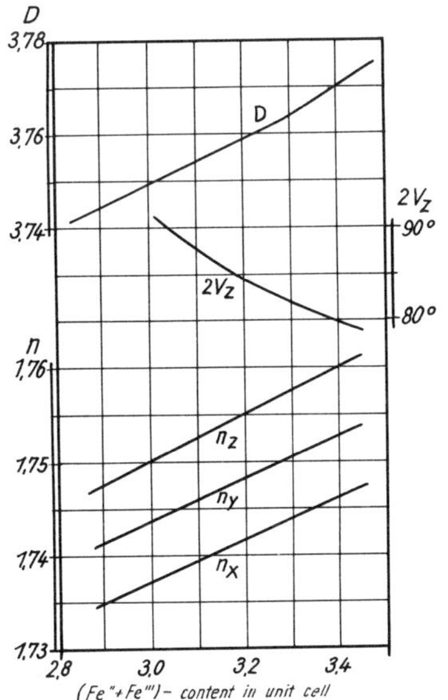


134-2. Staurolite

JUURINEN 1956 (Ann. Acad. Sci. Fenn., A III. 47. 33)

See also GRIFFEN & RIBBE 1974 (Am. J. Sci., COOPER Vol.)

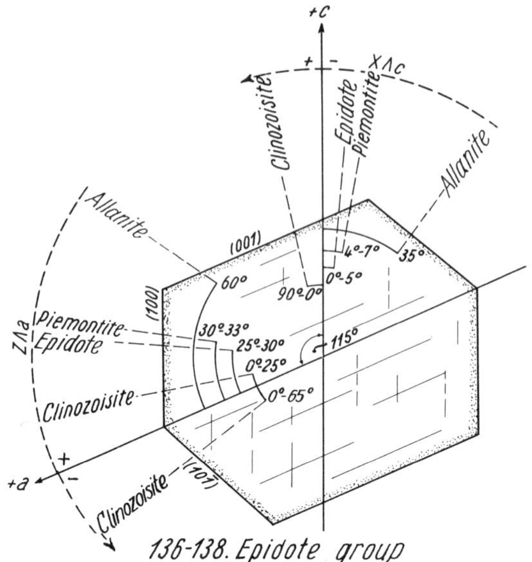


136-138. Epidote group
section // (010)

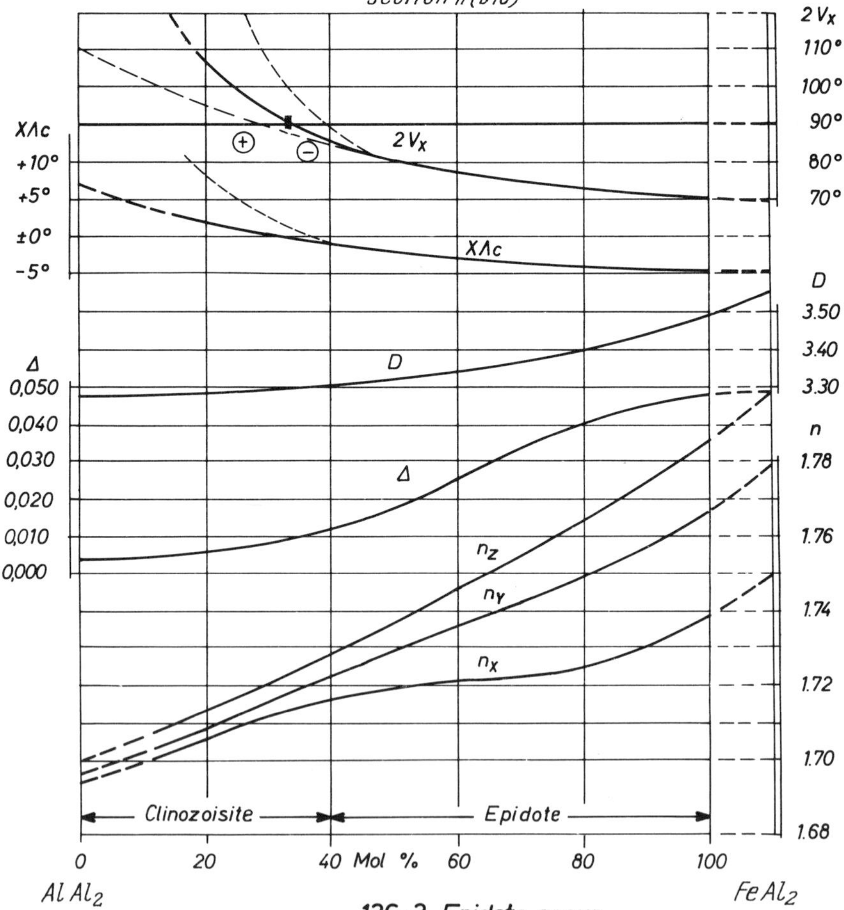


**136-3. Epidote group**

TRÖGER 1952 corrected TROCHIM 1969

after STRENS 1966 (Min. Mag. 35, 928-944)
and HÖRMANN & RAITH 1971 (N. Jb. Miner. Abh. 116, 41-60)

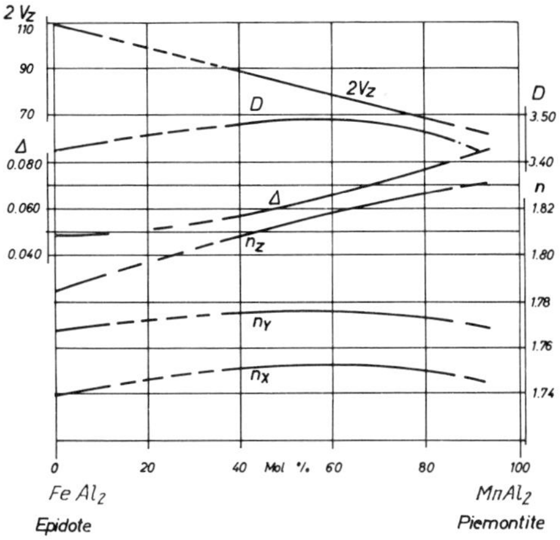


137-2. Piemontite group
MARMO & al. 1959 (Bull. Comm. géol. Finlande 184, 11-20)

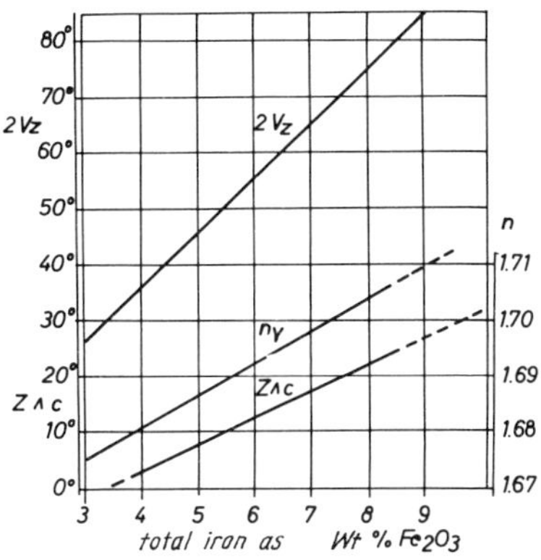


141a-3. Pumpellyite
COOMBS 1953 (Min. Mag. 30, 113-135)

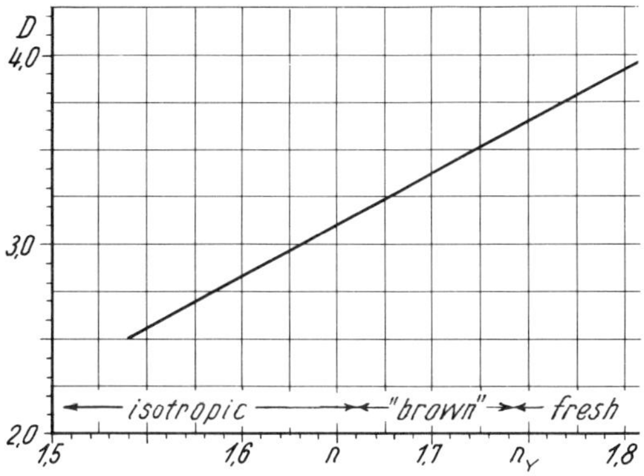


138-2. Allanite (Orthite)
TRÖGER 1950

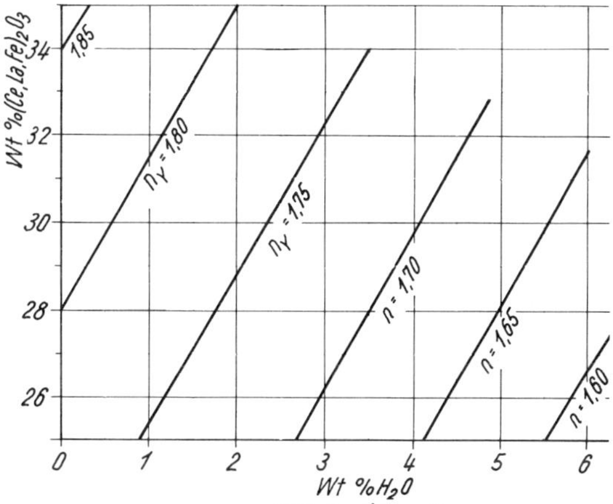


138-3. Allanite
TRÖGER 1950

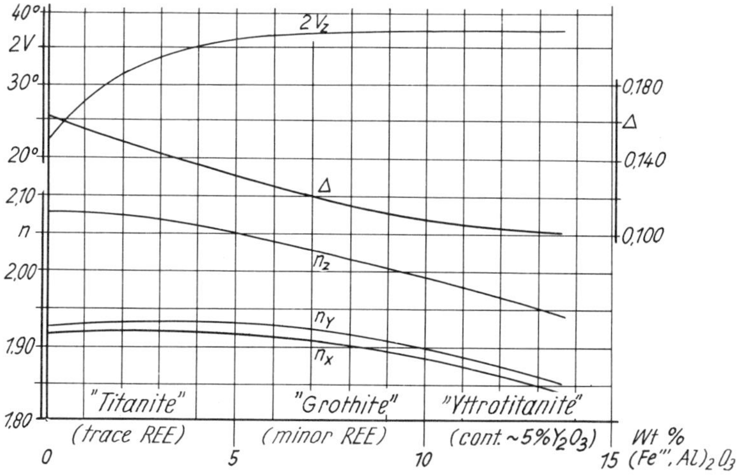


142-3. Titanite (Sphene)
TRÖGER 1955 (Handb. Mikr. Techn., 4, 1, 207)
REE = rare earth elements

| No. | Name<br>Symmetry<br>Chem. Composition | Common Crystal Forms, Twinning<br>Crystal Habit<br>Aggregates | Cleavage<br>Translation Gliding<br>Deform. Twinning<br>Hardness, Density | Optic Orientation | Refractive Indices<br>Birefringence |
|---|---|---|---|---|---|
| 141a | Pumpellyite mcl<br>$Ca_4(Mg,Fe^{3+})Al_4$<br>$[(OH)_2/(SiO_3OH)$<br>$(SiO_4)_5] \cdot 2H_2O$ | {100},{001};<br>β = 97½°;<br>◈ (001)<br>◈ (100) rare;<br>intergrown ‖ (010),<br>bladed ‖ b, granular | # (100) perf. } 97½°;<br># (001) dist. }<br>H = 5½ D = 3.16± | X Λ a = +11 to +40°<br>Y ‖ b, O.A.P. (010)<br>Z Λ c = −4° to −32°, bi;<br>l = (+) (−)<br>fig. p. 59 | $n_x$ = 1.665 – 1.702<br>$n_y$ = 1.670 – 1.709<br>$n_z$ = 1.683 – 1.722<br>⊕Δ = 0.018 – 0.020 |
| 142 | Titanite mcl<br>(Sphene)<br>$CaTi[O/SiO_4]$ | {111},{100},{001},<br>{110};<br>β = 119¾°<br>◈ (100);<br>lenticular,<br>tabular ‖ (001),<br>prismatic, granular | # {110} distinct,<br># {111} poor,<br>parting ~ ‖ {552};<br>H = 5 – 5½ D = 3.51±<br>(yttrotitanite:<br>D = 3.65) | X Λ a = −21°<br>Y ‖ b, O.A.P. (010)<br>Z Λ c = +51° (sphene), bi;<br>+59° (yttrotit.)<br>Z ≈ ⊥ (102)<br>fig. p. 59 | sphene yttrotit.<br>$n_x$ = 1.921–1.920–1.885<br>$n_y$ = 1.927–1.932–1.896<br>$n_z$ = 2.081–2.050–1.993<br>⊕Δ = 0.160–0.130–0.108 |
| 143 | Woehlerite mcl<br>$Ca_2NaZr$<br>$[(F,OH,O)_2/Si_2O_7]$ | {100},{010},{hk0},<br>{011};<br>β = 109°;<br>◈ (100) lam.;<br>tabular ‖ (100) | # (010) poor,<br>very brittle;<br>H = 5½ – 6 D = 3.45± | X Λ c = −43° to −45°, bi;<br>Y Λ a = +62° to +64°,<br>Z ‖ b, O.A.P. ⊥ (010)<br>fig. p. 68 | $n_x$ = 1.681 – 1.700<br>$n_y$ = 1.690 – 1.716<br>$n_z$ = 1.704 – 1.726<br>⊖Δ = 0.023 – 0.026 |
| 144 | Låvenite mcl<br>$Na_3CaZr$<br>$[(F,OH,O)_2/Si_2O_7]$ | {110},{100},{101},<br>{111};<br>β = 110¼°;<br>◈ (100) lam.;<br>columnar ‖ c,<br>tabular ‖ (100), granular | # (100) distinct;<br>H = 6 D = 3.53± | X Λ c = −19° to −21°, bi;<br>Y ‖ b, O.A.P. (010)<br>Z Λ a = +40° to +41°<br>l' = (−)<br>fig. p. 68 | $n_x$ = 1.690 – 1.698<br>$n_y$ = 1.707 – 1.723<br>$n_z$ = 1.720 – 1.745<br>⊖Δ = 0.030 – 0.047 |
| 145 | Guarinite trcl<br>(= Hjortdahlite)<br>$Ca_2NaZr$<br>$[(F,O)_2/Si_2O_7]$ | {100},{010},{hk0},<br>{011},{1̄11};<br>always ◈ [001], lam.,<br>comp. plane (100):<br>pseudorhombic; tabular<br>‖ (100), elongated ‖ c | # (03̲2) dist. } ~ 90°;<br># (03̄2) dist. }<br>very brittle;<br>H = 5 – 5½ D = 3.26± | in (100): Y' Λ c = 25°<br>in (010): Y' Λ c = 15°<br>O.A.P. ~ (011)<br>stereogr. p. 52<br>fig. p. 68 | $n_x$ = 1.652 – 1.704<br>$n_y$ = 1.658 – 1.716<br>$n_z$ = 1.665 – 1.727<br>Δ = ⊕ 0.013 – ⊖0.023 |
| 146 | Hainite trcl<br>Ca, Na, Ti, Zr-Silicate | {100},{010},{hk0};<br>◈ (100);<br>acicular, poikilitic<br>flakes | # (010) good,<br># (100) poor;<br>brittle;<br>H = 5 D = 3.18± | in (100): X' Λ c ~ 0°<br>in (010): X' Λ c = −4°<br>⊥ to [001]: Y' Λ a = −5°<br>Z ~ ⊥ (010) bi;<br>l' = (−) fig. p. 68 | $n_y$ ~ 1.7<br>⊕Δ = 0.012 |
| 147 | Rosenbuschite trcl<br>$(Ca,Na)_6ZrTi$<br>$[(F,O)_2/Si_2O_7]_2$ | {100},{010},{12̄0},<br>{h0l};<br>acicular ‖ c,<br>stellate | # (100) perf.,<br># (12̄0) distinct,<br># (010) poor;<br>H = 5 – 6 D = 3.31± | X ~ ‖ c<br>Y Λ b' = −25°<br>O.A.P. Λ ⊥ (100) = 29°<br>Z Λ a = +47°, bi;<br>l = (−)<br>fig. p. 68 | $n_x$ = 1.678<br>$n_y$ = 1.668<br>$n_z$ = 1.705<br>⊕Δ = 0.027 |
| 148 | Rinkite trcl<br>$(\underline{Na},\underline{Ca},Ce)_3$<br>$(\underline{Ti},Ce,Zr)$<br>$[(OH,F,O)_2/Si_2O_7]$ | {100}, {hk0}, {102},<br>{102̄};<br>◈ (100) lam.;<br>tabular ‖ (100),<br>elongated ‖ c | # (100) perf.;<br>H = 5 D = 3.46± | X ‖ b, O.A.P. ⊥ (010)<br>Y Λ c = −8°<br>Z Λ a = +9°, bi;<br>l' = (−)<br>fig. p. 68 | $n_x$ = 1.665<br>$n_y$ = 1.668<br>$n_z$ = 1.681<br>⊕Δ = 0.016 |
| 149 | Mosandrite trcl<br>(Johnstrupite)<br>$(\underline{Na},\underline{Ca},Ce)_3$<br>$(\underline{Ti},Ce,Zr)$<br>$[(OH,F,O)_2/Si_2O_7]$ | {100}, {hk0}, {h0l};<br>β = 93°, pseud.-mcl;<br>◈ (100) lam.;<br>tabular ‖ (100) and<br>elongated ‖ c | # (100) good,<br>very brittle;<br>irregular cracks;<br>H = 4 D = 3.0 – 3.3 | X Λ c = −2°<br>Y ‖ b, O.A.P. (010)<br>Z Λ a = +5°, bi;<br>l' = (−)<br>fig. p. 69 | $n_x$ = 1.646 – 1.661<br>$n_y$ = 1.649 – 1.666<br>$n_z$ = 1.658 – 1.673<br>⊕Δ = 0.012 – 0.012 |
| 150 | Lamprophyllite mcl<br>$Na_3Sr_2Ti_3$<br>$[(O,OH,F)_2/Si_2O_7]$ | {100}, {hk0}, {201},<br>{3̄02};<br>β ~ 90°, pseud.-orh.;<br>◈ (100), lam.;<br>tabular ‖ (100),<br>elongated ‖ c,<br>stellate | # (100) perf.,<br>resembling mica;<br># {011}? poor;<br>H = 2 – 3 D = 3.47± | X ‖ b<br>Y Λ a = +5° to +13°<br>O.A.P. (010)<br>Z Λ c = −5° to −13°, bi;<br>l' = (+) (−)<br>fig. p. 69 | $n_x$ = 1.744 – 1.745<br>$n_y$ = 1.747 – 1.754<br>$n_z$ = 1.773 – 1.780<br>⊕Δ = 0.029 – 0.035 |
| 151 | Astrophyllite trcl<br>$(\underline{K}_2,\underline{Na}_2,Ca)$<br>$(\underline{Fe},Mn)_4(\underline{Ti},Zr)$<br>$[OH/Si_2O_7]_2$ | {100},{hkl}, {h0l},<br>{001};<br>β = 94°, pseud.-mcl.;<br>elongated ‖ c or b,<br>stellate | # (100) perf.,<br>resembling mica;<br>brittle;<br>percussion figure<br>{021}: 83°;<br>H = 3 – 4 D = 3.35± | X Λ a = +4° to +6°<br>Y ~ ‖ c, O.A.P. ⊥ (010)<br>Z ~ ‖ b, bi;<br>l' = (+)<br>fig. p. 69 | $n_x$ = 1.676 – 1.708<br>$n_y$ = 1.700 – 1.726<br>$n_z$ = 1.731 – 1.758<br>⊕Δ = 0.055 – 0.050 |

## Nesosilicates (cont.) and Sorosilicates (cont.) No. 141a–151

| Angle and Dispersion of Optic Axes | Color in Thin Section Pleochroism | Chemical Reactions Special Remarks | Distinguishing Features of Similar Minerals | Occurrences | |
|---|---|---|---|---|---|
| $2V_z = 10° - 85°$<br>$r \ll v$ very strong | X = colorless,<br>Y = colorless to blue-green,<br>Z = colorless | insoluble in HCl;<br>abnormal interference colors:<br>XY = leather brown,<br>YZ = blue;<br>usually xenomorphic | epidote: } X-ray<br>zoisite: }<br>lawsonite: sm. 2V,<br>usually idiomorphic;<br>jadeite: gr. n | low grade to burial metam. in green schists and in glaucophane rocks; in metam. basaltic rocks | 141a |
| pure $AlFe^{3+}$<br>$2V_z = 23° - 40° - (50°)$<br>$r \gg v$ strong | usually colorless;<br>otherwise<br>X = almost colorless,<br>Y = pale greenish to yellow,<br>Z = pale brown to reddish,<br>X < Y < Z | soluble in conc. $H_2SO_4$;<br>yttrotitanite may have pleochroic haloes; leucoxene: alteration product of ilmenite | baddeleyite: better #,<br>$2V_x$; monazite: sm. n,<br>sm. 2V; epidote, allanite:<br>sm. n, sm. $\Delta$, gr. 2V | magmatic with hornblende in syenites etc.;<br>metam. in marbles, amphibolites, skarns;<br>hydrothermal in fissures | 142 |
| $2V_x = 71° - 79°$<br>$r < v$ weak,<br>horizontal disp. | X = yellowish to colorless,<br>Y = very light yellow,<br>Z = wine yellow,<br>$X \leqq Y < Z$ | soluble in HCl;<br>differing $n_z' - n_x'$ for twin-halves ⊥ X and ⊥ Y | guarinite: 2 #;<br>låvenite: gr. $\Delta$;<br>melinophane: sm. n,<br>uniaxial | with zircon and sphene deuteric in nepheline foyaites and in foyaite-pegmatites | 143 |
| $2V_x = 80° - 85°$<br>$r < v$ distinct | X = pale yellow to colorless,<br>Y = yellowish green to colorless,<br>Z = orange to colorless,<br>X < Y < Z | slightly soluble in HCl;<br>because of # (100) only one opt. axis visible | woehlerite: sm. $\Delta$;<br>guarinite: 2 #; rosenbuschite: $\oplus$, sm. 2V;<br>epidote: diff. pleochroism; diaspore: $\oplus$,<br>different paragenesis | late magmatic with fluorite, sphene, mosandrite, rinkite in alkaline syenites, foyaites and their pegmat.;<br>pneumatolytic in sanidinites | 144 |
| $2V = 90° \pm 5°$<br>$r < v$ distinct | X = colorless,<br>Y = pale yellowish,<br>Z = pale to wine yellow,<br>$X \lesssim Y < Z$ | gelatinizes with HCl | woehlerite: one #;<br>låvenite: gr. $\Delta$;<br>rinkite: better #,<br>sm. $2V_x$; mosandrite:<br>sm. n; augite: O.A.P. halves angle of #, sm. $2V_z$ | late magmatic to pegmatitic in foyaites; pneumatolytic in sanidinite-volcanic ejecta | 145 |
| $2V_z$ = large<br>$r > v$ strong | X = colorless,<br>Y = pale yellow,<br>Z = wine yellow,<br>$X \lesssim Y < Z$ | easily soluble in acids | guarinite: sm. n;<br>rinkite: sm. n, 2V, X ‖ b;<br>mosandrite: sm. n,<br>sm. 2V; rosenbuschite:<br>gr. $\Delta$ | deuteric to pneumatolytic in phonolites and tinguaites | 146 |
| $2V_z = 60° - 80°$<br>$r > v$ weak | X = colorless,<br>Y = pale yellowish,<br>Z = pale yellow,<br>$X \lesssim Y < Z$ | soluble in HCl;<br>optics appear monoclinic; needles with often transverse fractures | epidote: $\ominus$, O.A.P. ⊥ elongation; zoisite: sm. $\Delta$;<br>augite: 2 #, $\sphericalangle$ 90°;<br>låvenite: $\ominus$, gr. $\Delta$; hainite,<br>rinkite: sm. $\Delta$ | deuteric in foyaitic plutonites and related dike rocks; with låvenite, mosandrite, eucolite in syenite- and foyaite-pegmatites | 147 |
| $2V_z = 60° - 80°$<br>$r > v$ weak | X = colorless,<br>Y = pale yellowish,<br>Z = pale yellow,<br>$X \lesssim Y < Z$ | easily soluble in diluted acids | mosandrite: O.A.P. (010);<br>rosenbuschite: gr. $\Delta$,<br>acicular; guarinite: gr. 2V;<br>woehlerite: gr. 2V, gr. n and $\Delta$ | with låvenite, astrophyllite, and violet fluorite in foyaites, foyaite-pegmatites, alkali trachytes and phonolites; in skarns | 148 |
| $2V_z = 70° - 75°$<br>$r > v$ strong,<br>oblique dispersion | yellowish to colorless;<br>no conspicuous pleochroism | easily soluble in HCl;<br>turbid along cracks due to weathering | rinkite: O.A.P. ⊥ (010),<br>sm. 2V; guarinite: gr. 2V,<br>O.A.P. oblique; rosenbuschite: gr. $\Delta$, acicular;<br>hainite: gr. n, 2V | deuteric with rosenbuschite, woehlerite and violet fluorite in sodic syenites, foyaites and their pegmatites; in corresp. volcanics | 149 |
| $2V_z = 23° - 50°$<br>$r > v$ strong | X = yellow,<br>Y = pale yellow,<br>Z = orange-yellow,<br>$Y \leqq X < Z$ | slightly soluble in acids | biotite: sm. n, $\ominus$;<br>astrophyllite: gr. 2V,<br>gr. $\Delta$, sm. n | deuteric with eudialyte, aegirine and rinkolite in foyaites and their pegmatites | 150 |
| $2V_z = 70° - 85°$ (Na)<br>$r \ll v$ | X = orange to dark red,<br>Y = lemon-yellow,<br>Z = orange to lemon-yellow,<br>$Y < Z < X / Z < Y \ll X$ | soluble only in HF;<br>often turbid due to weathering | biotite: absorption<br>X < Z, $\ominus$, sm. 2V;<br>staurolite: # less perf.,<br>sm. $\Delta$, diff. parag.;<br>lamprophyllite: sm. 2V,<br>absorption X < Z;<br>diaspore: colorless | deuteric with sphene, pectolite, aegirine in nepheline-syenitic plutonites and their pegmatites;<br>hydrotherm. in natrolite veins; metam. in appropriate gneisses | 151 |

(40, 48)

| No. | Name<br>Symmetry<br>Chem. Composition | Common Crystal Forms, Twinning<br>Crystal Habit<br>Aggregates | Cleavage<br>Translation Gliding<br>Deform. Twinning<br>Hardness, Density | Optic Orientation | Refractive Indices<br>Birefringence |
|---|---|---|---|---|---|
| 152 | Axinite trcl<br>$Ca_2(\underline{Fe},\underline{Mn},Mg)$<br>$AlAl[BO_3OH/Si_4O_{12}]$ | {010},{$\bar{1}$20},{011},<br>{$\bar{1}$11};<br>cryst. setting: PEACOCK;<br>no ⟐;<br>lenticular ~ ‖ ($\bar{1}$21),<br>flaky, granular | # (100) distinct;<br>splintery;<br>H = 6½ – 7 D = 3.3± | X ~ ⊥ ($\bar{1}$11)<br>O.A.P. ~ (111), one opt. axis ~ ⊥ ($\bar{1}$20);<br>stereogr. p. 71,<br>fig. p. 69 | Mn only (Fe,Mn)<br>$n_x$ = 1.678 – 1.687<br>$n_y$ = 1.685 – 1.695<br>$n_z$ = 1.688 – 1.698<br>⊖Δ = $0.009_5$ – $0.010_5$<br>diagr. p. 70 |
| 153 | Kornerupine orh<br>$(\underline{Mg},Fe)_4Al_6$<br>$[(O,OH)_2/BO_4/$<br>$(SiO_4)_4]$ | {110},{100},{010};<br>no ⟐;<br>prismatic ‖ c,<br>acicular, stellate | # {110} good: 81½°;<br>parting (001);<br>H = 6½ – 7 D = {3.27(Mg), 3.45 (Mg,Fe)} | X ‖ c, bi;<br>Y ‖ a, O.A.P. (100)<br>Z ‖ b<br>l = (–)<br>fig. p. 69 | pure Fe-cont.<br>$n_x$ = 1.665 – 1.682<br>$n_y$ = 1.677 – 1.696<br>$n_z$ = 1.677 – 1.699<br>⊖Δ = 0.012 – 0.017<br>diagr. p. 71 |
| 154 | Serendibite trcl<br>$(\underline{Ca},\underline{Mg},Fe)_5(AlO)_5$<br>$[BO_3/(SiO_4)_3]$ | tabular with hexagon-like outline;<br>⟐ lam. ‖ to elongated direction; granular<br>fig. p. 69 | no #;<br>H = 6½ – 7 D = 3.42± | extinction in sections ⊥ ⟐ and ~ ⊥ Z = 35 – 40°<br>fig. p. 69 | $n_x$ = 1.701 – 1.738<br>$n_y$ = 1.703 – 1.740<br>$n_z$ = 1.706 – 1.743<br>⊕Δ = 0.005 – 0.005 |
| 155 | Grandidierite orh<br>$(Mg,Fe)\ Al_3$<br>$[O/BO_4/SiO_4]$ | {100},{010},{hk0};<br>acicular, massive | # (100) perf.<br># (010) good > 90°<br>H = 7½ D = 2.99± | X ‖ a, bi;<br>Y ‖ c, O.A.P. (001)<br>Z ‖ b<br>l = (+) (–)<br>fig. p. 69 | $n_x$ = 1.577 – 1.6018<br>$n_y$ = 1.630 – 1.6360<br>$n_z$ = 1.632 – 1.6385<br>⊖Δ = 0.035 – 0.0367 |
| 156 | Cordierite orh<br>(= Dichroite)<br>$\alpha$-$(\underline{Mg},Fe)_2Al_3$<br>$[AlSi_5O_{18}]$ | {110},{010},{001},<br>{100},{111};<br>⟐ {110} and {130}:<br>penetration triplets,<br>rarely lamellar; pseudo-hex.; short columns ‖ c,<br>granular | # (010) poor,<br>(001) parting, poor;<br>irregular cracks;<br>H = 7 – 7½<br>D = 2.57 – 2.76 | X ‖ c, bi;<br>Y ‖ a, O.A.P. (100)<br>Z ‖ b<br>l = (–)<br>fig. p. 69 | $n_x$ = 1.530 – 1.560<br>$n_y$ = 1.535 – 1.574<br>$n_z$ = 1.538 – 1.578<br>⊕⊖Δ = 0.008 – 0.018<br>diagr. p. 70 |
| 157 | Bertrandite orh<br>$Be_4[(OH)_2/SiO_4/SiO_3]$ | {001},{110},{100},<br>{010},{0kl};<br>⟐ (021) heart-shaped,<br>⟐ (011);<br>tabular ‖ (001), (010) | # (001) perf.,<br># {110} good, 121°;<br># (010) distinct,<br>H = 6 – 6½ D = 2.60± | X ‖ a, bi;<br>Y ‖ b, O.A.P. (010)<br>Z ‖ c<br>l = (–)<br>fig. p. 69 | $n_x$ = 1.584 – 1.591<br>$n_y$ = 1.603 – 1.605<br>$n_z$ = 1.611 – 1.614<br>⊖Δ = 0.027 – 0.023 |
| 157a | Euclase mcl<br>$AlBe[SiO_4/OH]$ | {130},{120},{011},<br>{021};<br>columnar ‖ c,<br>tabular ‖ (010) | # (010) perf.,<br># (110), (001) poor;<br>H = 7½ D = 3.0± | Y ‖ b, O.A.P. (010)<br>Z ∧ c = +41° | $n_x$ = 1.651 – 1.653<br>$n_y$ = 1.655 – 1.656<br>$n_z$ = 1.671 – 1.673<br>⊕Δ = 0.020 – 0.020 |
| 158 | Prehnite orh<br>$Ca_2Al_2[(OH)_2/Si_3O_{10}]$ | {001},{110},{010},<br>{031};<br>⟐ (100) not recogn. optically;<br>tabular ‖ (001),<br>short columns ‖ c,<br>stellate | # (001) perf.,<br># {110} poor, 80°;<br>H = 6 – 6½ D = 2.91± | X ‖ a<br>Y ‖ b, O.A.P. (010)<br>Z ‖ c, bi;<br>l = (±)<br>fig. p. 69 | $Fe^{3+}$-free 20% Fe<br>$n_x$ = 1.611 – 1.630<br>$n_y$ = 1.617 – 1.641<br>$n_z$ = 1.632 – 1.669<br>⊕Δ = 0.021 – 0.039<br>diagr. p. 71 |
| 159 | Rhodonite trcl<br>$CaMn_4[Si_5O_{15}]$ | {010},{100},{001},<br>{110},{1$\bar{1}$0},{101},<br>{10$\bar{1}$};<br>⟐? no specifications;<br>tabular ‖ (010),<br>granular | # {110} and {1$\bar{1}$0}<br>perf., 92°;<br># (001) poor;<br>H = 5½ – 6½ D = 3.6 | Y ∧ b ~ 20°<br>Z' ∧ c in (010) ≈ 35°<br>Z = bi;<br>l = (±)<br>fig. p. 74 | $n_x$ = 1.711 – 1.738<br>$n_y$ = 1.716 – 1.741<br>$n_z$ = 1.723 – 1.752<br>⊕Δ = 0.012 – 0.014 |
| 160 | Babingtonite trcl<br>$Ca_2Fe^{2+}Fe^{3+}$<br>$[OH/Si_5O_{14}]$ | {110},{100},{110},<br>{001}; cryst. setting:<br>RICHMOND;<br>no ⟐;<br>short, slightly bladed prisms ‖ (100) | # (001) perf.,<br># {1$\bar{1}$0} good,<br># {110} poor;<br>H = 6 D = 3.36± | O.A.P. ≈ ‖ (1$\bar{1}$0)<br>opt. axis A ~ 10° – 20° oblique to (001);<br>opt. axis B ≈ 10° oblique to (110),<br>fig. p. 74 | $n_x$ = 1.700 – 1.720<br>$n_y$ = 1.710 – 1.731<br>$n_z$ = 1.725 – 1.753<br>⊕Δ = 0.025 – 0.033 |
| 161 | Wollastonite trcl<br>$Ca_3[Si_3O_9]$ | {100},{001},{10$\bar{1}$},<br>{101},{1$\bar{1}$0};<br>⟐ (100) lam.;<br>acicular ‖ b,<br>fibrous | #(100) perf. } 70° } 84°<br>#(102) good<br>#(001) good<br># {101} distinct,<br># {101} distinct;<br>H = 4½ – 5 D = 2.91± | X ∧ c = –28° to –34°, bi;<br>Y ∧ b = +3° to 5°<br>O.A.P. ~ ⊥ l<br>Z ∧ a = +34° to 39°<br>l = (±)<br>fig. p. 74, 75 | $n_x$ = 1.618 – 1.622<br>$n_y$ = 1.630 – 1.634<br>$n_z$ = 1.632 – 1.636<br>⊖Δ = 0.014 – 0.014 |

| Angle and Dispersion of Optic Axes | Color in Thin Section Pleochroism | Chemical Reactions Special Remarks | Distinguishing Features of Similar Minerals | Occurrences | |
|---|---|---|---|---|---|
| $2V_X = 55^o - 81^o$<br>$r < v$<br>diagr. p. 70 | colorless to pale brown-violet, almost no pleochroism: ($Y \geqq X \geqq Z$) | slightly soluble in acids, boron reactions! gelatinizes after melting in HCl; acute-angled rhombic cross-sections; somewhat abnormal interf. colors; strong pleochroism in ultra-violet light | quartz: sm. n, uniaxial$\oplus$; sapphirine: sm. $\Delta$; serendibite: ◈ lam., $\oplus$; zoisite: sm. $\Delta$, $\oplus$; | with datolite, epidote, diopside e.g. contact-pneumatolytic in calc-silicate rocks, in marbles, skarns; in fissures of diabases, granites; hydrotherm. in quartz-chlorite veins of metam. rocks | 152 |
| $2V_X = 3^o - 48^o$<br>$r < v$ very weak<br>diagr. p. 71 | usually colorless; also:<br>X = pale red-yellow,<br>Y = pale brownish yellow,<br>Z = pale yellowish green,<br>$X < Y < Z$ | slightly soluble in acid; rimmed by a fine reticulate, mica-like alteration product | andalusite: sm. n, gr. 2V; topaz: sm. n, $\oplus$, Z ‖ c; sillimanite: gr. $\Delta$, Z ‖ c; tremolite: oblique extinction, sm. n, gr. $\Delta$ | metam. in Al-rich, quartz-poor paragneisses and their albite-pegmatoide mobilisates with cordierite, sapphirine, sillimanite, gedrite, corundum; in placers | 153 |
| $2V_Z \sim 90^o$<br>$r < v$ strong | X = pale yellowish green,<br>Y = almost colorless to dark blue-green<br>Z = dark greenish blue,<br>$Y \leqq X \ll Z$ | slightly soluble in HF; plagioclase-like ◈ lam.; strong pleochroism | grandidierite: sm. $2V_X$; sm. n, gr. $\Delta$, no ◈; sapphirine: sm. $2V_X$, no ◈; lazulite: sm. $2V_X$, sm. n, gr. $\Delta$ | contact-metasom. in calc-silicate rocks, marbles, and skarns with spinel, diopside, grossularite etc. | 154 |
| $2V_X = 30^o$<br>$r < v$ strong | X = blue-green,<br>Y = colorless,<br>Z = bluish green,<br>$Y \ll X < Z$ | insoluble in acids; alteration to sericite; poikilitically intergrown with quartz; pleochr.: ⊥ l: deep colors, ‖ l: colorless | lazulite: gr. 2V; dumortierite: gr. n, X>Z; sapphirine: gr. n, sm. $\Delta$, gr. 2V; tourmaline: $O \gg E$, uniaxial; serendibite: gr. n, sm. $\Delta$, gr. $2V_Z$, ◈ lam. | deuteric in pegmatites and aplites with almandine, pleonaste and andalusite | 155 |
| $2V_X = 42^o - 104^o$<br>$r < v$<br>diagr. p. 70 | colorless; also<br>X = pale orange,<br>Y = light blue-violet,<br>Z = pale blue-violet,<br>$X < Z < Y$ | slightly soluble in acids; pinitic alteration; yellow pleochroic haloes, poikiloblastically in contact-metam. schists | quartz: uniaxial $\oplus$, no alteration; plagioclase: usually ◈ lam.; high-cordierite: $n_E = 1.534, n_O = 1.539$, $\Delta = 0.005$, $\ominus$ uniaxial | in magmatites due to resorption; contact-metam. in schists, hornfelses; regional-metam. with anthophyllite, sillimanite, garnet in paragneisses, granulites and associated pegmatitic veins | 156 |
| $2V_X = 65^o - 85^o$<br>$r < v$ weak | colorless | insoluble in acids, even HF | muscovite: one #; tremolite: different #; amblygonite: sm. $\Delta$ | secondary from beryl in pegmatites, probably also in granites, aplites; pneumatolytic to hydrotherm. | 157 |
| $2V_Z = 50^o - 46^o$<br>$r > v$ | colorless, light green | only soluble in hot, conc. HF | diopside: different paragenesis | in granite-pegmatites, gneisses, in alpine fissures also alteration of beryl | 157a |
| $2V_Z = 65^o - 69^o$<br>($2V\ 0^o - 30^o$, $r \ll v$, abnormal interference colors due to submicroscopic twinning | colorless | slightly decomposed by HCl, precipitation of $SiO_2$-powder; divergent or "parquet"-like subindividua; abnormal interference colors | lawsonite: gr. n; danburite: sm. $\Delta$; datolite: no #; thomsonite: sm. n; chondrodite: stainable; pargasite: amphibole-# | hydrothermal in fissures and vesicles; secondary at the expense of plagioclase and analcime in calc-silicate rocks, zeolite- to green schist facies | 158 |
| $2V_Z = 74^o - 58^o$<br>$r < v$,<br>asymmetric disp.<br>diagr. p. 75 | colorless; thick sections:<br>~ ‖ X = orange,<br>~ ‖ Y = pink,<br>~ ‖ Z = pale orange | slowly decomposed by HCl, hardly stainable, weathering to Mn-oxides | bustamite: sm. n and D, $\ominus$; augite: gr. $\Delta$; kyanite: coarser #; staurolite: color | with quartz and rhodochrosite in cherts, with spessartine and magnetite contact-metasom. in calc-silicate rocks | 159 |
| $2V_Z = 58^o - 74^o$<br>$r > v$ strong,<br>asymmetric disp. | X = dark green,<br>Y = light purplish grey,<br>Z = orange-brown,<br>$X > Z > Y$ | not decomposed by acids; characteristic absorption, dispersion of bisectrices | pyroxene: weaker pleochroism; hornblende: $\ominus$ | with epidote, prehnite, calcite; hydrothermal in fissures in diabase and pegmatites, in skarns | 160 |
| $2V_X = 42^o - 36^o$<br>$r > v$,<br>distinct oblique disp.<br>diagr. p. 75 | colorless | gelatinizes with HCl, then stainable $SiO_2$-coating, frequently ◈ lam. | parawollastonite: in volcanic ejecta, very similar, but Y ‖ b; pectolite: l = (+), gr. $\Delta$; tremolite: O.A.P. ‖ l | with diopside, grossularite and epidote contact-metam. in marbles; product of resorption in nepheline rocks | 161 |

(48, 54)

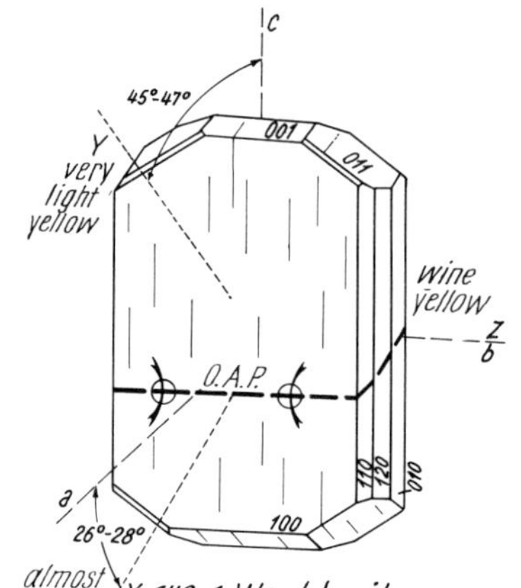


143-1. Woehlerite

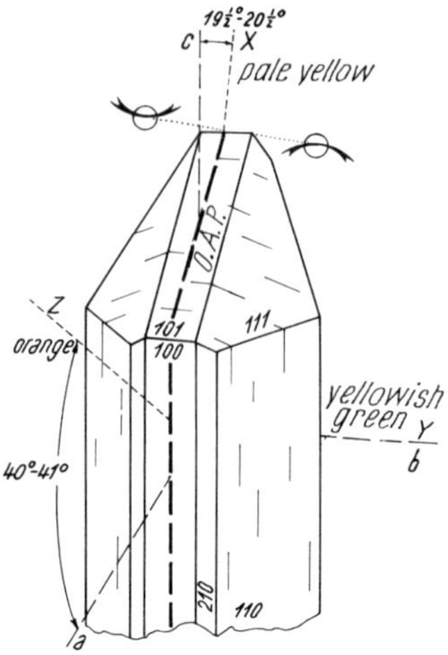


144-1. Låvenite

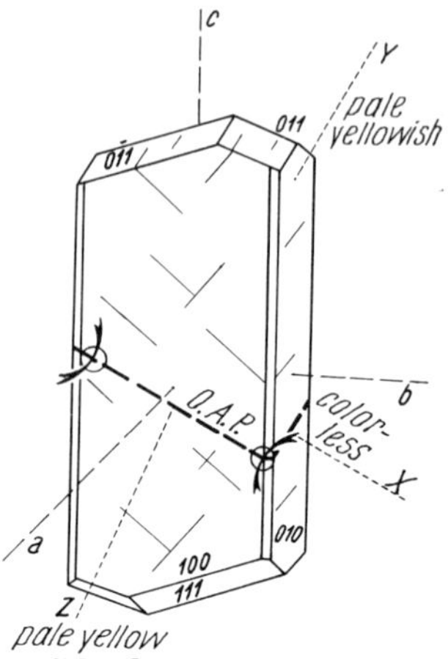


145-1. Guarinite
(= Hjortdahlite)

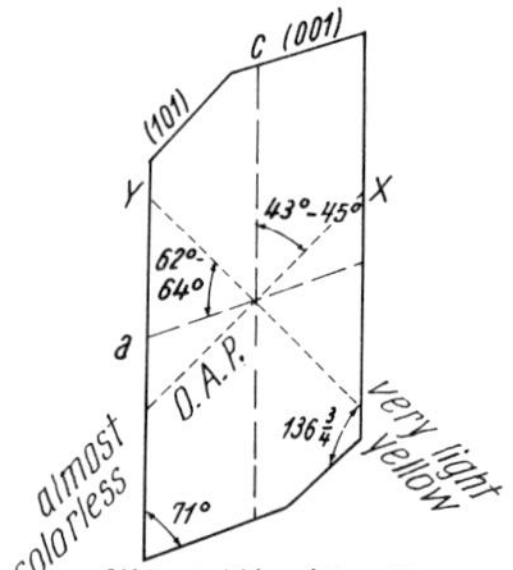


143-2. Woehlerite
section ∥(010)

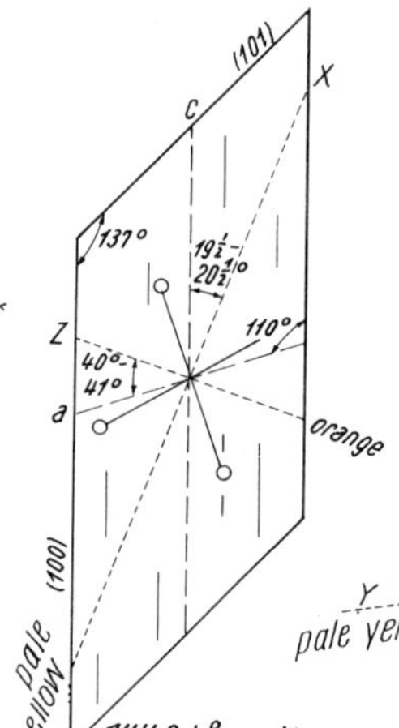


144-2. Låvenite
section ∥(010)

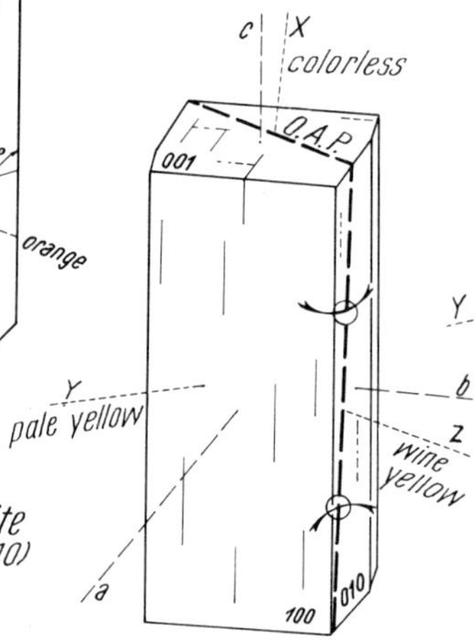


146-1. Hainite

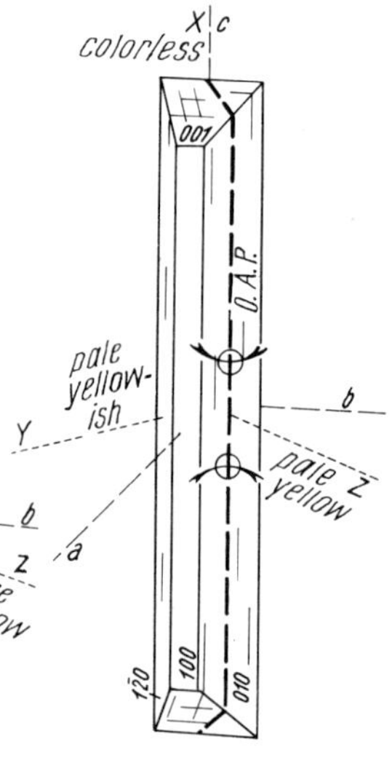


147-1. Rosenbuschite

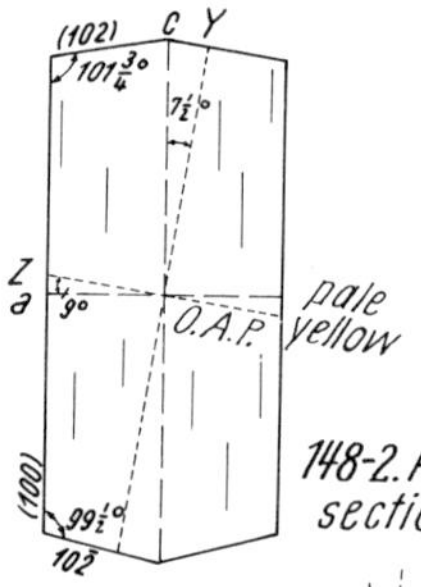


148-2. Rinkite
section ∥(010)

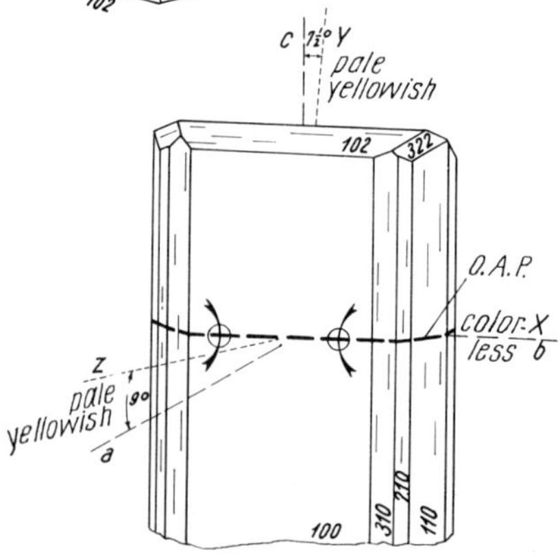


148-1. Rinkite

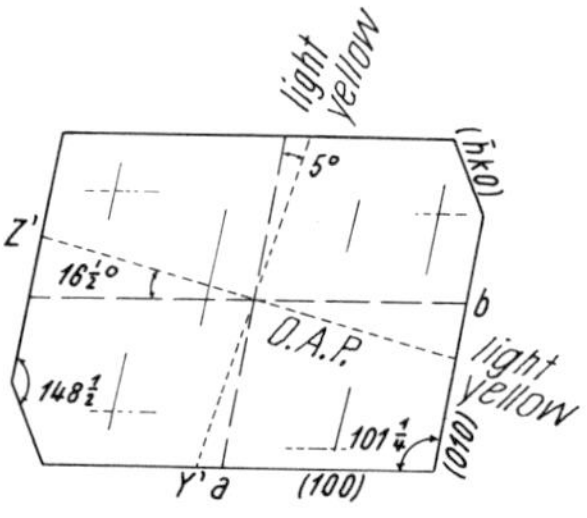


146-2. Hainite
section ⊥c

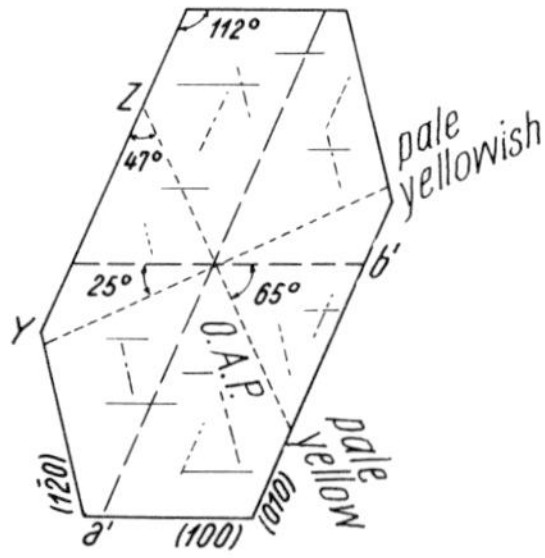


147-2. Rosenbuschite
section ⊥c

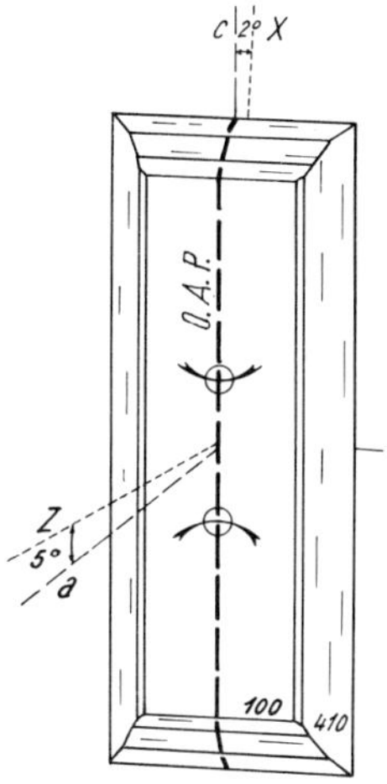


149-1. Mosandrite

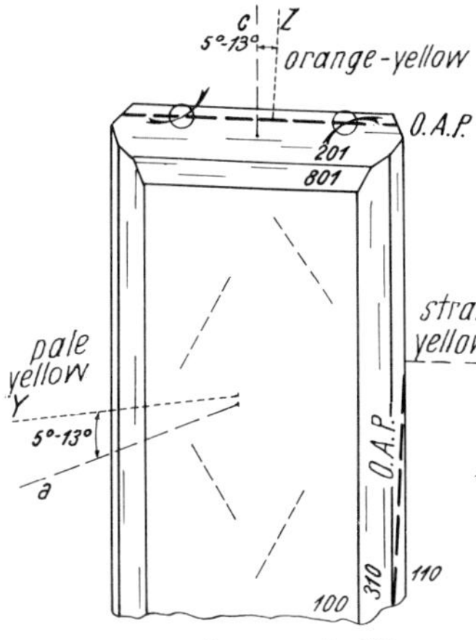


150-1. Lamprophyllite

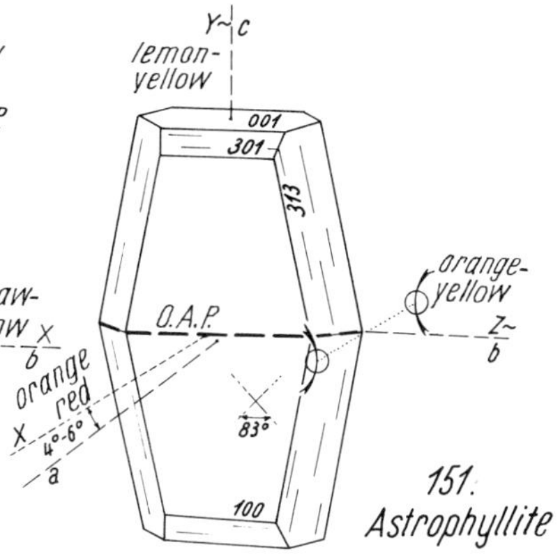


151. Astrophyllite

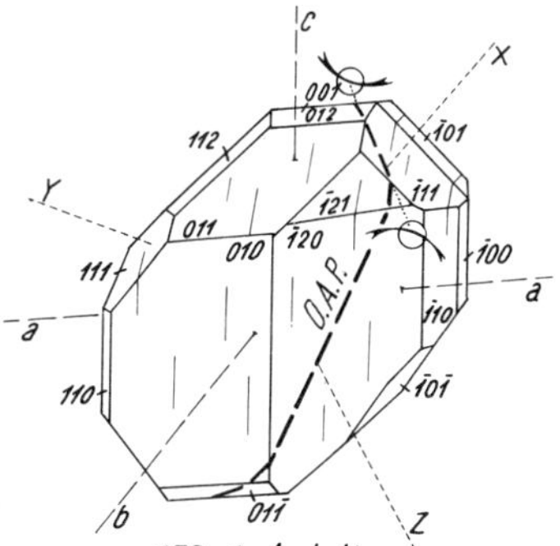


152-1. Axinite

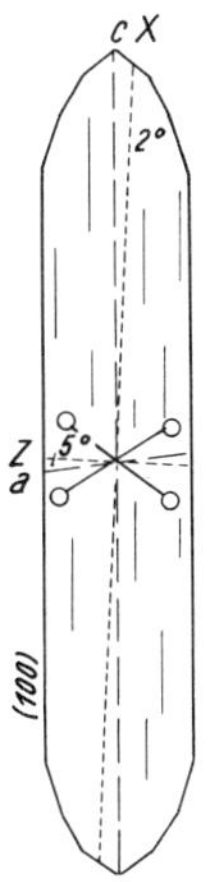


149-2. Mosandrite
section ‖ (010)

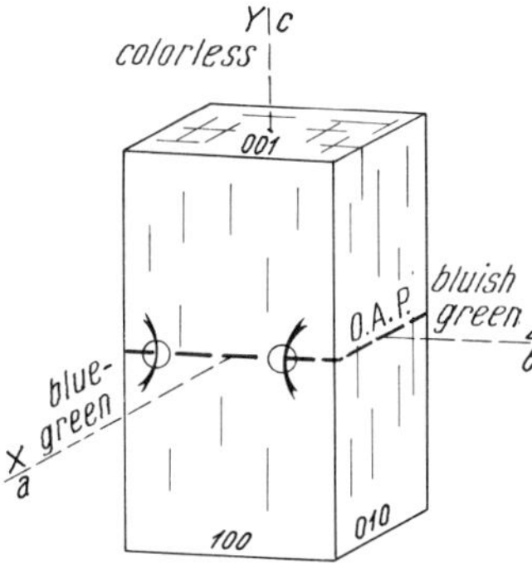


155. Grandidierite

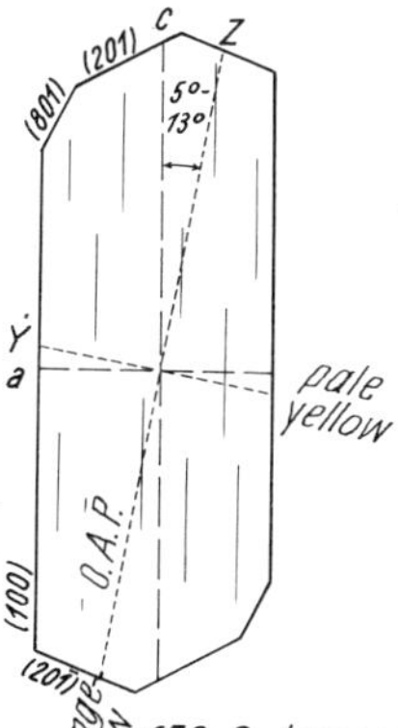


150-2. Lamprophyllite
section ‖ (010)

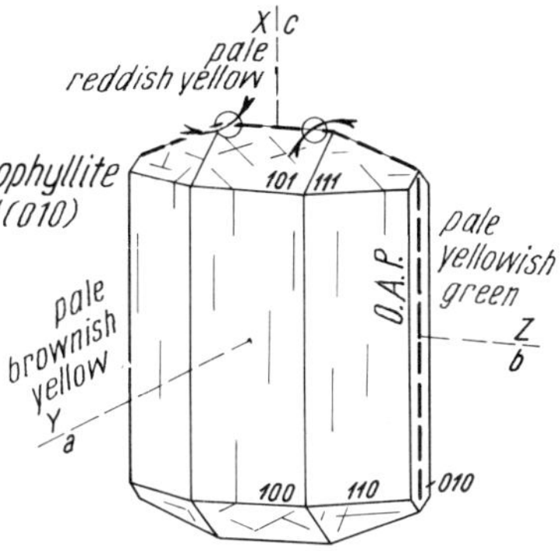


153-1. Kornerupine

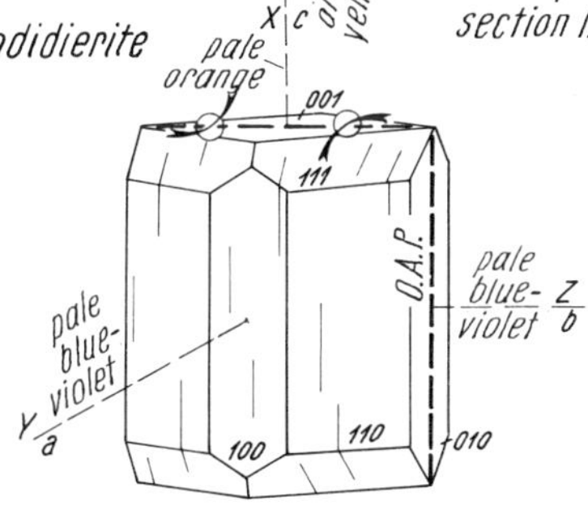


156-1. Cordierite

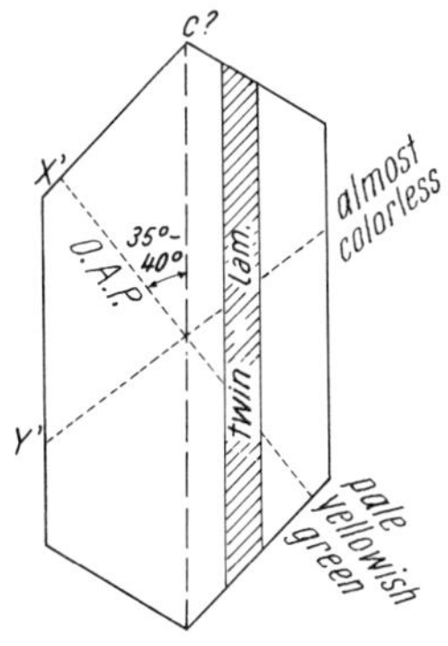


154. Serendibite
section ⊥ ◊ and ~⊥Z

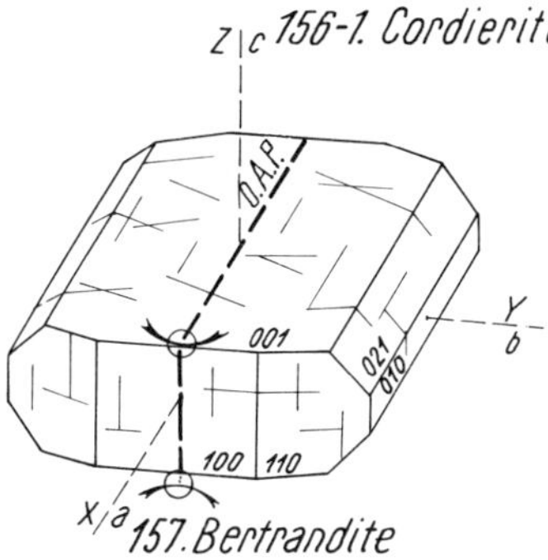


157. Bertrandite

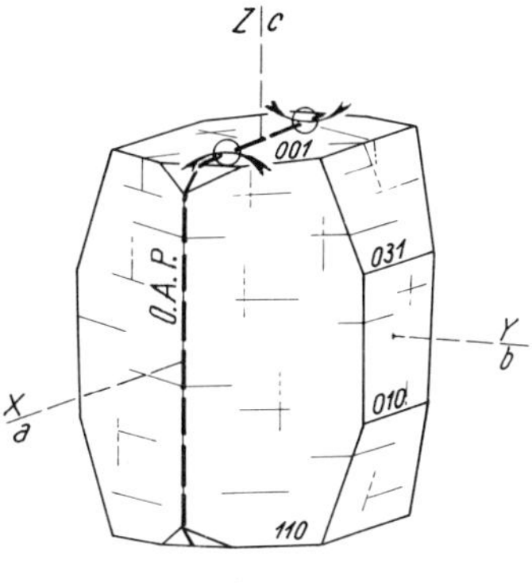


158-1. Prehnite

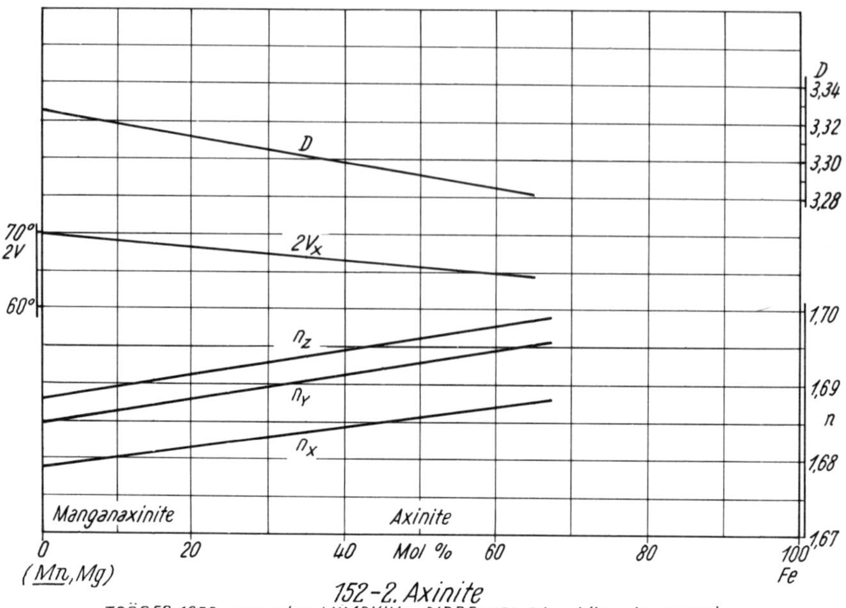


152-2. Axinite
TRÖGER 1950, see also LUMPKIN & RIBBE 1978 (Am. Min., in press)

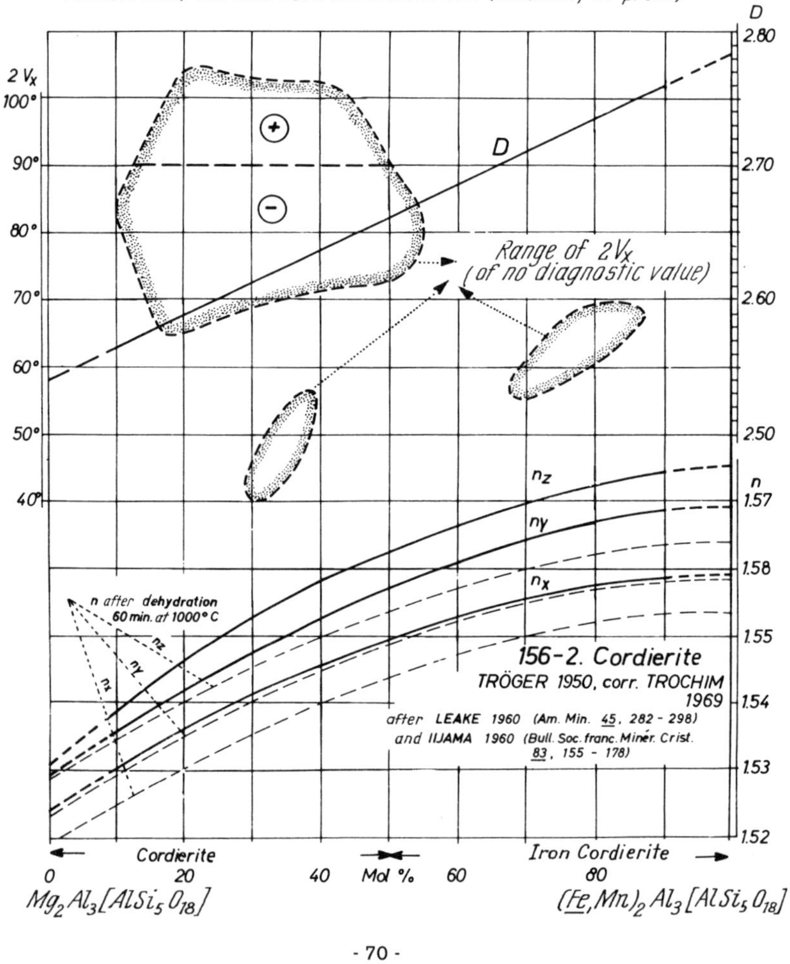


156-2. Cordierite
TRÖGER 1950, corr. TROCHIM 1969
after LEAKE 1960 (Am. Min. 45, 282 - 298)
and IIJAMA 1960 (Bull. Soc. franc. Minér. Crist. 83, 155 - 178)

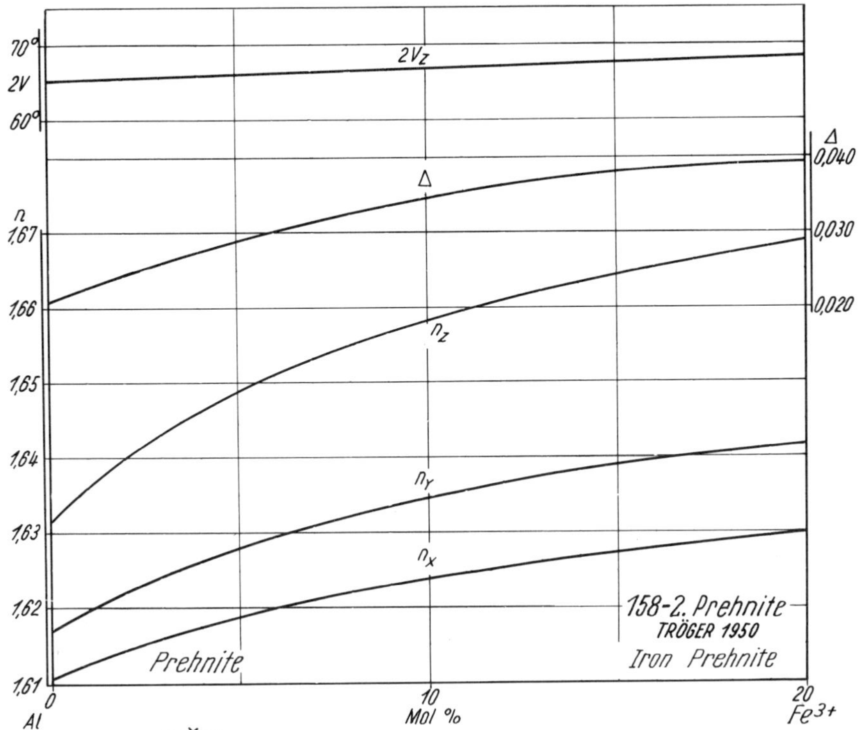


158-2. Prehnite
TRÖGER 1950

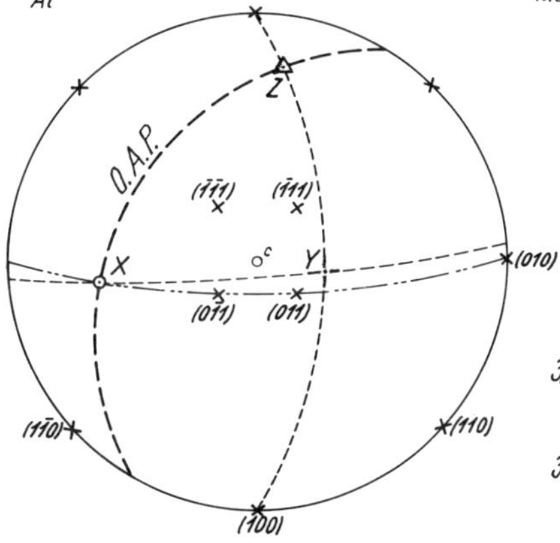


145-2. Guarinite (=Hjortdahlite)

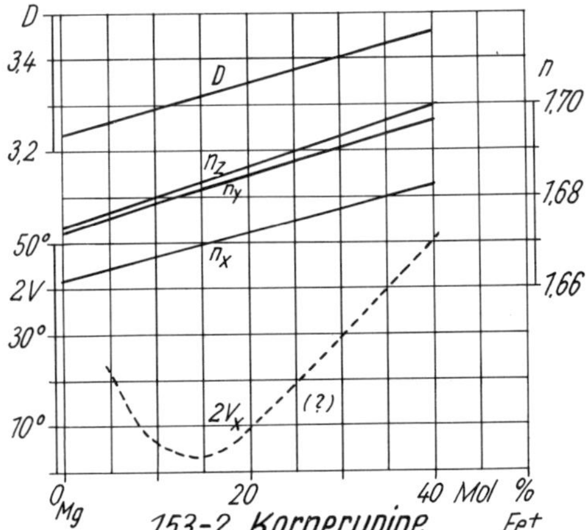


153-2. Kornerupine
GIRAULT 1952 (Am.Min. 37, 537)

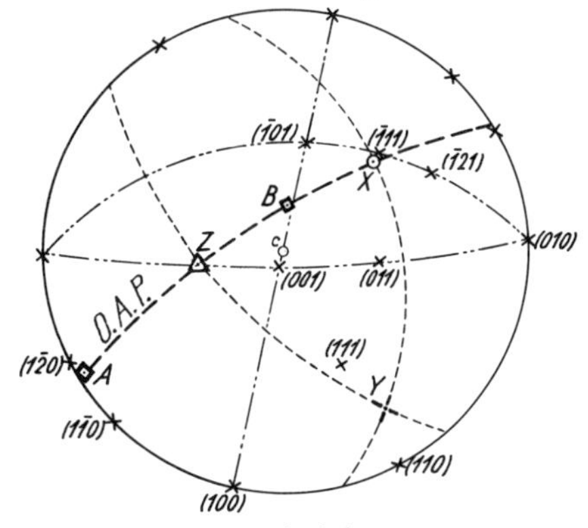


152-3. Axinite

| No. | Name<br>Symmetry<br>Chem. Composition | Common Crystal Forms, Twinning<br>Crystal Habit<br>Aggregates | Cleavage<br>Translation Gliding<br>Deform. Twinning<br>Hardness, Density | Optic Orientation | Refractive Indices<br>Birefringence |
|---|---|---|---|---|---|
| 161a | Scawtite mcl<br>$Ca_6[Si_3O_9]_2 \cdot CaCO_3$<br>$2H_2O$ | {100};<br>◊ (100)? ;<br>tabular | # (010) good,<br># (100) poor;<br>H = 4½ D = 2.77 | Y ∥ b<br>Z ∧ c = 28° | $n_x$ = 1.595 – 1.603<br>$n_y$ = 1.605 – 1.609<br>$n_z$ = 1.621 – 1.618?<br>⊕Δ = 0.026 – 0.015 |
| 162 | Bustamite trcl<br>(Mn,<u>Ca</u>)$_3$ [$Si_3O_9$] | {010},{100},{001},<br>{110}; xenomorphic;<br>◊ (100) and (001);<br>columnar ∥ [010],<br>tabular ∥ (010), granular | # (001) perf. } 84°<br># (100) good.<br># {101} dist. } 86°<br># {10$\bar{1}$} poor<br>H = 4½ D = 3.33 | Z ∧ b ~ 35°<br>X = bi<br>X ∧ c ~ 30°<br>fig. p. 74, 75 | $n_x$ = 1.662 – 1.689<br>$n_y$ = 1.675 – 1.701<br>$n_z$ = 1.677 – 1.704<br>⊖Δ = 0.015 – 0.015 |
| 163 | Pectolite trcl<br>$Ca_2Na[Si_3O_8/OH]$ | {100},{001}, {hk0},<br>{11$\bar{1}$}; cryst. setting:<br>PEACOCK;<br>◊ (100);<br>acicular ∥ b, stellate | # (100) perf.,<br># (001) perf.;<br>H = 4½ – 5 D = 2.87± | Z' ∧ b in (100) ≈ 0°<br>Z' ∧ b in (001) ≈ 2°<br>O.A.P. ~ (100)<br>l' = (+)<br>fig. p. 74, 75 | $n_x$ = 1.595 – 1.610<br>$n_y$ = 1.604 – 1.614<br>$n_z$ = 1.633 – 1.642<br>⊕Δ = 0.038 – 0.032 |
| 164 | Orthopyroxene orh<br>Enstatite-Bronzite<br>-Hypersthene<br>- Eulite-Ferrosilite<br>(<u>Mg</u>,Fe)$_2$ [$Si_2O_6$] | {100},{010},{210},<br>{h0l};<br>◊ (011), c : <u>c</u> ~ 61°,<br>◊ (023) and (043);<br>tabular ∥ (100),<br>elongated ∥ c, also<br>tabular ∥ (010), granular | # {210} good, 92°<br># {100} perf.,<br>parting;<br>T = (100), t = [001];<br>H = 5 – 6<br>D = 3.20 – 3.90 | X ∥ b, bi; (hyp.)<br>Y ∥ a, O.A.P. (010)<br>Z ∥ c, bi; (enst.)<br>l = (+)<br>fig. p. 76 | enst. hyp. eul.<br>$n_x$ = 1.657–1.712–1.766<br>$n_y$ = 1.659–1.724–1.770<br>$n_z$ = 1.665–1.727–1.788<br>Δ=⊕0.008 ⊖0.015 ⊕0.022 |
| 165 | Clinoenstatite group mcl<br>Clinoenstatite<br>Clinohypersthene<br>Clinoferrosilite<br>(<u>Mg</u>,Fe)$_2$ [$Si_2O_6$] | {100},{010},{110},<br>{111},{001};<br>β = 104½°<br>◊ (100), lam.;<br>columnar ∥ c,<br>tabular ∥ (100) | #{110} good 92°– 91°,<br>(001) parting;<br>D = {3.18 en. / 3.90 fs.}<br>H = 6 | X ∥ b<br>Y ∧ a = –8° to 20°,<br>O.A.P. ⊥ (010)<br>Z ∧ c = {+22° en / +42° / +34° fs} bi<br>fig. p. 76 | en fs<br>$n_x$ = 1.651 – 1.762<br>$n_y$ = 1.653 – 1.764<br>$n_z$ = 1.660 – 1.794<br>⊕Δ = 0.009 – 0.032 |
| 166 | Pigeonite mcl<br>(<u>Mg</u>,Fe,Ca)$_2$ [$Si_2O_6$] | {110},{100},{010},<br>{111},{22$\bar{1}$};<br>β = 108½°;<br>◊ (100), lam.;<br>columnar ∥ c,<br>tabular ∥ (100) | # {110} good, 92°<br># (100) and (010) rare,<br>(001) parting frequent;<br>H = 6 D = 3.30–3.46 | usually:<br>X ∥ b, O.A.P. ⊥ (010)<br>Y ∧ a = –21°–28°<br>Z ∧ c = 37 – 44°, bi;<br>rarely:<br>X ∧ a = –21° – 28°<br>Y ∥ b, O.A.P. (010)<br>Z ∧ c = +37° – 44°, bi;<br>fig. p. 76 | en-rich fs-rich<br>$n_x$ = 1.682 – 1.722<br>$n_y$ = 1.684 – 1.722<br>$n_z$ = 1.705 – 1.751<br>⊕Δ = 0.023 – 0.029<br>diagr. p. 80 |
| 167 | Diopside group mcl<br>Diopside-Salite-<br>Hedenbergite<br>Ca(Mg,Fe) [$Si_2O_6$] | {100},{010},{110},<br>{001},{111},{22$\bar{1}$};<br>β = 105½°;<br>◊ (100) lam.;<br>acicular ∥ c,<br>granular | # {110} good, 93°,<br># (010) distinct,<br># (100) rare,<br>(100),(001) parting;<br>$K_1$=(001), $K_2$=(100),<br>T = (100)<br>H = 5½–7<br>D = 3.25–3.62 | X ∧ a = –22° – 32°<br>Y ∥ b, O.A.P. (010)<br>Z ∧ c = +38 – 48°, bi;<br>(r > v)<br>fig. p. 76 | diops. sal. hed.<br>$n_x$ = 1.665–1.699–1.727<br>$n_y$ = 1.672–1.709–1.735<br>$n_z$ = 1.696–1.728–1.756<br>⊕Δ = 0.031–0.029–0.029<br>diagr. p. 82 |
| 167a | Johannsenite mcl<br>CaMn [$Si_2O_6$] | ◊ (100) lam.;<br>granular | # (110) good;<br>H = 6± D = 3.6 | Y ∥ b<br>Z ∧ c = +48°<br>O.A.P. (010) | $n_x$ = 1.700<br>$n_y$ = 1.708<br>$n_z$ = 1.729<br>⊕Δ = 0.029 |
| 168 | Augite group mcl<br>Augite-Ferroaugite<br>Ca(<u>Mg</u>,<u>Fe</u>, Al)<br>[(Si,Al)$_2$O$_6$] | {100},{010},{110},<br>{111},{001};<br>β = 105¾°;<br>◊ (100), {$\bar{1}$22},<br>and (001);<br>short prisms ∥ c, grains | # {110} good, 93°<br># (010) distinct,<br>parting ∥ (100):<br>"diallage";<br>$K_1$=(001), $K_2$=(100),<br>T = (100); H = 5½ – 6<br>D = 3.35–3.45 | X ∧ a = –23° to –31°<br>Y ∥ b, O.A.P. (010)<br>Z ∧ c = +39° to<br>+47°, bi;<br>fig. p. 76 | diops. sal. hed.<br>$n_x$ = 1.670–1.693–1.743<br>$n_y$ = 1.676–1.698–1.750<br>$n_z$ = 1.694–1.720–1.772<br>⊕Δ = 0.024–0.027–0.029 |
| 169 | Fassaite group mcl<br>Ca(Mg,$Fe^{2+}$,$Fe^{3+}$,Al)<br>[(Si,Al)$_2$O$_6$] | {110},{$\bar{1}$11},{22$\bar{1}$},<br>{021};<br>β = 105¾°;<br>◊ (100);<br>pyramids, grains | # {110} good, 93°<br># (010) distinct;<br>H = 5½ – 6<br>D = 2.97 – 3.31 | X ∧ a = –26° to –29°<br>Y ∥ b, O.A.P. (010)<br>Z ∧ c = +36° to +47°,bi;<br>(r > v)<br>fig. p. 76 | diops. fs. 20<br>$n_x$ = 1.676 – 1.693<br>$n_y$ = 1.685 – 1.699<br>$n_z$ = 1.702 – 1.719<br>⊕Δ = 0.026 – 0.026 |
| 170 | Titanaugite mcl<br>Ca(<u>Mg</u>,<u>Fe</u>,Al,Ti)<br>[(Si,Al)$_2$O$_6$] | {100},{110},{010},<br>{111}; β = 106°;<br>◊ (100) lam.,<br>◊ (101) and {122},<br>penetration ◊; short<br>columnar, tabular | # {110} good to poor,<br>93° partly conchoidal<br>fractures<br>H = 5½–6 D = 3.36± | X ∧ a = –16° to –39°<br>Y ∥ b, O.A.P. (010)<br>Z ∧ c = +32° to +55°<br>(v > r), bi;<br>changing in zones<br>fig. p. 76 | $n_x$ = 1.695 – 1.741<br>$n_y$ = 1.700 – 1.746<br>$n_z$ = 1.728 – 1.762<br>⊕Δ = 0.033 – 0.021 |
| 171 | Aegirine-augite group mcl<br>m Na$Fe^{3+}$[$Si_2O_6$] +<br>n Ca(Mg,Fe)[$Si_2O_6$] | {110},{100},{010},<br>{461},{$\bar{1}$11},{101};<br>β = 106°;<br>◊ (100); columnar ∥ c,<br>rarely granular | # {110} perf., 93°<br># (010) distinct,<br>(001) parting;<br>H = 5½ – 6<br>D = 3.3 – 3.6 | X ∧ c = 8°–0° – 50°,bi;<br>Y ∥ b,O.A.P. (010)<br>Z ∧ a = +46° to +16°<br>l' = (–)<br>fig. p. 77 | di-rich ac-rich<br>$n_x$ = 1.690 – 1.774<br>$n_y$ = 1.700 – 1.812<br>$n_z$ = 1.724 – 1.826<br>⊕Δ = 0.034 – 0.052 |
| 172 | Chloromelanite mcl<br>≈ 1 di + 1 ja + 1 ac | xenomorphic;<br>β = 106°;<br>◊ (100);<br>prismatic ∥ c,<br>granular, fibrous | # {110} good, 93°<br>H = 5½–6 D ≈ 3.4 | Z ∧ c ≈ 55°– 63°, bi;<br>Y ∥ b, O.A.P. (010)<br>X ∧ a ≈ 39°– 47° | jad.-rich di-rich<br>$n_x$ = 1.690 – 1.703<br>$n_y$ = 1.700 – 1.717<br>$n_z$ = 1.715 – 1.734<br>⊕Δ = 0.025 – 0.031 |

| Angle and Dispersion of Optic Axes | Color in Thin Section Pleochroism | Chemical Reactions Special Remarks | Distinguishing Features of Similar Minerals | Occurrences | |
|---|---|---|---|---|---|
| $2V_Z = 74^\circ - 78^\circ$ | colorless | effervescent and soluble in HCl | prehnite: insoluble in HCl; X-ray | contact zones between limestones and basaltic magm. | 161a |
| $2V_X = 36^\circ - 48^\circ$<br>$r < v$<br>diagr. p. 75 | colorless, also<br>X = pale orange,<br>Y = pinkish red,<br>Z = pale orange | gelatinizes in HCl; then stainable; weathering to Mn-oxides | rhodonite: gr. n, D;<br>piemontite: gr. n, Δ;<br>johannsenite: gr. Δ | with ilvaite, spessartine contact-metasomatic in marbles with other Mn-minerals | 162 |
| $2V_Z = 60^\circ - 45^\circ$<br>$r > v$ weak | colorless | gelatinizes in HCl; then stainable | wollastonite: O.A.P. ⊥ l;<br>rosenbuschite: l' = (–);<br>katapleite: different habit | postmagmatic in basic volcanics and their joints | 163 |
| $2V_Z = \geq 54^\circ$ enst. ↑ $r < v$<br>$2V = 90^\circ$ brz.<br>$2V_X = > 45^\circ$ hyp. ↓ $r > v$<br>$2V = 90^\circ \pm$ eul. $r < v$<br>diagr. p. 78 | enst.-brz: almost colorless; hyp.-eul.:<br>X = pale reddish brown,<br>Y = pale yellow-green to ocre,<br>Z = light green to grey-green<br>$X \gtreqless Y < Z$ | slowly decomposed by cold HF; unmixing: ilmenite lam. ∥ (100) in hyp.; lamellae of clino-pyroxene ∥ (010) or ∥ (110); frequently serpentinization | clinopyroxene: gr. Δ;<br>zoisite: usually l = (–) and sm. Δ; andalusite: l = (–) and sm. n;<br>amphibole: diff. #;<br>kyanite: oblique extinction | enstatite-bronzite: magmatic in gabbroic to ultrabas. rocks and meteorites; hyperst. magmatic in gabbroic rocks, metam. in hornfelses and schists; eulite: medium grade metam. in eulysite | 164 |
| $2V_Z \sim 30^\circ$ Ca-free; smaller if Ca-cont.<br>$r < v$ | clinoenstatite: colorless<br>clinohypersthene:<br>X = light reddish br.,<br>Y = light yellowish green,<br>Z = light green,<br>$Y < X < Z$ | lamellar intergrowth with enstatite | orthopyroxene: straight extinction; pigeonite: gr. Z∧c and sm. 2V;<br>diopside group: gr. Δ | in meteorites, especially in chondrites; in kimberlites; rarely in magmatic rocks; synthetic in slags | 165 |
| $2V_Z$ usually = $30^\circ - 0^\circ$ in ⊥ (010); $r < v$,<br>$2V_Z$ rarely = $0^\circ - 15^\circ$ in (010); $r > v$<br>diagr. p. 81 | colorless to<br>X = pink to greenish brown,<br>Y = pink to greenish brown,<br>Z = light yellow to light green<br>$X \sim Y \sim Z$ | not decomposed by HCl, lamellar intergrowth ∥ (100) with enstatite and diopside | enstatite: sm. Δ and O.A.P. (100);<br>olivine: gr. Δ;<br>diopside: gr. 2V;<br>orthopyr.: straight extinction | magmatic in basaltic rocks, rarely in gabbroic plutonites; not metam. | 166 |
| $2V_Z = 56^\circ - 63^\circ$<br>$r > v$,<br>opt. axis A (near a) stronger dispersed than opt. axis B<br>diagr. p. 82 | diopside: colorless<br>sal.-hed.:<br>X = light blue-green to pale yellow,<br>Y = pale yellowish green<br>Z = pale green to pale bluish green<br>$X \geqq Z \geqq Y$ | insoluble in HCl, soluble in HF;<br>hed.: in sections ∥ (010) distinctly abnormal interf. colors | augite: more vivid colors;<br>pigeonite: sm. 2V;<br>bronzite: sm. Δ;<br>olivine: poor #<br>others: see Vol. 2 | diopside: magmatic in acid and intermed. rocks; contact- and regional-metam. in calc-silicate rocks; sal., hed.: predom. in Fe-skarns and metam. limestones | 167 |
| $2V_Z = 70^\circ$ | X = pale green,<br>Y = yellowish green,<br>Z = green | insoluble in HCl, soluble in HF | bustamite: sm. $n_y$,<br>sm. Z∧c;<br>rhodonite: sm. Δ | in skarn deposits; metasomatic Mn-ores; in metam. Mn-ore deposits | 167a |
| $2V_Z = 42^\circ - 60^\circ$<br>$r > v$ | X = pale green to pale blue-green,<br>Y = yellowish green to green,<br>Z = olive to grey-green,<br>$Y < X \lesssim Z$ | slowly soluble by HCl; exsolution lamellae of ilmenite ∥ (100), elongated ∥ b and orthopyrox.-lamellae ∥ (100), (001) | orthopyroxene: straight extinct.; diopside: somewhat gr. Δ and paler absorption color<br>see Vol. 2 | magmatic in basaltic rocks of calc-alkali provinces; medium to high grade metam. in amphibole and pyroxene rocks | 168 |
| $2V_Z = 58^\circ - 65^\circ$<br>$r \gg v$,<br>opt. axis B (near c) stronger dispersed than opt. axis A | usually colorless; otherwise<br>X = pale green,<br>Y = pale yellowish gr.,<br>Z = pale green | not decomposed by HCl, abnormally leather-brown and bluish grey interference colors, especially in (010) | augite: usually no ext. dispersion; diopside-augite, titanaugite: absorption color | contact-metamorphic in calc-silicate rocks and marbles with spinel | 169 |
| $2V_Z = 42^\circ - 65^\circ - (0^\circ)$<br>$r \gg v$ strongly inclined; opt. axis B (near c) stronger dispersed than opt. axis A | X = pale brownish yellow (⊥ (100)),<br>Y = brownish lilac,<br>Z = brownish yellow to violet-brown (∥ c)<br>$X < Y \gtrsim Z$ | corroded by hot HCl; zonal structure, hourglass structure, abnormal interference colors in (010) | schorlite (tourmaline): uniaxial;<br>glaucophane: sm. n, ⊖, diff. # | magmatic in basic to intermediate rocks of alkaline provinces; not metamorphic | 170 |
| $2V_Z = 60^\circ$ ↑ di-rich<br>$2V = 90^\circ$ $r > v$<br>$2V_X = 60^\circ$ ↓ aeg-rich $r < v$<br>diagr. p. 83 | X = green<br>Y = pale green<br>Z = pale green-yellow to pale brownish yell.,<br>acmite: { X = br., Y = p. br.<br>$X > Y > Z$ | slightly decomposed by HCl, easily fusible with blowpipe; zonal structure | pyroxenes: slightly green and ⊕;<br>green hornblende: diff. # | magmatic in intermediate rocks of alkaline provinces; rim around titanoaugite in basic alkaline rocks | 171 |
| $2V_Z = 80^\circ - 84^\circ$<br>$r \gg v$ | acmite-rich:<br>$X \approx Y$ emerald green,<br>Z = yellowish green,<br>(if acmite-poor only pale green)<br>$X \approx Y > Z$ | often fractured cataclastic, acmite-rich members strongly pleochroic, characteristic paragenesis | augite: sm. 2V and diff. colors; aegirine-augite: gr. n and Δ; omphacite, jadeite: sm. 2V, sm. Z∧c | low grade metamorphic in serpentinites, in alpine eclogites and low grade metam. basic to ultrabasic rocks | 172 |

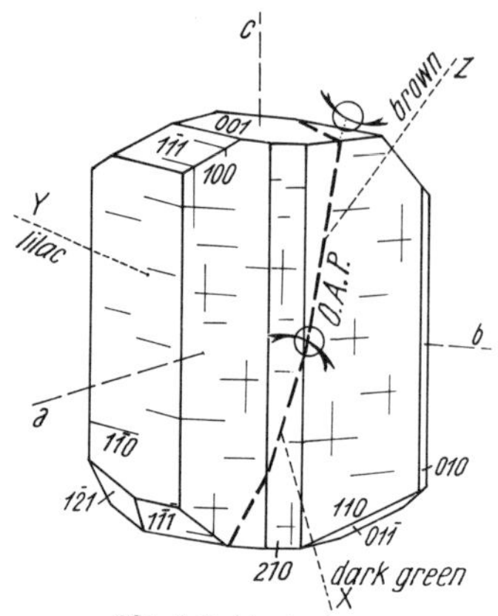


*160-1. Babingtonite*
(modif. after RICHMOND 1937)

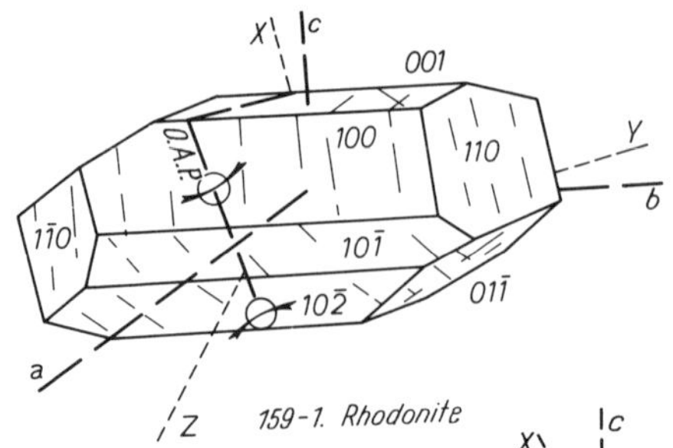


*159-1. Rhodonite*

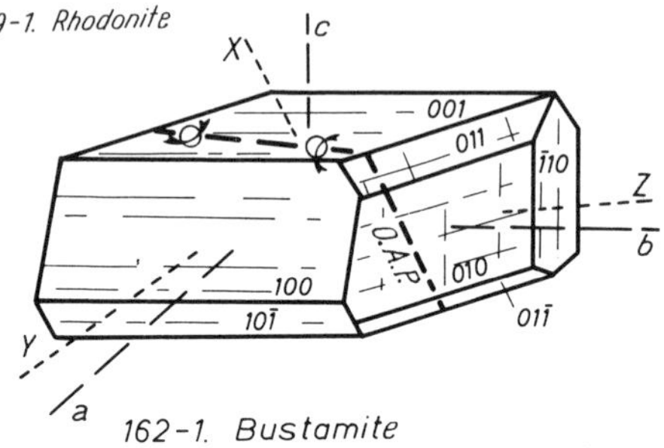


162-1. Bustamite
*(idealized sketch, crystals xenomorphic)*

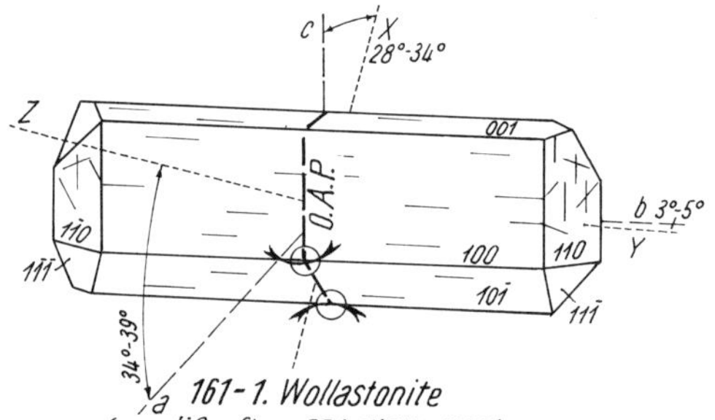


*161-1. Wollastonite*
*(modif. after PEACKOK 1935)*

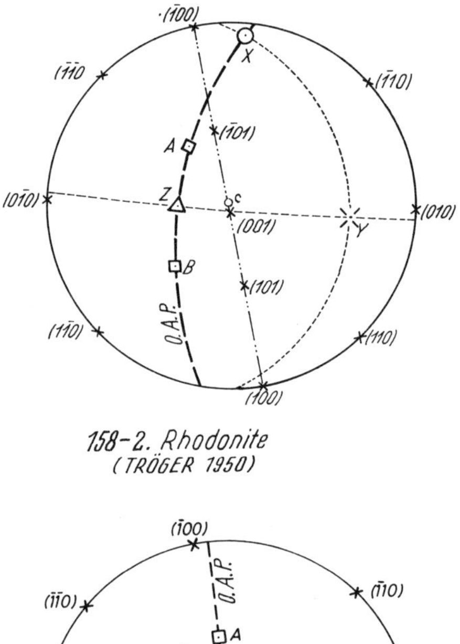


*158-2. Rhodonite*
*(TRÖGER 1950)*

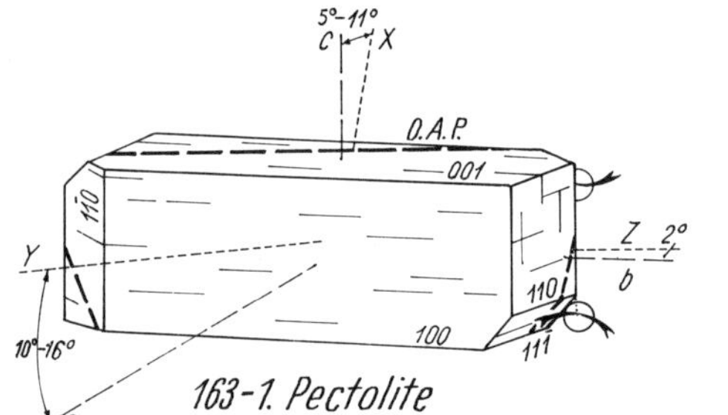


*163-1. Pectolite*
*(PEACKOK 1935)*

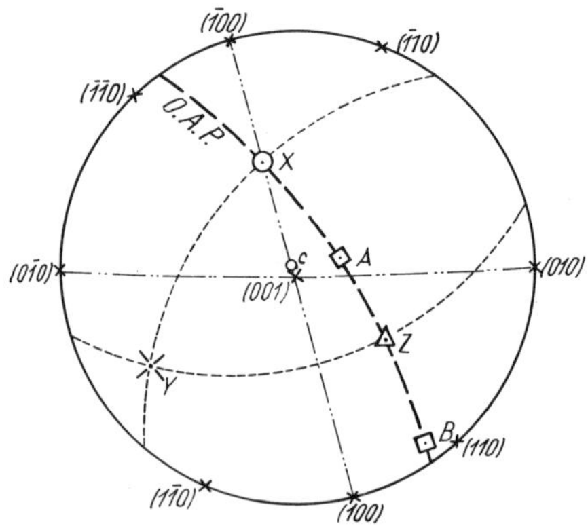


*160-2. Babingtonite*
*(RICHMOND 1937)*

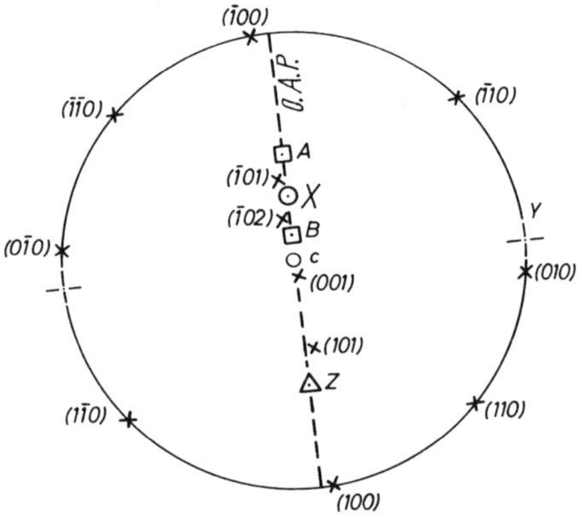


161-2. Wollastonite
(BRAUNER 1960)

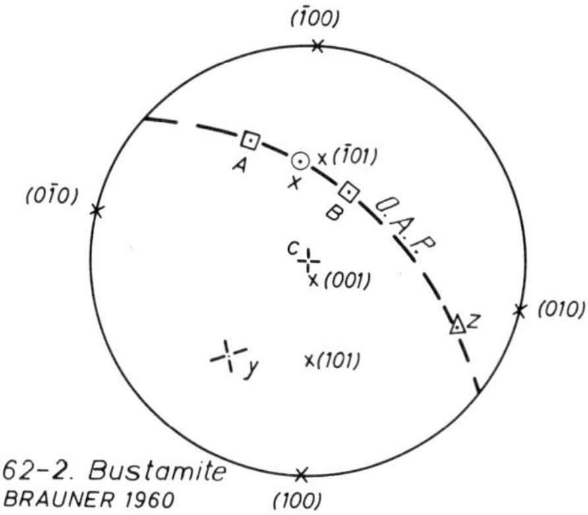


162-2. Bustamite
BRAUNER 1960

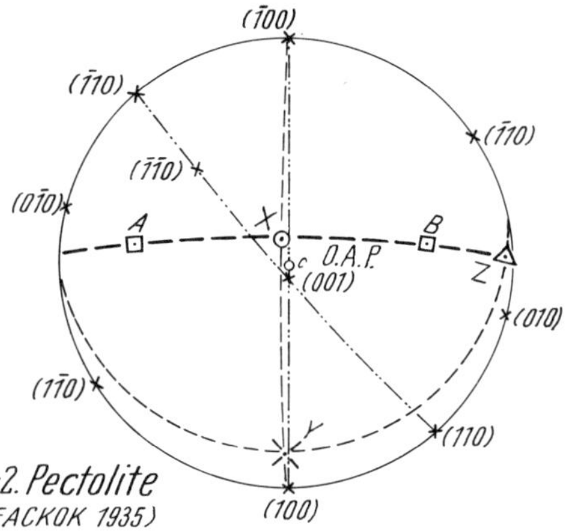


163-2. Pectolite
(PEACKOK 1935)

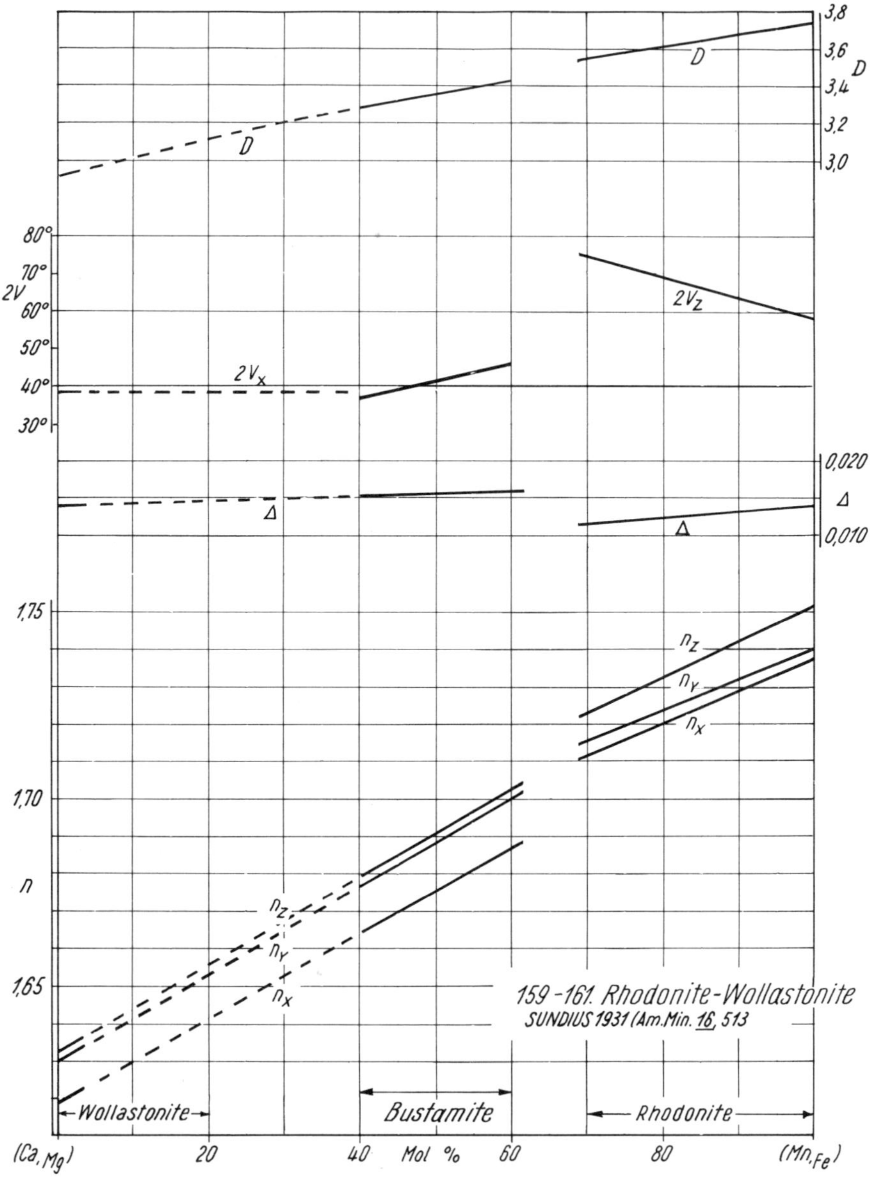


159-161. Rhodonite-Wollastonite
SUNDIUS 1931 (Am.Min. 16, 513

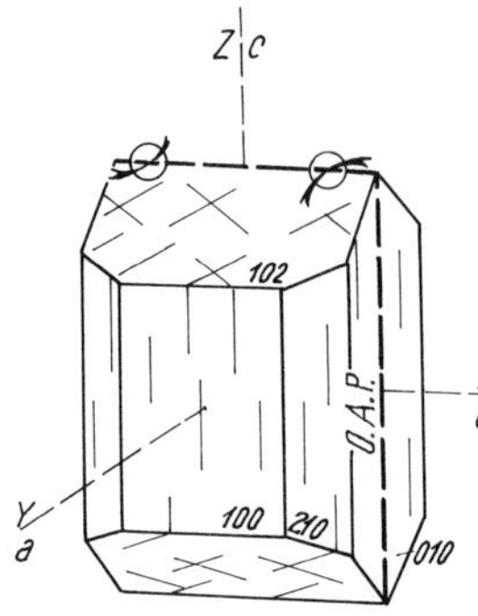


164-1. Enstatite

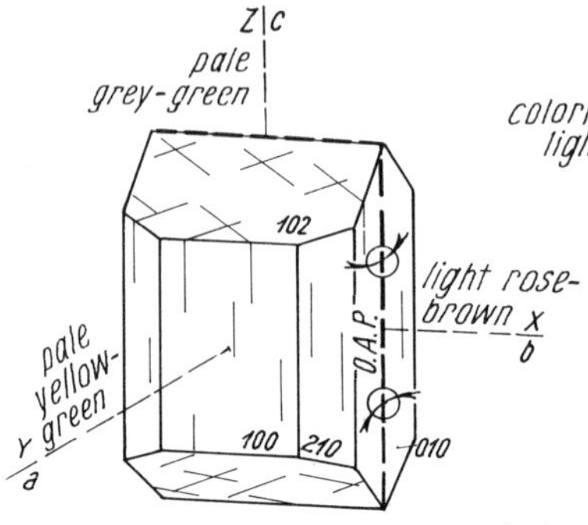


164-2. Hypersthene

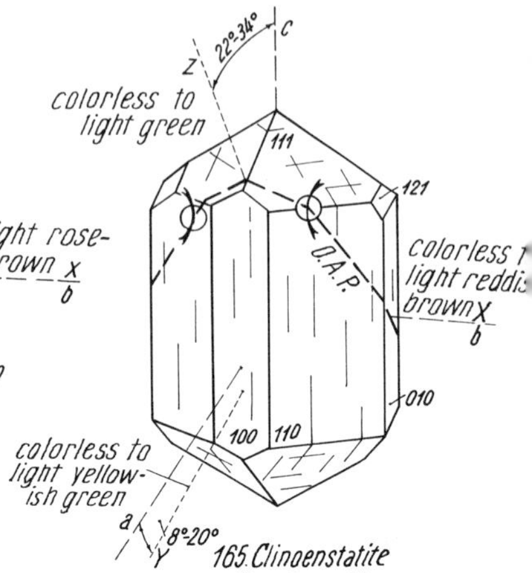


165. Clinoenstatite
Clinoferrosilite

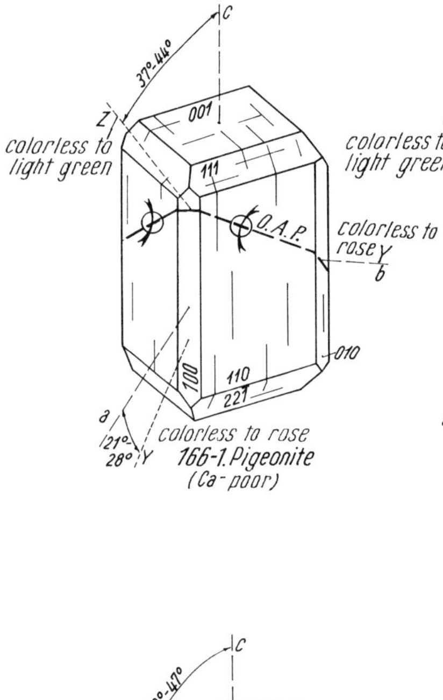


166-1. Pigeonite
(Ca-poor)

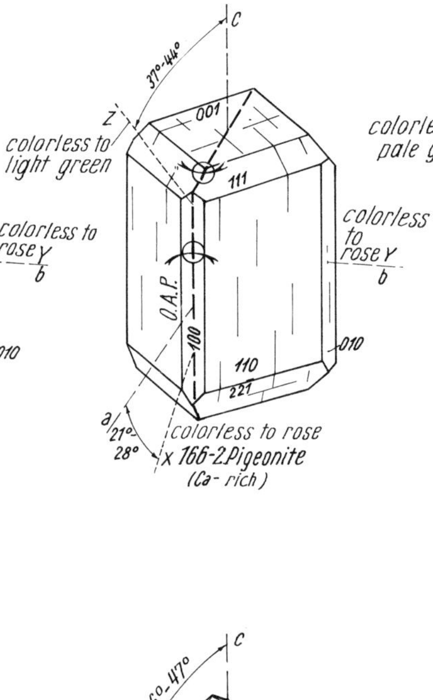


166-2. Pigeonite
(Ca-rich)

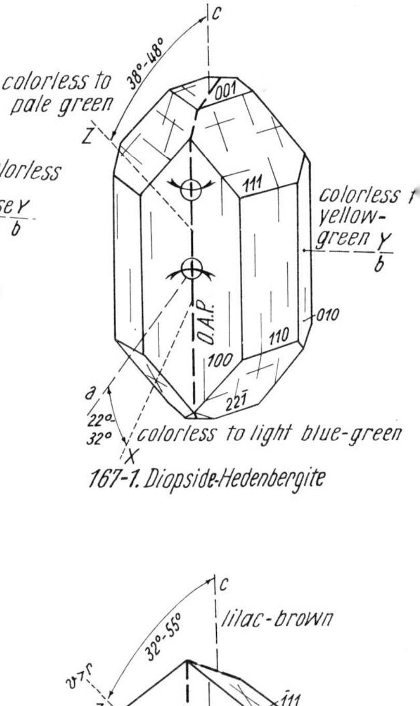


167-1. Diopside-Hedenbergite

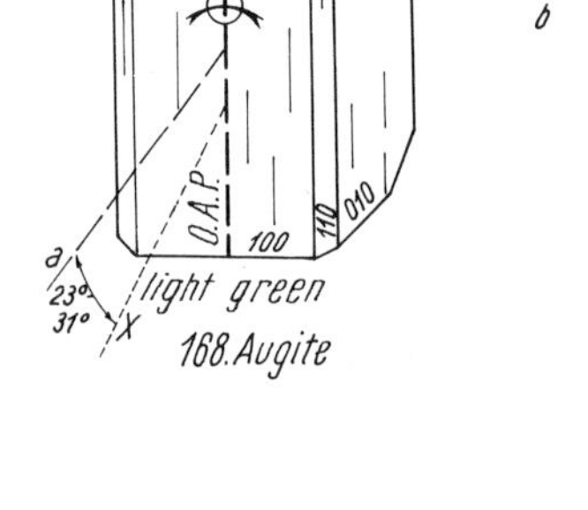


168. Augite

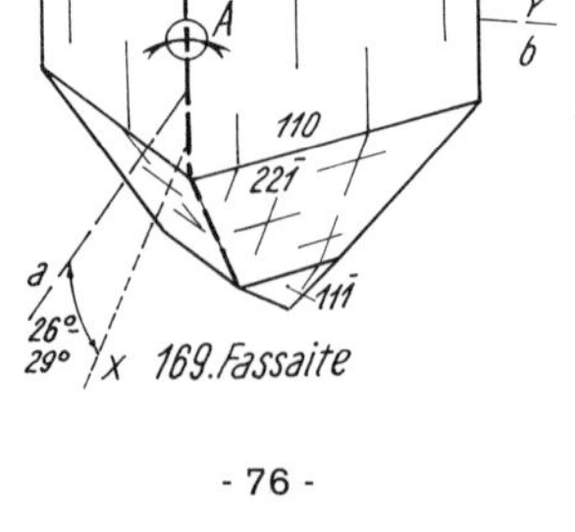


169. Fassaite

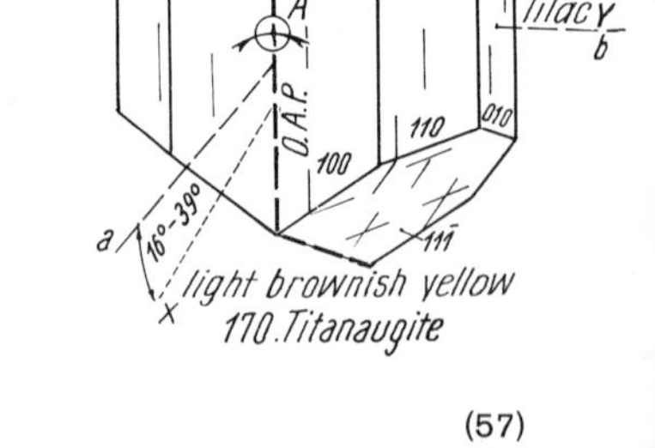


170. Titanaugite

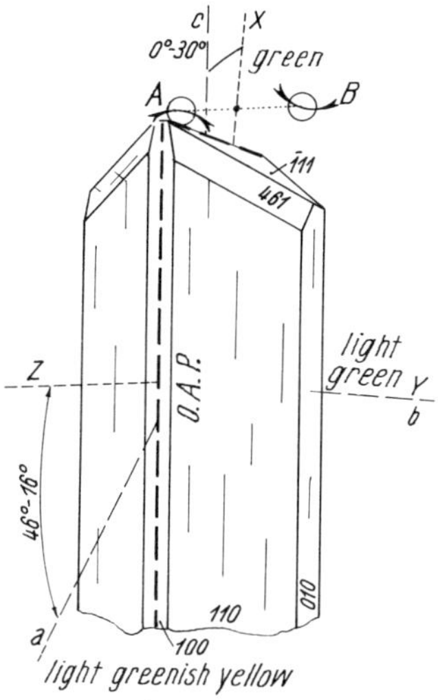


171-1. Aegirine-Augite

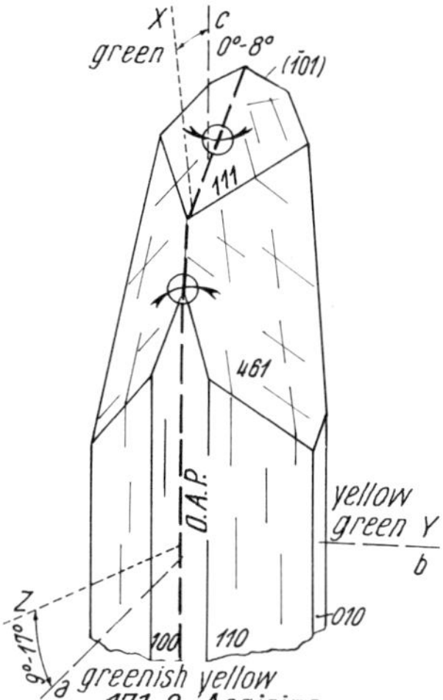


171-2. Aegirine

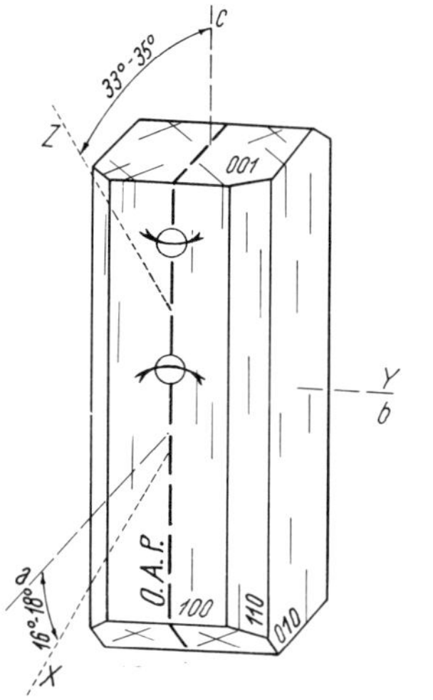


174. Jadeite

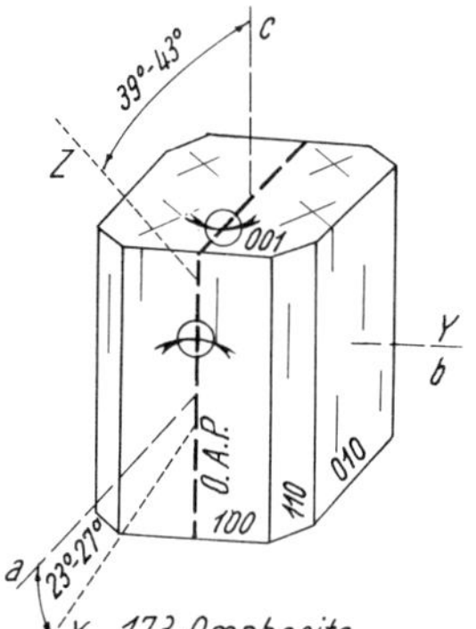


173. Omphacite

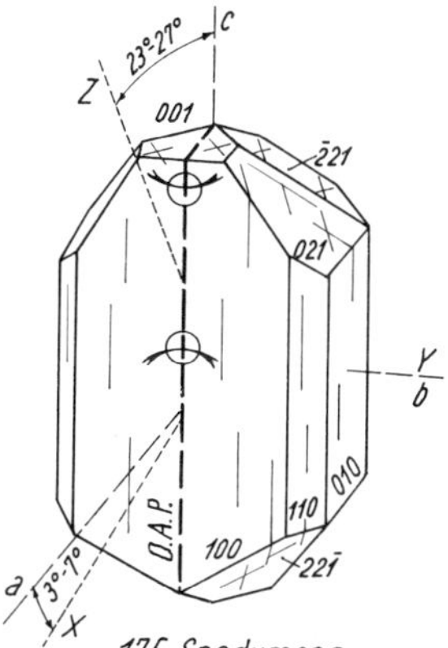


175. Spodumene

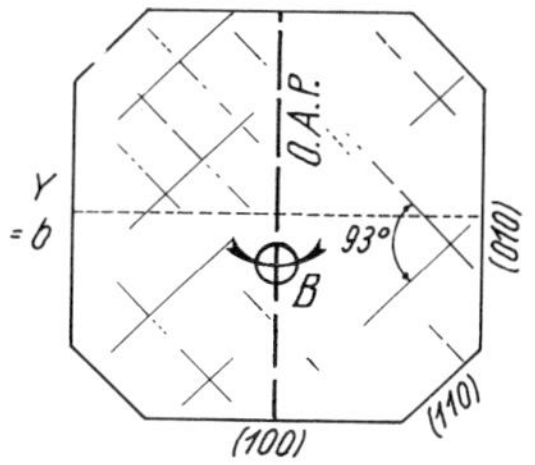


166 bis 175-1. Clinopyroxenes, section ⊥ c

## 165 to 175-2. Clinopyroxenes

Average extinction angles
Z ∧ c
in section ∥ (010)

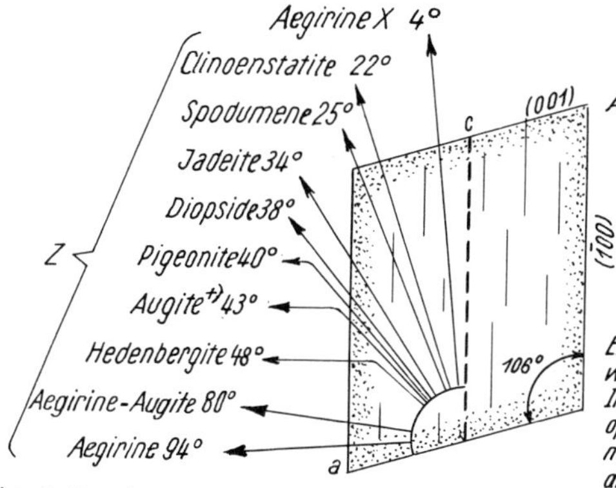


+) including Fassaite, Omphacite, Titanaugite

Exact data obtainable only with universal stage.
In a twin pair, measure optic axes in a section nearly ⊥c or extinction angles in a section nearly ∥ (010).
(BAMBAUER, 1959, Z. Krist. 111, 421)

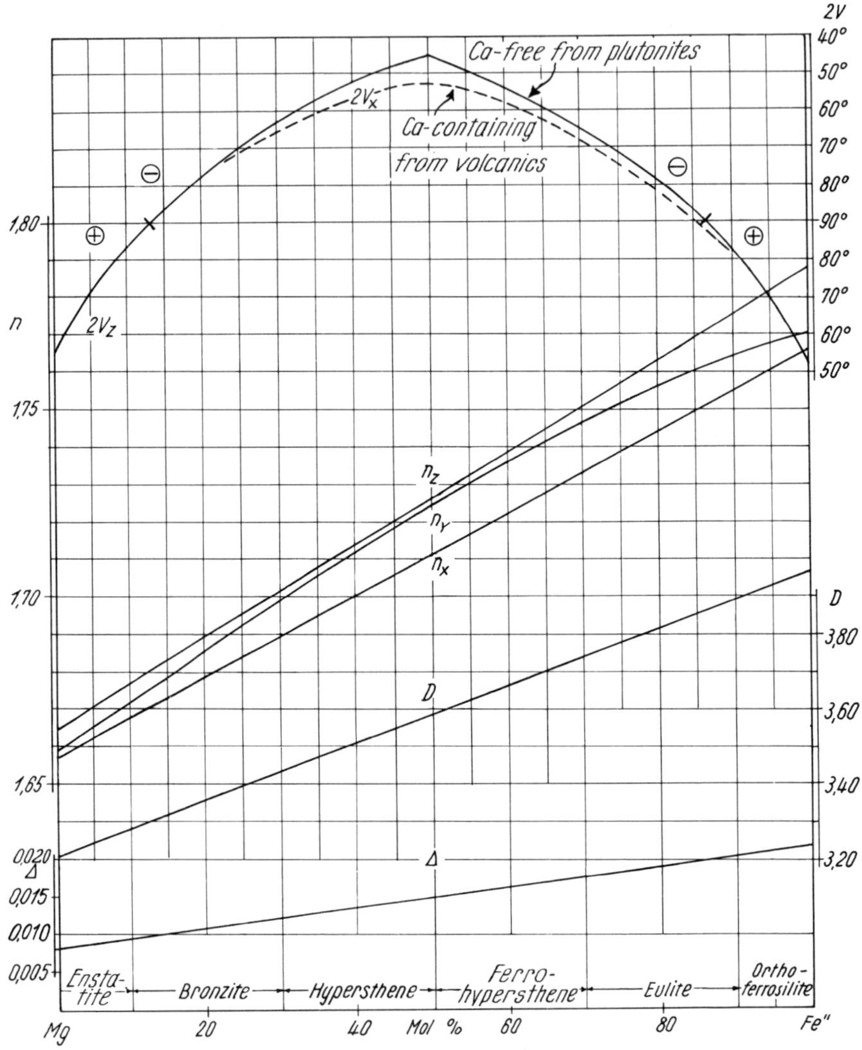


164-3. Orthopyroxenes

H.H.HESS 1952, Amer. J.Sci., BOWEN-VOL., 180.

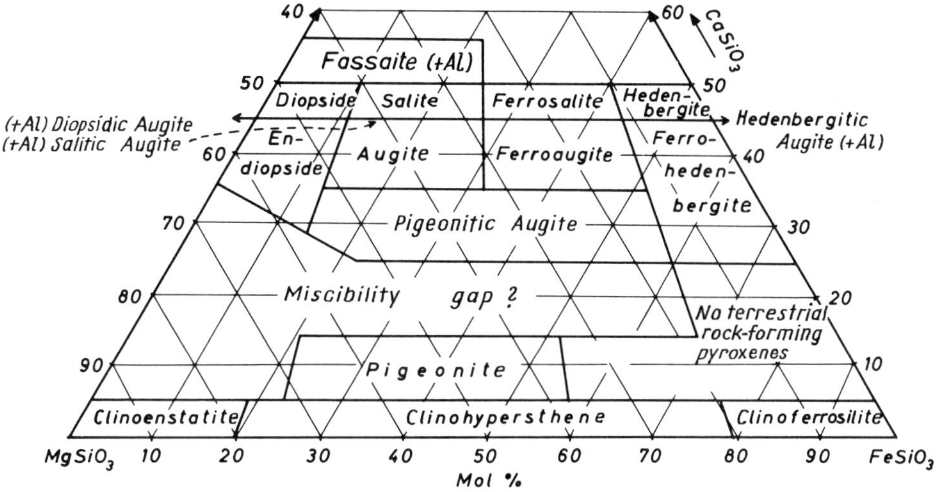


*165–169. Chemical variation of clinopyroxenes in the system $CaSiO_3 - MgSiO_3 - FeSiO_3$*
*TRÖGER 1951, (N. Jb. Miner. Mh. 1951, 136) after HESS 1941 and 1949 corrected after MORIMOTO et al. Z. Krist. 114, 120, 1960*

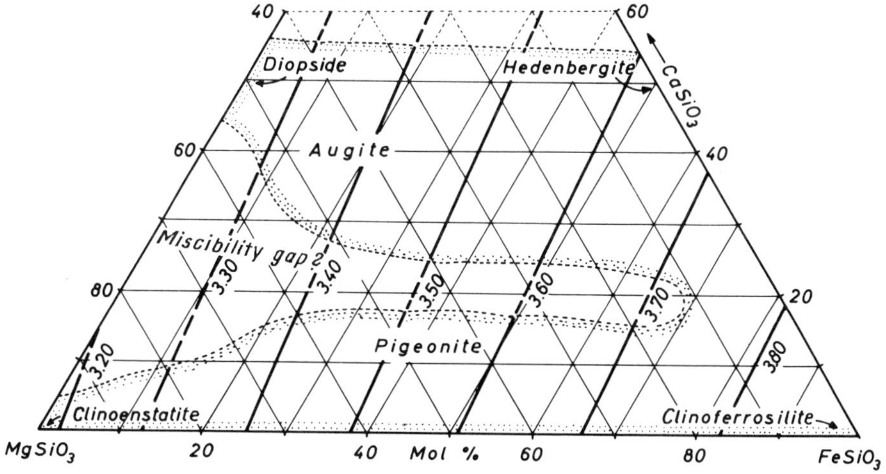


*165–169. Densities of clinopyroxenes in the system $CaSiO_3 - MgSiO_3 - FeSiO_3$*
*TRÖGER 1951 (N. Jb. Min., Mh. 1951, corrected after MORIMOTO et al. Z. Krist. 114, 120, 1960)*

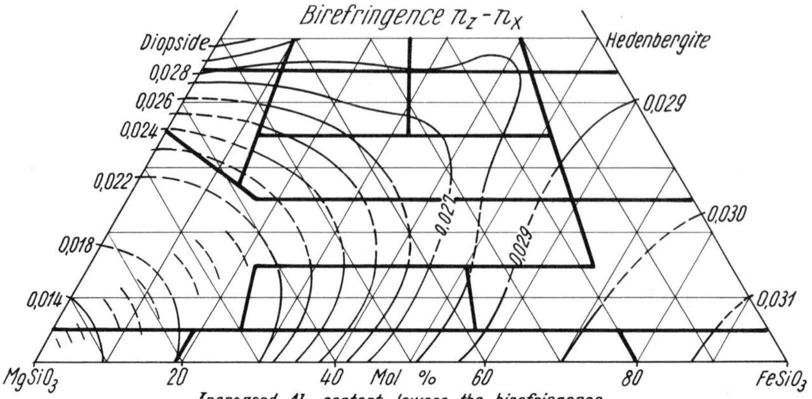


Increased Al-content lowers the birefringence,
exolution of hypersthene lamellae raises the birefringence

**165-167. Clinopyroxenes in the system $CaSiO_3 - MgSiO_3 - FeSiO_3$**
HESS 1949 (Am. Min. 34, 639)

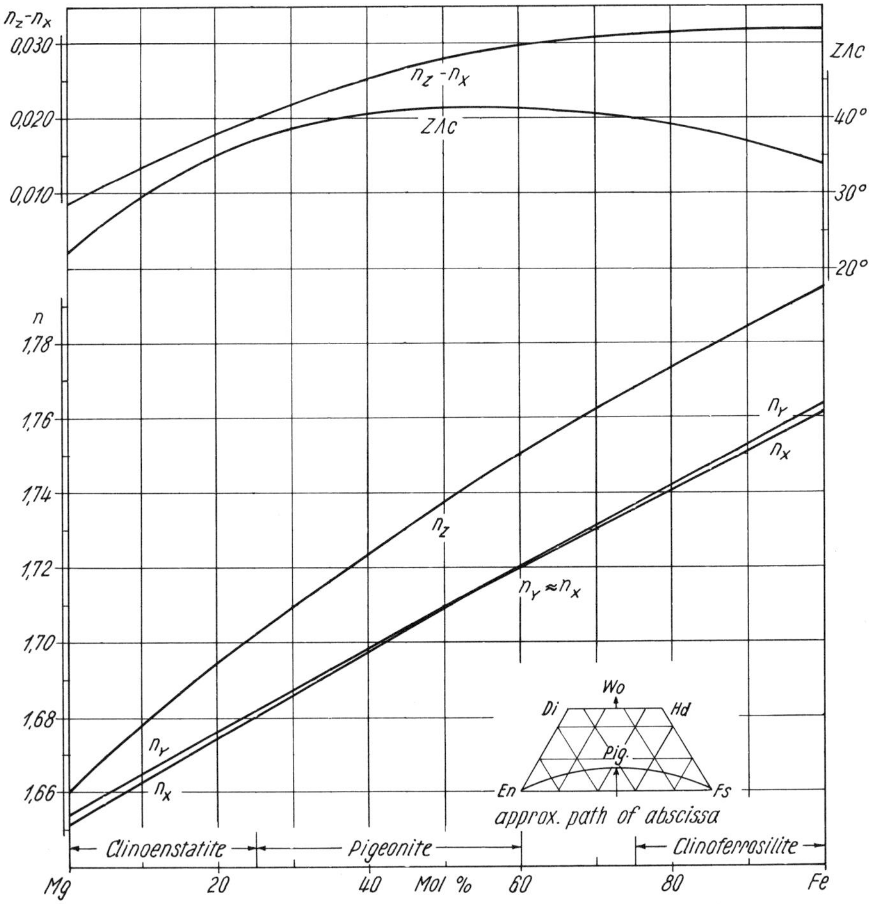


165-166-3. Pigeonite
HESS 1949 (Am. Min. 34, 643)

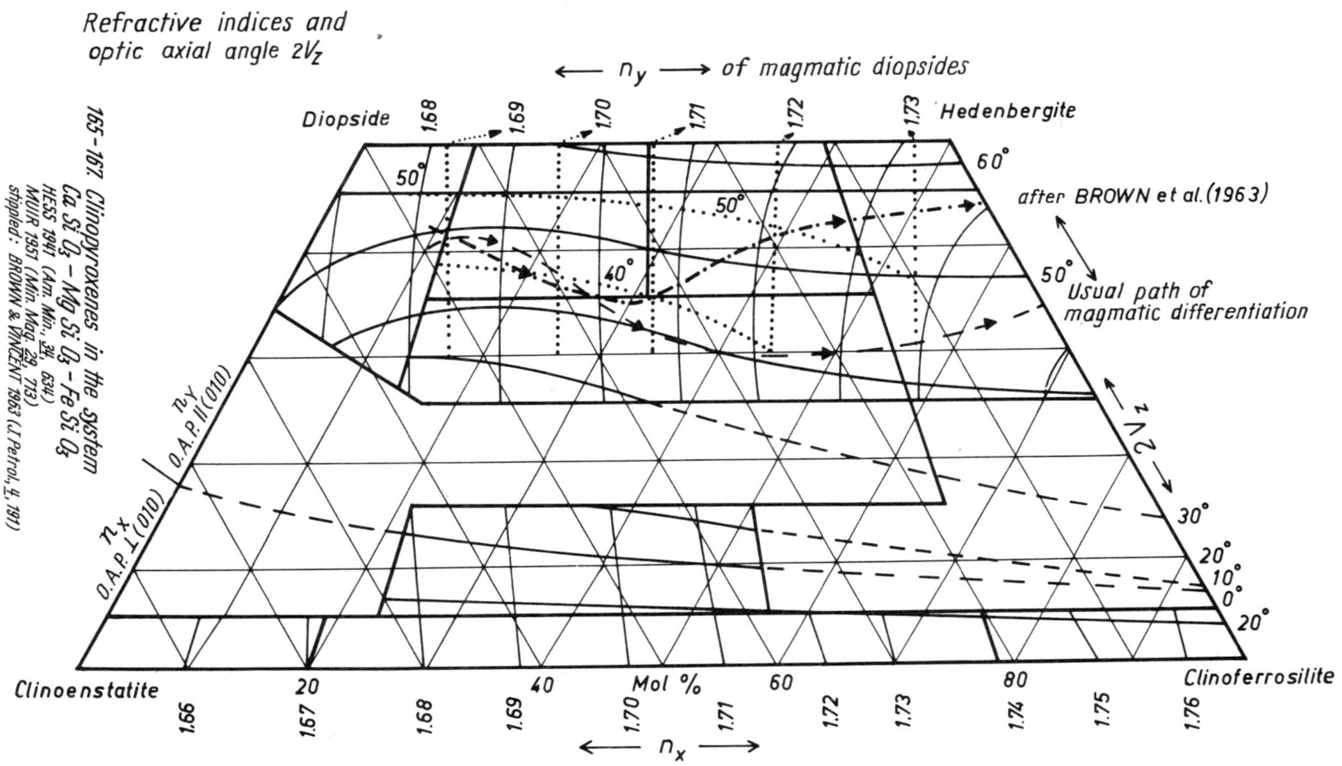


*165–167. Clinopyroxenes in the system*
*$CaSiO_3 - MgSiO_3 - FeSiO_3$*
*HESS 1941 (Am. Min., 34, 634)*
*MUIR 1951 (Min. Mag. 29, 713)*
*stippled: BROWN & VINCENT 1963 (J. Petrol., 4, 191)*

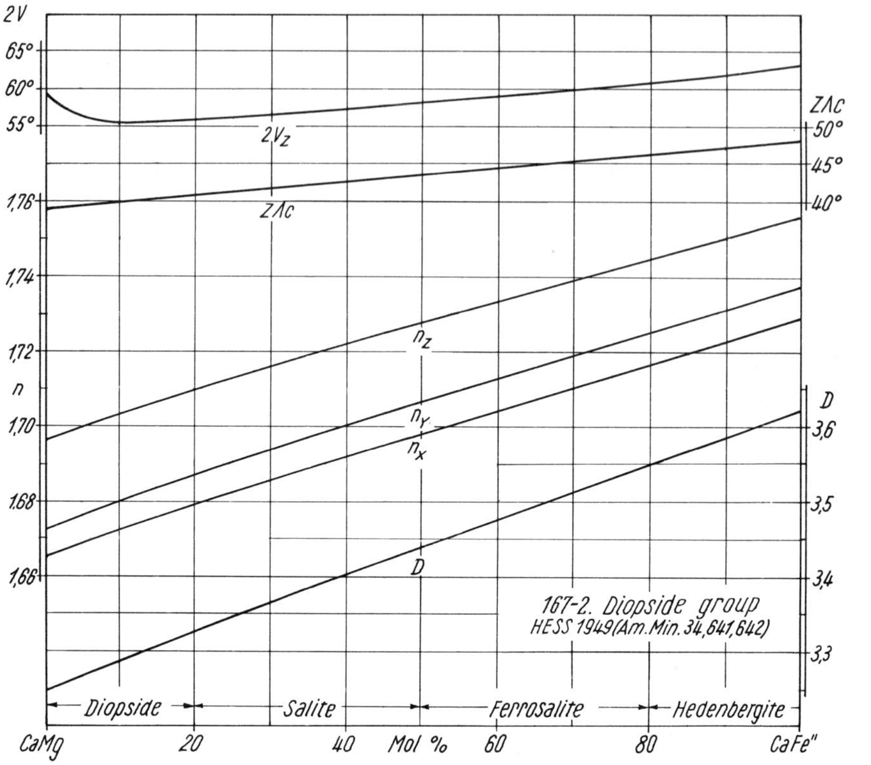


167-2. Diopside group
HESS 1949 (Am. Min. 34, 641, 642)

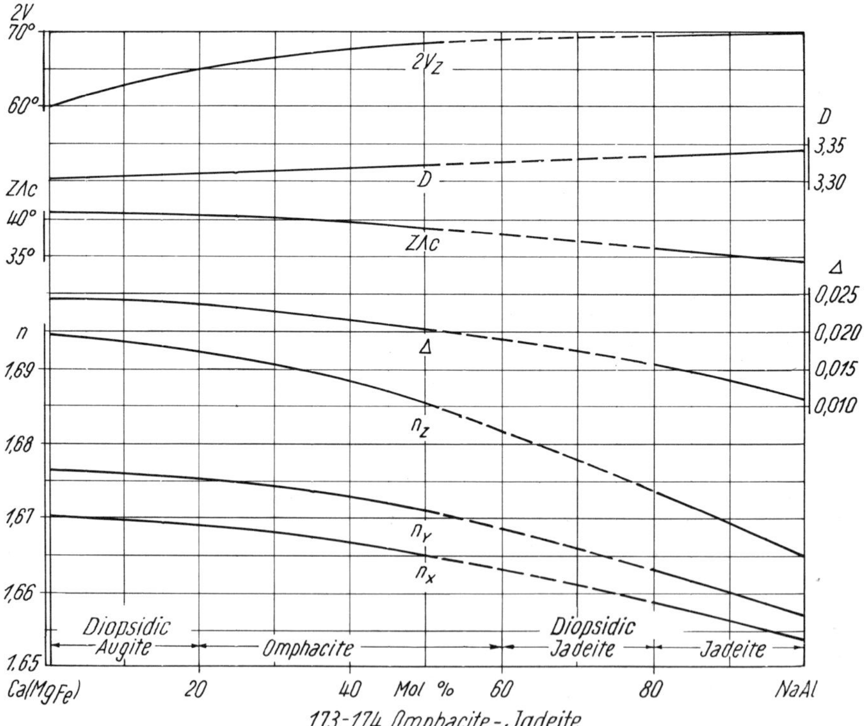


173-174. Omphacite - Jadeite
TRÖGER 1951 (N. Jb. Min. Mh. 1951, 138)
corrected after TRÖGER 1961, Tscherm. Mitt. 8, 24-35

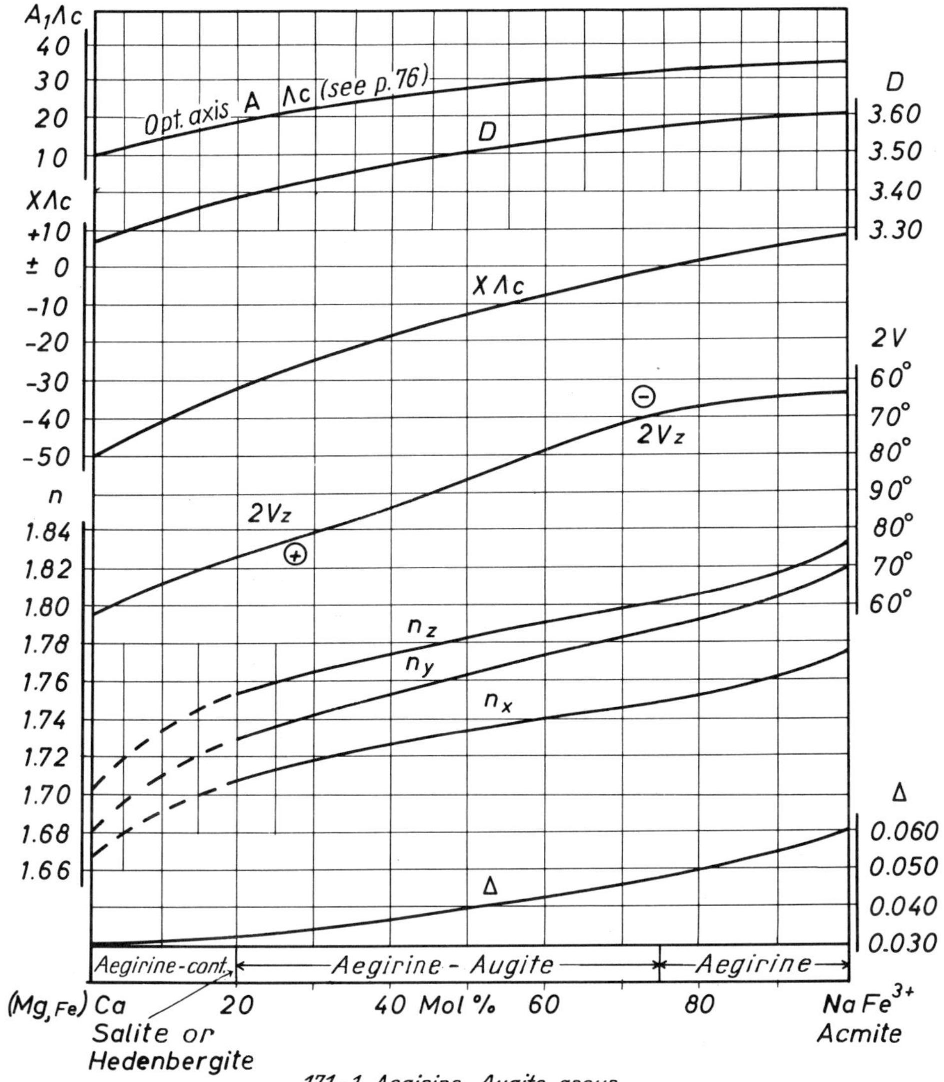


171-1. Aegirine-Augite group

TRÖGER 1950 from LARSEN 1942.
SABINE 1950, KING 1962.

| No. | Name<br>Symmetry<br>Chem. Composition | Common Crystal Forms, Twinning<br>Crystal Habit<br>Aggregates | Cleavage<br>Translation Gliding<br>Deform. Twinning<br>Hardness, Density | Optic<br>Orientation | Refractive Indices<br>Birefringence |
|---|---|---|---|---|---|
| 173 | Omphacite mcl<br>≈ di. 80–40<br>jad. 20–60 | xenomorphic;<br>β = 106°;<br>◈ (100), lam.;<br>short columns,<br>granular | # {110} perf. 92°;<br>(100) parting;<br>H = 5–6 D ≈ 3.31 | XΛa = −23° to −27°<br>Y ∥ b, O.A.P. (010)<br>ZΛc = +43° to +39°, bi;<br>fig. p. 77 | $n_x$ = 1.669 – 1.667<br>$n_y$ = 1.676 – 1.674<br>$n_z$ = 1.693 – 1.689<br>⊕Δ = 0.024 – 0.022<br>diagr. p. 82 |
| 174 | Jadeite mcl<br>$NaAl[Si_2O_6]$ | xenomorphic;<br>β = 107.2°;<br>◈ (100) and (001) lam.<br>(110); flakes ∥ c;<br>reticulated, granular,<br>massive | # {110} good 92°;<br>(100) parting;<br>H = 6½ D = 3.35 | XΛa = −16°–18° (pure)<br>to −22° – 30°<br>Y ∥ b, O.A.P. (010)<br>ZΛc = +33° to +36°, bi;<br>fig. p. 77 | pure di-rich<br>$n_x$ = 1.654 – 1.658<br>$n_y$ = 1.657 – 1.663<br>$n_z$ = 1.665 – 1.674<br>⊕Δ = 0.011 – 0.016<br>diagr. p. 82 |
| 175 | Spodumene mcl<br>$LiAl[Si_2O_6]$ | {100},{010},{110},<br>{0kl}, {221},{001};<br>β = 110½°;<br>◈ (100) rare;<br>tabular, bladed,<br>flaky, acicular ∥ c | # {110} perf. 94°;<br>(100) parting;<br>H = 6½ – 7 D ≈ 3.2 | XΛa = −3° to −7°, bi;<br>Y ∥ b, O.A.P. (010)<br>ZΛc = +23° to +27°, bi;<br>fig. p. 77 | $n_x$ = 1.651 – 1.668<br>$n_y$ = 1.660 – 1.671<br>$n_z$ = 1.676 – 1.681<br>⊕Δ = 0.025 – 0.013 |
| 176 | Anthophyllite orh<br>$(Mg, Fe)_7[(OH)_2/Si_8O_{22}]$ | {210},{100},{010};<br>no ◈;<br>acicular ∥ c, flaky | # {210} perf. 54°,<br># (010) distinct,<br># (100) poor,<br>(001) parting;<br>H = 5½ D = 3.1± | X ∥ a, bi (Mg-rich)<br>Y ∥ b, O.A.P. (010)<br>Z ∥ c, bi (Fe-rich)<br>l = (+)<br>fig. p. 88 | Mg-rich Fe-rich<br>$n_x$ = 1.598 – 1.647<br>$n_y$ = 1.616 – 1.651<br>$n_z$ = 1.623 – 1.664<br>Δ = ⊖ 0.026–⊕0.017<br>diagr. p. 92 |
| 177 | Gedrite orh<br>$(Mg,Fe)_6Al[(OH)_2/AlSi_7O_{22}]$ | {210},{100},{010};<br>no ◈;<br>acicular ∥ c,<br>fibrous, flaky | # {210} perf. 54°,<br># (010) distinct,<br># (100) poor,<br>(001) parting;<br>H = 5½ D = 3.3± | X ∥ a, bi (Mg-rich)<br>Y ∥ b, O.A.P. (010)<br>Z ∥ c, bi (Fe-rich)<br>l = (+)<br>fig. p. 88 | Mg-rich Fe-rich<br>$n_x$ = 1.642 – 1.674<br>$n_y$ = 1.651 – 1.681<br>$n_z$ = 1.658 – 1.691<br>Δ = ⊖ 0.016 –⊕0.017<br>diagr. p. 92 |
| 178 | Cummingtonite group mcl<br>Cummingtonite-Grunerite<br>$(Fe,Mg)_7[(OH)_2/Si_8O_{22}]$ | {hk0}, {010};<br>β = 109½°;<br>◈ (100), fine lam.;<br>divergent, acicular ∥ c,<br>flaky, asbestiform:<br>amosite | # {110} perf.<br>54½ – 56°.<br># (100) poor,<br>(001) parting;<br>H = 5 – 6<br>D = 3.30 – 3.60 | XΛa = 5° – 9° (bi), grun.,<br>0° – 5° cum.,<br>ZΛc = 10° – 15° grun.,<br>15° – 20° (bi) cum.,<br>l' = (+)<br>fig. p. 88 | cum. grun.<br>$n_x$ = 1.644 – 1.685<br>$n_y$ = 1.657 – 1.709<br>$n_z$ = 1.614 – 1.728<br>Δ = ⊕ 0.030 – ⊖0.043 |
| 179 | Actinolite group mcl<br>Tremolite-Actinolite-Ferroactin.<br>$Ca_2(Mg,Fe)_5[(OH)_2/Si_8O_{22}]$ | {110},{010},{011};<br>β = 106°;<br>◈ (100) lam.,<br>◈ (001) lam.;<br>acicular-fibrous ∥ c,<br>flaky | # {110} perf. 55¾°,<br># (010) poor,<br>(001) parting;<br>H = 5 – 6<br>D = 3.0 – 3.5 | XΛa = −1° to +6°, bi;<br>Y ∥ b, O.A.P. (010)<br>ZΛc = 15° – 10°<br>l' = (+)<br>fig. p. 88 | trem. act. fe-act.<br>$n_x$ = 1.608–1.647–1.688<br>$n_y$ = 1.618–1.659–1.699<br>$n_z$ = 1.630–1.667–1.704<br>⊖Δ = 0.022–0.020–0.016<br>diagr. p. 92 |
| 179a | Smaragdite mcl<br>Cr-bearing<br>actinolitic hornbl. | {110},{010},{011};<br>β = 106°;<br>◈ (100) lam.;<br>acicular | # {110} perf.;<br>H = 5 – 6 D = 3.4± | | like tremolite<br>⊖Δ about like tremolite |
| 180 | Na-Tremolite-Richterite mcl<br>$Na_2Ca(Mg,Fe,Mn)_5[(OH)_2/Si_8O_{22}]$ | {110}, {hk0}, {010},<br>{011};<br>β = 104¾°;<br>columnar to acicular<br>∥ c | # {110} perf. 56°;<br>H = 5 – 6 D = 2.90± | XΛa = −3° to 5°, bi;<br>Y ∥ b, O.A.P. (010)<br>ZΛc = +16° to 26°<br>l' = (+)<br>fig. p. 88 | $n_x$ = 1.606 – 1.628<br>$n_y$ = 1.613 – 1.638<br>$n_z$ = 1.623 – 1.644<br>⊖Δ = 0.017 – 0.016 |
| 181 | Glaucophane group mcl<br>Glaucophane-Crossite-Mg-Riebeckite<br>$Na_2(Mg,Fe^{2+})_3(Al,Fe^{3+})_2[(OH)_2/Si_8O_{22}]$ | {110},{010},{10$\bar{1}$},<br>{011};<br>β = 104¼°;<br>◈ (100) rare;<br>acicular ∥ c,<br>fibrous, flaky | # {110} perf. 55½°<br>~ (001) parting;<br>H = 5 – 6<br>D = 3.07 – 3.33 | {ZΛc = +7° to 5°, l' = (+)<br>{Y=b, O.A.P. (010): glauc.,<br>YΛc = +5° to 9°: cross.,<br>Z=b XΛc = 3° to 15°:<br>Mg-riebeckite,<br>l = (+) to (±)<br>fig. p. 88 | glauc. cross. Mg.-rieb.<br>$n_x$ = 1.595–1.642–1.668<br>$n_y$ = 1.614–1.656–1.676<br>$n_z$ = 1.620–1.657–1.678<br>⊖Δ = 0.025–0.015–0.010<br>diagr. p. 94 |
| 182 | Riebeckite Osannite mcl<br>$Na_2(Mg,Fe)_3(Al,Fe^{3+})_2[(OH)_2/Si_8O_{22}]$ | {110},{010},{$\bar{1}$01};<br>β = 103½° – 107½° (?)<br>acicular ∥ c,<br>reticulated, fibrous | # {110} perf. 56–55°,<br># (010) good,<br># (001) distinct;<br>H = 5½ – 6<br>D = 3.38± | XΛc = 0° to −3° (−9°)<br>bi; r < v<br>Y ∥ b, O.A.P. (010): rieb.,<br>Z ∥ b, O.A.P. ⊥ (010):<br>osannite,<br>l' = (−)<br>fig. p. 89 | $n_x$ = 1.685 – 1.695<br>$n_y$ = 1.687 – 1.697<br>$n_z$ = 1.689 – 1.699<br>⊖Δ = 0.004 – 0.004 |
| 183 | Arfvedsonite group mcl<br>Mg-Arfvedsonite-Fluotaramite-Arfveds.<br>$Na_3(Mg,Fe^{2+})_4(Fe^{3+},Al)[(OH)_2/Si_8O_{22}]$ | {110},{010}, rarely<br>idiomorphic;<br>β = 104¼°;<br>◈ (100);<br>columnar ∥ c, granular,<br>flaky ∥ (010) | # {110} perf. 56°;<br># (010) distinct;<br>H = 5 – 6<br>D = 3.00 – 3.40 | XΛc = −10° (−25°) – 0°, bi;<br>−25° to −53°: Mg-arf.,<br>{Y ∥ b, O.A.P. (010): Mg-arf.,<br>{Z ∥ b, O.A.P. ⊥ (010):<br>fluot.-arf.,<br>l' = (±) to (−) fig. p. 89 | Mg-arf. fluot. arf.<br>$n_x$ = 1.633–1.660–1.702<br>$n_y$ = 1.639–1.671–1.705<br>$n_z$ = 1.642–1.672–1.707<br>⊖Δ = 0.009–0.012–0.005<br>diagr. p. 95 |

| Angle and Dispersion of Optic Axes | Color in Thin Section Pleochroism | Chemical Reactions Special Remarks | Distinguishing Features of Similar Minerals | Occurrences | |
|---|---|---|---|---|---|
| $2V_z = -(60^o) - 66^o - (74^o)$<br>r > v<br>diagr. p. 82 | almost colorless | not decomposed by HCl, often inclusions of rutile, often intergrowths with smaragdite: (010) ‖ (010), c ‖ c | diopside, augite: sm. 2V, different paragenesis; jadeite: sm. Δ; gedrite: diff. # | high grade metamorphic in eclogitic rocks together with garnet, smaragdite, kyanite, rarely in glaucophane rocks | **173** |
| $2V_z = 72^o - 68^o$<br>r < v | usually colorless; rarely:<br>X = pale greenish,<br>Y = colorless,<br>Z = pale yellowish | insoluble in HCl, easily fusible before the blowpipe: flame with Na-color | actinolite: sm. Z∧c; diopside, augite: sm. 2V and gr. Z∧c; omphacite: gr. n, gr. Δ; chloromelanite: gr. n, gr. Δ | low grade /high pressure metam. in basic alkaline rocks and limestones, together with glaucophane; also in certain eclogites | **174** |
| $2V_z = 55^o - 80^o$<br>r < v, slightly oblique; also:<br>$2V_x = 60^o$ to $80^o$ ⊖<br>in $n_y$ slightly higher! (1.665 – 1.676) | almost colorless; in thick sections lilac (kunzite) or green (hiddenite), then: X > Y > Z, pale | not decomposed by acids; flame with Li-color! | clinoenstatite: not in pegmatites; diopside: gr. Z∧c, sm. 2V; petalite: sm. n and Δ | with lepidolite, tourmaline and triphyline in granite-pegmatites and -aplites | **175** |
| $2V_x = 65^o$, $2V = 90^o$, $2V_z = 58^o$ — Mg:Fe 95:5, 83:17, 60:40<br>r ≷ v weak<br>fig. p. 92 | colorless | decomposed only by HF; easily altered to talc and antigorite | zoisite: gr. n, sm. Δ; carpholite: gr. n; sillimanite: {110} ~ 90°; monoclin. amph.: oblique extinction | with clinoamph. in metamorphic schists, especially in serpentines, not primary magmatic | **176** |
| $2V_x = 80^o$, $2V = 90^o$, $2V_z = 70^o$ — Mg:Fe 80:20, 66:34, 40:60<br>r ≷ v weak<br>fig. p. 92 | X = pale yellow-brown to greenish yellow,<br>Y = pale brownish yel. to greenish yellow,<br>Z = yellowish brown to greyish green,<br>X ≈ Y < Z | decomposed only by HF; inclusions of biotite and spinel | anthophyllite: chem. anal. necessary<br>zoisite: gr. n, sm. Δ; cummingtonite: Z∧c ~ 15°; pyroxenes: # ~ 90° | medium grade metam. with garnet and cordierite in amphibolites, eclogites, cordierite gneisses; in hornfelses | **177** |
| $2V_z = 80^o - 90^o$ cum.,<br>$2V_x = 90^o - 84^o$ grun.,<br>r < v weak<br>fig. p. 92 | colorless, with higher Fe-content also:<br>X = pale yellowish,<br>Y = very pale brownish,<br>Z = pale brownish to yellow or green,<br>X < Y < Z | only soluble in HF; turns black upon heating to red heat (if Fe-rich); sometimes pleochroic haloes, ◊lam. mostly fibrous | anthophyllite: Z ‖ c; actinolite: ⊖, sm. n; chrysotile: sm. n, sm. Δ | contact-metam. and metam. in medium grade mica schists, marly rocks, basic metam. rocks; with grunerite in iron ores | **178** |
| $2V_x$ 85° tremolite, 80° actinolite, 74° ferroact.<br>r < v weak<br>fig. p. 93 | tremolite: colorless<br>otherwise<br>X = pale yellow-green,<br>Y = light green,<br>Z = light bluish green,<br>X < Y < Z | soluble only in HF; becomes red upon heating to red heat; pleochroic haloes. nephrite: massive actinolite, reticulated structure. smaragdite: contains Cr, Z = green | aegirine: l' = (–); wollastonite: sm. 2V; carpholite: sm. Z∧c; hornblende: green; orthoamphibole: Z ‖ c | low grade metam. with chlorite, epidote, calcite in greenschists and limestones; in hornfelses; with ferroactinolite in skarns | **179** |
| $2V_x$ ~ 85° | X = light yellow-green,<br>Y = light green,<br>Z = light green | soluble only in HF; finely reticulated pseudomorph. after omphacite and diopside | distinguished from tremolite and actinolite only by chem. anal.; Cr-diopside: diff. # ⊕ and sm. 2V | low to medium grade metam. in serpentine-schists, in uralite gabbros; in eclogites | **179a** |
| $2V_x = 60^o - 82^o$<br>r < v | Na-tremolite-richterite<br>X = colorless<br>Y = colorless to green yellowish orange,<br>Z = colorless to bluish green to greenish orange<br>X < Z ≶ Y | easily fusible; turns dark upon heating to red heat | actinolite: gr. 2V; richterite, gedrite: Z ‖ c, smaller Na, Mn-contents | with rhodonite and schefferite in skarns around Mn-Fe deposits, Na-tremolite in metam. dolomit. limestones | **180** |
| $2V_x$ = 60° – 10° glauc. r < v, 0° – 40° crossite r < v, 40° – 70° Mg-rieb. r ≫ v<br>fig. p. 94 | glaucophane:<br>X = colorless to yellowish green,<br>Y = lavender blue,<br>Z = ultramarine; X<Y<Z<br>Mg-riebeckite:<br>X = dark blue; Y = purple,<br>Z = violet-blue; X<Y≈Z<br>see fig. 181 – 2 | easily fusible; slightly soluble in acids, may have pleochroic haloes, often alteration to chlorite, „rhodusite" = asbestiform Mg-riebeckite | tourmaline: absorp. ⊥ l ≫ ‖ l; arfvedsonite: l' = (–); dumortierite: gr. Δ; riebeckite: diff. paragenesis; chloritoid: polysynth. ◊ | with garnet, epidote-minerals, lawsonite in low grade/high pressure metamorphites; in eclogites | **181** |
| $2V_x = 80^o - 90^o$<br>(rarely $2V_z$ ~ 80°),<br>inclined dispersion<br>r > v | riebeckite: X = dark blue, Y = light blue, Z = greenish yellow; X ~ Y > Z<br>osannite: X = dark steel-blue, Y = pale brownish to yellow-green, Z = dark blue-opaque; Z>X>Y | easily fusible; insoluble in acids; alteration to siderite + limonite; crocidolite ≈ asbestiform riebeckite; silicified: tiger-eye | tourmaline: absorp. ⊥ l ≫ ‖ l; dumortierite: gr. Δ; glaucophane: l' = (+); arfvedsonite: usually gr. X∧c; chloritoid: polysynth. ◊ | with aegirine ± quartz magmatic in felsic alkaline rocks; as relict in related gneisses; rarely contact-metamorphic | **182** |
| $2V_x$ = 80° – 15° Mg-arf. v > r; 0° – 50° fluot.; 50° – 100° arf. r > v<br>fig. p. 95 | Mg-arfvedsonite: X = blue-green, Y = light blue-green, Z = light yellow-green; Y<X>Z<br>fluotaramite: X = blue-green, Y = greenish blue, Z = dark green; Y<X~Z.<br>arfvedsonite: X = dark blue-green, Y = greyish blue, Z = greenish blue-grey; Y<X<Z | easily fusible; strong dispersion of the extinction r ≪ v (3° – 20°) | tourmaline: absorp. ⊥ l ≫ ‖ l; riebeckite: sm. X∧c | with aegirine only magmatic in foid-bearing sodic rocks, especially in plutonites and related pegmatites; Mg-arfvedsonite also hydrothermal metasomatic | **183** |

(54, 66)

| No. | Name<br>Symmetry<br>Chem. Composition | Common Crystal Forms, Twinning<br>Crystal Habit<br>Aggregates | Cleavage<br>Translation Gliding<br>Deform. Twinning<br>Hardness, Density | Optic Orientation | Refractive Indices<br>Birefringence |
|---|---|---|---|---|---|
| 183a | Eckermannite mcl<br>$(\underline{Na},Ca)_3(\underline{Mg},Fe^{3+})_4$<br>$(Fe^{3+},Al)$<br>$[(\underline{OH}),F)_2/AlSi_7O_{22}]$ | {110},{010};<br>◇ (100) lam.;<br>columnar | # {110} perf.;<br>H = 5 – 6<br>D = 3.0 – 3.4 | Y ∥ b, O.A.P. (010)<br>XΛc ≈ –25° to (–53°)<br>ZΛc ≈ +65° to (+37°) | $n_x$ = 1.614 – 1.636<br>$n_y$ = 1.625 – 1.644<br>$n_z$ = 1.630 – 1.649<br>⊖Δ = 0.016 – 0.013 |
| 184 | Hastingsite group mcl<br>Magnesiohastingsite, Barkevikite-Hastingsite<br>$NaCa_2(Mg,Fe^{2+})_4$<br>$(\underline{Fe}^{3+},Al)$<br>$[(OH)_2/Al_2Si_6O_{22}]$ | {110},{010},{011},<br>{10$\bar{1}$};<br>β = 105¾°;<br>short to long prisms,<br>acicular ∥ c | # {110} perf. 56°,<br>~(100), (001) partings;<br>H = 5 – 6<br>D = 3.1 – 3.6 | XΛa = 0° to 5°, bi;<br>Y ∥ b, O.A.P. (010)<br>ZΛc = +20° to +15°<br>l' = (±) to (+)<br>fig. p. 89 | Mg-hast. bark. hast.<br>$n_x$ = 1.646–1.680–1.700<br>$n_y$ = 1.658–1.696–1.719<br>$n_z$ = 1.662–1.700–1.722<br>⊖Δ = 0.016–0.020–0.022<br>diagr. p. 96 |
| 184a | Barkevikite mcl<br>$Ca_2(\underline{Na},K)(\underline{Fe}^{2+},Mg,$<br>$Mn,Fe^{3+})_5$<br>$[(\underline{OH}),F)_2/AlSi_7O_{22}]$ | {110},{010};<br>◇ (100) lam.;<br>acicular, columnar | # {110} perf.;<br>H = 5 – 6<br>D = 3.1 – 3.6 | XΛa = 5°, O.A.P. (010)<br>Y ∥ b<br>ZΛc = 15° | n: like hastingsite group<br>⊖Δ: like hastingsite group |
| 184b | Oxyhornblende mcl<br>$NaCa_2Mg_2Fe_3^{3+}$<br>$[O_2Al_2Si_6O_{22}]$ | {110},{010};<br>◇ (100)<br>columnar | # {110} perf.;<br>H = 5 – 6<br>D = 3.2 – 3.6 | Y ∥ b, O.A.P. (010)<br>ZΛc ~ 0° | n = hast. or kaersut. +<br>$n_x$ = +0.010 – 0.020<br>$n_y$ = +0.040 – 0.070<br>$n_z$ = +0.040 – 0.100<br>⊖Δ = +0.030 – 0.080 |
| 185 | Kaersutite mcl<br>$NaCa_2Mg_3Fe^{2+}(Ti,Fe^{3+})$<br>$[(OH)_2/Al_2Si_6O_{22}]$ | {110},{010},{011},<br>{10$\bar{1}$},{100};<br>β = 105¾°;<br>◇ (100) lam.;<br>columnar ∥ c | # {110} perf. 55°,<br>(001) parting;<br>H = 5 – 6<br>D = 3.1 – 3.3 | XΛc = –(90°)–80°<br>–(74°), bi;<br>Y ∥ b, O.A.P. (010)<br>ZΛc = +(0°) to 10° to (19°)<br>l' = (+) fig. p. 89 | Mg MgFe²⁺<br>$n_x$ = 1.667 – 1.690<br>$n_y$ = 1.680 – 1.708<br>$n_z$ = 1.691 – 1.717<br>⊖Δ = 0.024 – 0.027<br>diagr. p. 97 |
| 186 | Pargasite mcl<br>$NaCa_2(\underline{Mg},Fe^{2+})$<br>$(\underline{Al},Fe^{3+})$<br>$[(OH)_2/Al_2Si_6O_{22}]$ | {110},{010},{011},<br>rarely idiomophic;<br>β = 105¾°;<br>granular, rounded to subrounded | # {110} perf. 56°;<br>H = 5 – 6<br>D = 3.1 – 3.2 | XΛc = –62° to 65°<br>Y ∥ b, O.A.P. (010)<br>ZΛc = +30° to 25°, bi;<br>fig. p. 89 | Mg MgFe²⁺<br>$n_x$ = 1.611 – 1.660<br>$n_y$ = 1.615 – 1.673<br>$n_z$ = 1.630 – 1.681<br>⊕Δ = 0.019–⊖0.021<br>diagr. p. 97 |
| 186a | Karinthine mcl<br>~ Pargasite | {110},{010} | # {110} perf.;<br>H = 5 – 6 D = 3.2± | XΛc = –65°<br>Y ∥ b, O.A.P. (010)<br>ZΛc = +25° | n ≈ pargasite<br>⊖Δ ≈ pargasite |
| 187 | Tschermakite group mcl<br>$Ca_2(Mg,Fe^{2+})_3$<br>$(Al,Fe^{3+})_2$<br>$[(OH)_2/Al_2Si_6O_{22}]$ | {110},{010},{011};<br>β = 105¾°;<br>◇ (100) lam.;<br>elongated, acicular ∥ c | # {110} perf. 56°;<br>(100) and (001)partings;<br>H = 5 – 6<br>D = 3.1 – 3.4 | XΛc = –75° to 68°, bi;<br>Y ∥ b, O.A.P. (010)<br>ZΛc = +15° to 22°<br>l' = (+)<br>fig. p. 89 | Mg MgFe²⁺<br>$n_x$ = 1.640 – 1.673<br>$n_y$ = 1.649 – 1.690<br>$n_z$ = 1.658 – 1.696<br>⊖Δ = 0.018 – 0.023<br>diagr. p. 98 |
| 187a | Barroisite mcl<br>~ Tschermakite | {110},{010};<br>◇ (100) lam.;<br>columnar-acicular | # {110} perf.;<br>H = 5 – 6<br>D = 3.1 – 3.4 | ZΛc = 15° – 22°<br>Y ∥ b, O.A.P. (010) | n ≈ tschermakite<br>⊖Δ ≈ tschermakite |
| 188 | Katophorite mcl<br>$Na_2Ca(Mg,Fe^{2+})_4$<br>$(\underline{Fe}^{3+},Al)$<br>$[(OH)_2/AlSi_7O_{22}]$ | {110},{010},{011},<br>{$\bar{1}$01},{$\bar{1}$21};<br>short columns,<br>acicular ∥ c | # {110} perf. 55°;<br>(001) parting;<br>H = 5 – 6<br>D = 3.40 – 3.45 | Mg:<br>Y = c<br>ZΛc = –8° to –16°<br>XΛc = +82° to –74°<br>O.A.P. (010), r<v;<br>MgFe²⁺:<br>Z = c<br>YΛc = –16° to –65°<br>XΛc = +74° to 25°<br>O.A.P.⊥(010), r>v | Mg MgFe²⁺<br>$n_x$ = 1.640 – 1.681<br>$n_y$ = 1.658 – 1.688<br>$n_z$ = 1.660 – 1.690<br>⊖Δ = 0.020 – 0.009 |
| 189 | Aenigmatite trcl<br>(Cossyrite)<br>$Na_4Fe_{10}^{2+}Ti_2$<br>$[O_4/(Si_2O_6)_6]$ | {010},{100},{110},<br>{001},{1$\bar{1}$0};<br>β = 96°;<br>always ◇ lam. (010);<br>columnar ∥ c | # {110} } 66°<br># {1$\bar{1}$0}<br>H = 5½ D = 3.80±<br>fig. p. 99 | O.A.P. ~ (010)<br>Z'Λc in {(100) ~ 4°, (010) ~ 45°}<br>fig. p. 99 | $n_x$ = 1.81<br>$n_y$ ~ 1.82<br>$n_z$ = 1.88<br>⊕Δ = 0.07 |
| 190 | Rhoenite pseudo mcl<br>$Ca_4(\underline{Mg},Fe^{2+})_8Fe_2^{3+}Ti_2$<br>$[O_4(AlSiO_6)_6]$ | {100},{011},{0$\bar{1}$1};<br>β = 109°;<br>◇ (010) lam.;<br>columnar ∥ c,<br>tabular ∥ (1$\bar{1}$0) | # {201} and {$\bar{2}$01} good<br># {110} and {1$\bar{1}$0} dist.;<br>H < 5½ D ~ 3.58 | Z'Λc in (010) ~ 28 – 40°<br>Y'Λa in (010) ~ 9 – 21°<br>O.A.P. ⊥ (010)<br>X ~ b (0° – 6°)<br>fig. p. 99 | $n_x$ = 1.808<br>$n_y$ = 1.824<br>$n_z$ = 1.840<br>⊕Δ = 0.032 |

| Angle and Dispersion of Optic Axes | Color in Thin Section Pleochroism | Chemical Reactions Special Remarks | Distinguishing Features of Similar Minerals | Occurrences | |
|---|---|---|---|---|---|
| $2V_X = 74^o$<br>$r < v$ | X = bluish green,<br>Y = pale bluish green,<br>Z = yellowish green;<br>X>Y>Z | easily fusible | barroisite: sm. extinction, diff. paragenesis | with aegirine and pectolite metasomatic in sodic gneisses | **183a** |
| $2V_X$ = {90°–70° Mg-hast.; 53°–40° bark.; 50°–34° hast.}<br>r ≫ v to r < v<br>fig. p. 96 | Mg-hastingsite:<br>X=light green, Y=green,<br>Z=green;<br>barkevikite:<br>X=light brown-yellow,<br>Y=brown-red, Z=dark brown-green;<br>hastingsite:<br>X=yellow-grey, Y=olive-green,<br>Z=green; X<Y<Z variab. | easily fusible; alteration from barkevikite to aegirine + biotite | augite: diff. # diff. extinction;<br>hornblende: sm. n;<br>actinolite: similar optics, chem. anal. | usually magmatic in sodic rocks; åkerite, foyaite, sodic granite; hastingsite rarely in metam. rocks; in skarns; see Vol. 2 | **184** |
| $2V_X \approx 40^o - 55^o$ | X=light brownish yellow,<br>Y=red-brown,<br>Z=deep brown;<br>X<Y<Z | easily fusible; low $TiO_2$-contents | hastingsite: different absorption colors | mainly in foyaites and related dike rocks, rarely in trachytes, phonolites and tephrites | **184a** |
| $2V_X \approx 60^o$ | X≈Y light brown,<br>Z = dark brown | high $Fe^{3+}$-contents, low OH-contents, chem. anal. necessary | oxyhornblendes have unusually high $\Delta \sim 0.100$ | mostly in latites, dacites, andesites | **184b** |
| $2V_X = 85^o - 70^o$<br>r>v (rarely <v)<br>fig. p. 97 | X=light yellow-brown,<br>Y=reddish brown,<br>Z=dark red-brown to dark olive-brown;<br>X<Y≪Z | easily fusible; zonal coloring; effusive: resorption to augite + magnetite | barkevikite: gr. D and sm. 2V; karinthine: only in eclogites; aenigmatite: $Z' \wedge c \approx 45^o$; rhoenite: Z ~ opaque | exclusively in basic to intermediate magmas of alkaline series: essexite, theralite, monzonite, basanite, monchiquite | **185** |
| Mg MgFe²⁺<br>$2V_X = 124^o - 90^o - 75^o$<br>$r < v$ | usually colorless; at best:<br>Y=pale greenish,<br>Z=pale blue-green;<br>X<Y<Z | not easily fusible with blowpipe; usually rounded crystalloblasts | other colorless hornblendes: ⊖;<br>diopside: # ~ 90° | with chondrodite and wollastonite in high grade metam. gneisses and impure marbles | **186** |
| $2V_X = 85^o - 65^o$ | X=light grey-green,<br>Y=brownish green,<br>Z=grey-green to dark green; X≪Y>Z | | pargasite: gr. 2V | high grade metam. in eclogites, amphibolites and their open fissures | **186a** |
| $2V_X = 90^o - 65^o$<br>r ≷ v<br>fig. p. 98 | X=pale yellow-green to pale yellow-brown<br>Y=yellow-green to olive-green<br>Z=olive-green to brown-green;<br>X<Y≦Z | not decomposed by HCl, may show pleochroic haloes; barroisite: Z=greenish blue in amphibolites; uralite: pseudomorphic after augite, fibrous | actinolite: sm. n, sm. Δ;<br>hastingsite: gr. n;<br>augite: diff. # | with biotite and sphene magmatic in acid to intermediate (rarely in basic) rocks; medium grade metam. in Mg-Fe-rich schists | **187** |
| $2V_X = 90^o - 65^o$ | X=colorless to pale yellow,<br>Y=yellow-green to grey-green,<br>Z=bluish grey | | tschermakite: no bluish grey colors in Z | mainly in prasinitic schists | **187a** |
| $2V_X = (0^o) - 35^o - (65^o)$<br>r ≷ v | Mg MgFe²⁺<br>X=greyish red to yellowish red,<br>Y=dark brown to brownish red,<br>Z=greyish brown to dark red-brown;<br>X<Z<Y X<Y<Z | strong dispersion of 2V and extinction; high Ti-contents (=anophorite) | Ti-hornblende, barkevikite: strongest absorp. color ∥ c | with aegirine and arfvedsonite in syenites, foyaites, and volcanic equivalents, in sodic granites, in lamproites with diopside and phlogopite | **188** |
| $2V_Z = 32^o$<br>$r < v$<br>fig. p. 99 | X=light red-brown,<br>Y=dark chestnut-brown,<br>Z=very dark brown to opaque;<br>X<Y<Z | decomposed by acids; O.A.P. approx. halves the obtuse angle of the cleavages | rhoenite:<br>X=greenish brown;<br>titanhornblendes:<br>gr. Δ and Z near c | magmatic with aegirine and arfvedsonite in sodic syenites, sodic trachytes, phonolites | **189** |
| $2V_{Z,X}$ large, >70°<br>$r < v$<br>fig. p. 99 | X=brown,<br>Y=red-brown,<br>Z=dark red-brown to opaque<br>X≦Y<Z | angle of cleavages approx. halved by O.A.P. abnormal interference colors | ilmenite: no pleochr., no #;<br>aenigmatite: gr. $n_Z$, gr. Δ;<br>titanhornblendes: sm. n | in nepheline basanite, alkali basalt and nephelinites; only in volcanic rocks | **190** |

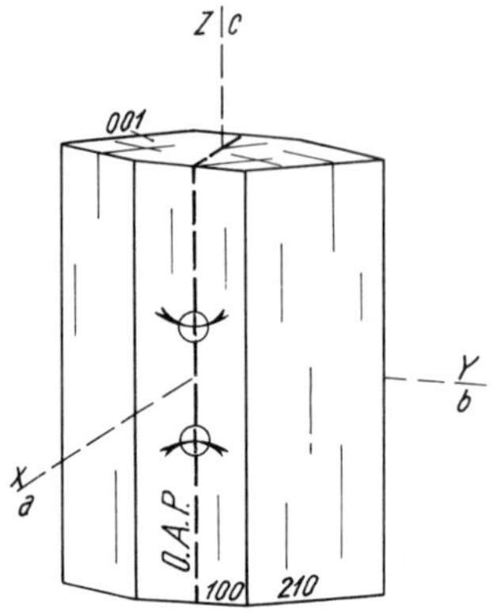


176 to 177-1. Anthoph. and Gedrite
Mg- rich

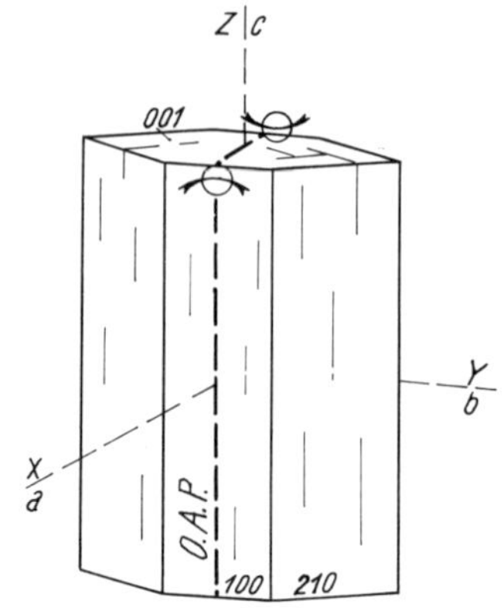


176 to 177-2. Anthoph. and Gedrite
Fe- rich

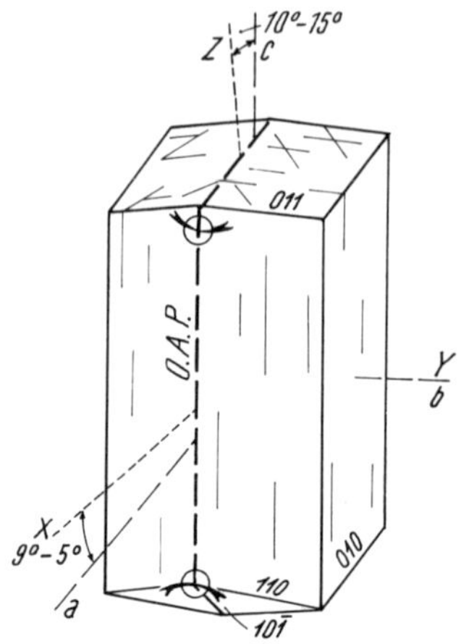


178-1. Grunerite

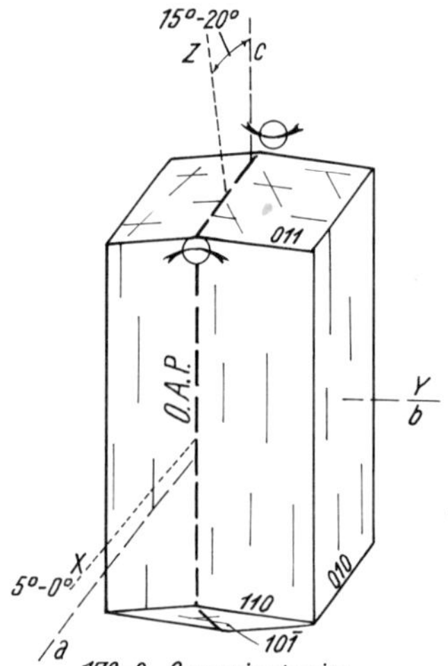


178-2. Cummingtonite

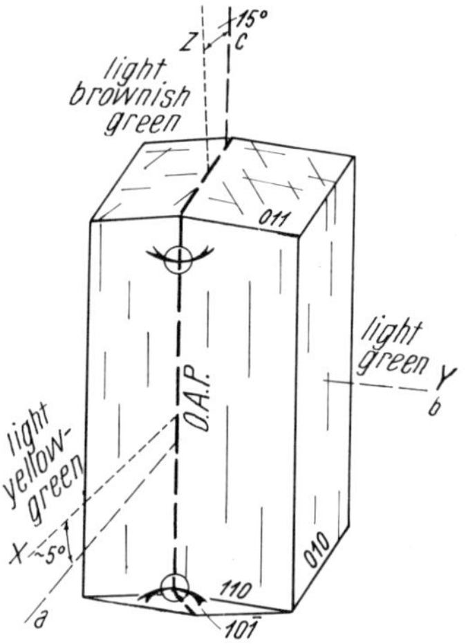


179-1. Actinolite group

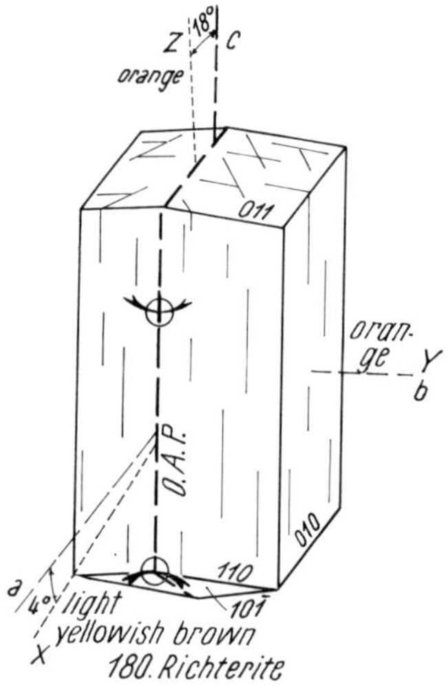


180. Richterite

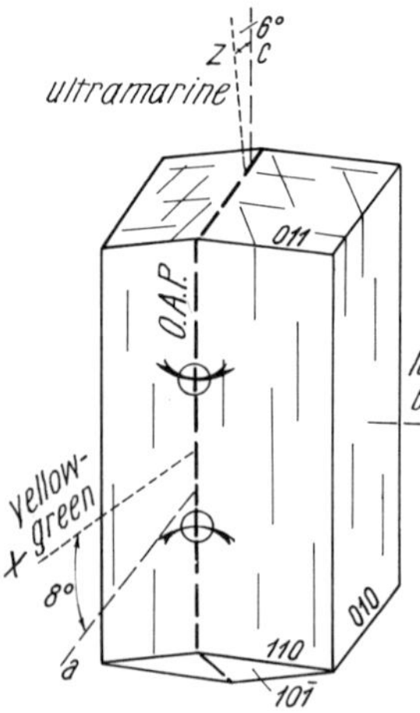


181-1. Glaucophane

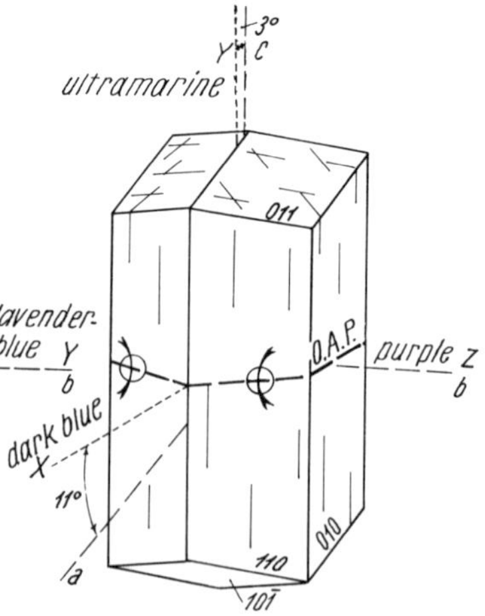


181-2. Crossite

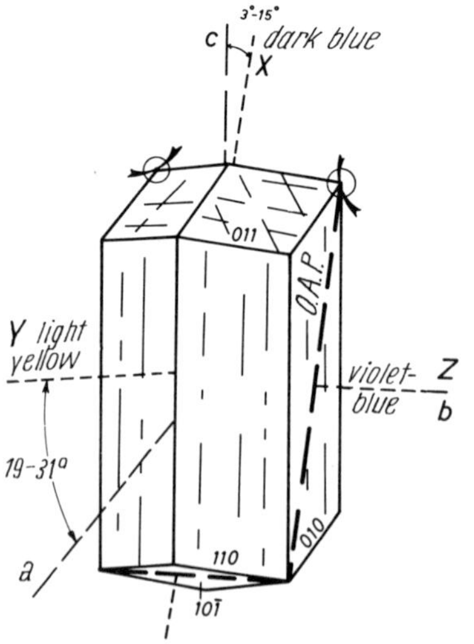


181-3. Magnesioriebeckite

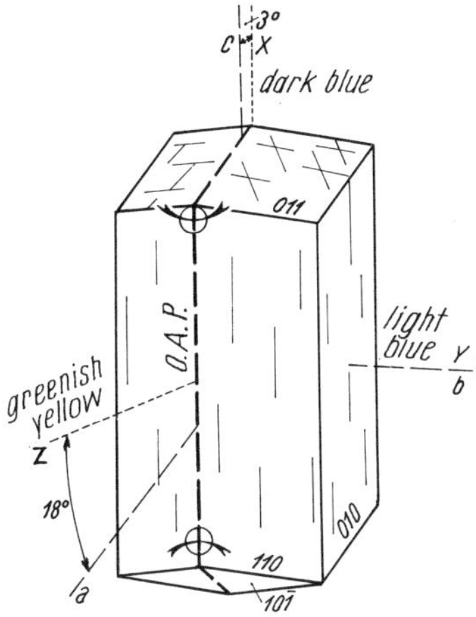


182-1. Riebeckite

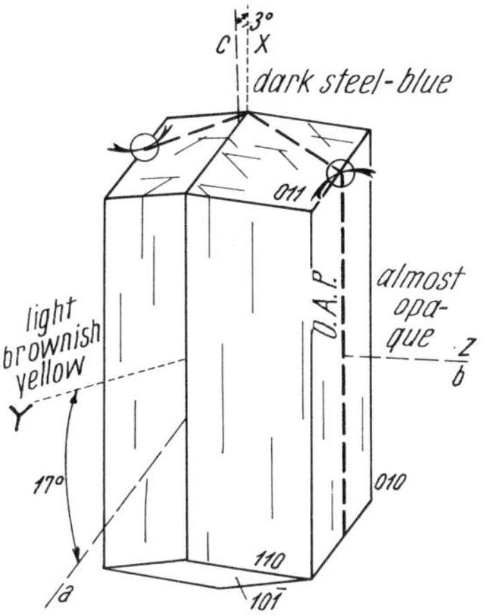


182-2. Osannite

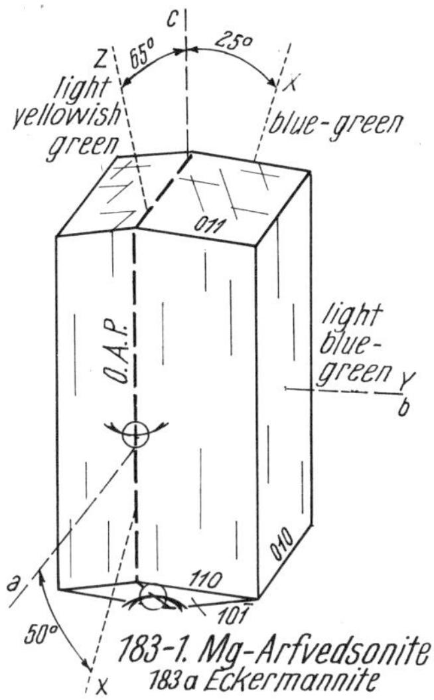


183-1. Mg-Arfvedsonite
183a Eckermannite

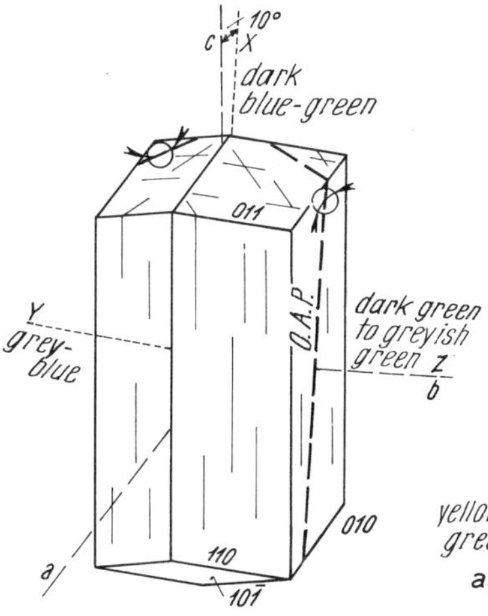


183-2. Fluotaramite-Arfvedsonite

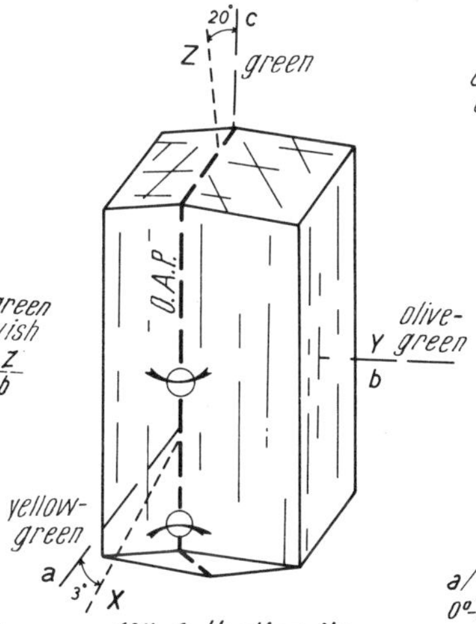


184-1. Hastingsite

184-2. Barkevikite

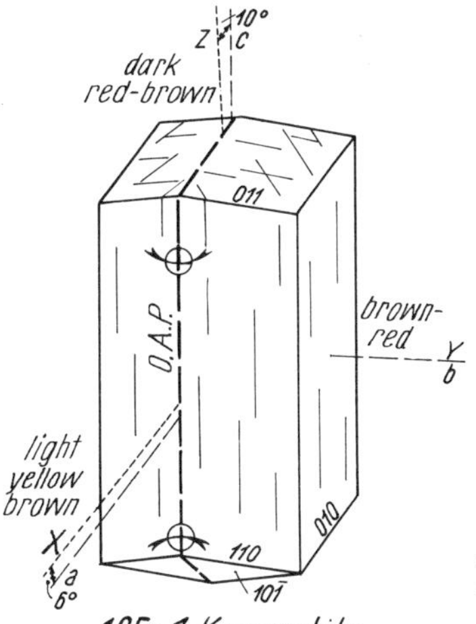


185-1. Kaersutite

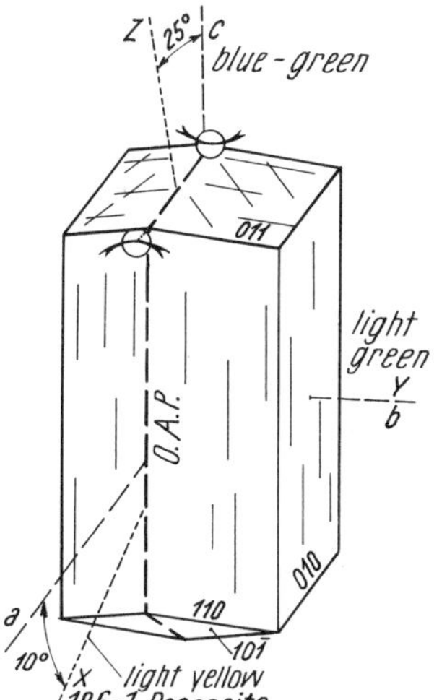


186-1. Pargasite

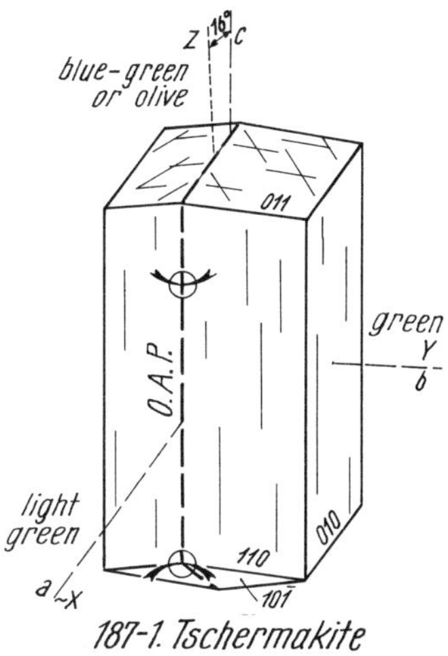


187-1. Tschermakite

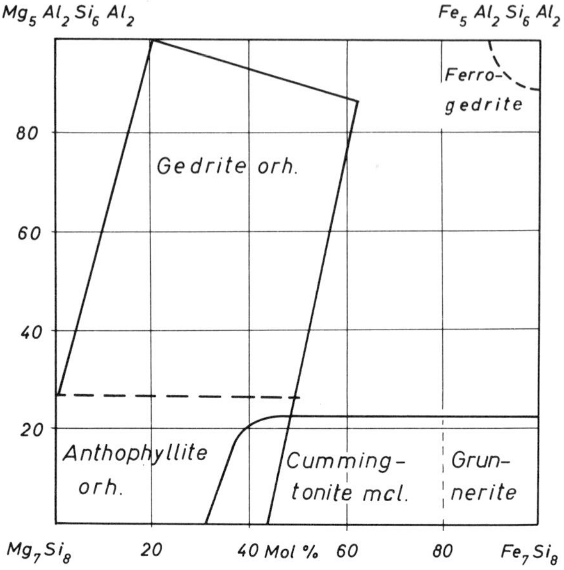


176-178. Chemical variation of the iron magnesium amphiboles

TRÖGER 1961 after SEKI & YAMASAKI, 1957

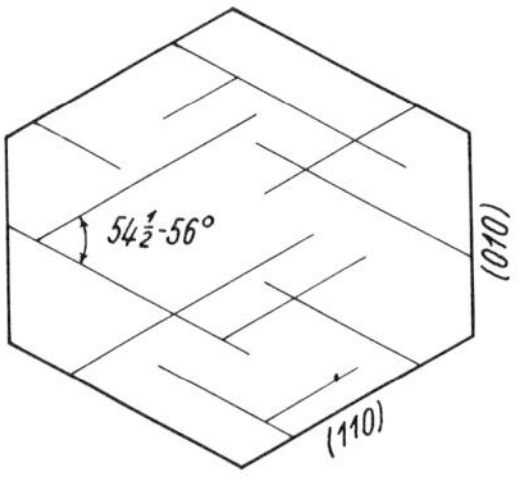


176 to 188-1. Amphiboles cleavage in section ⊥c

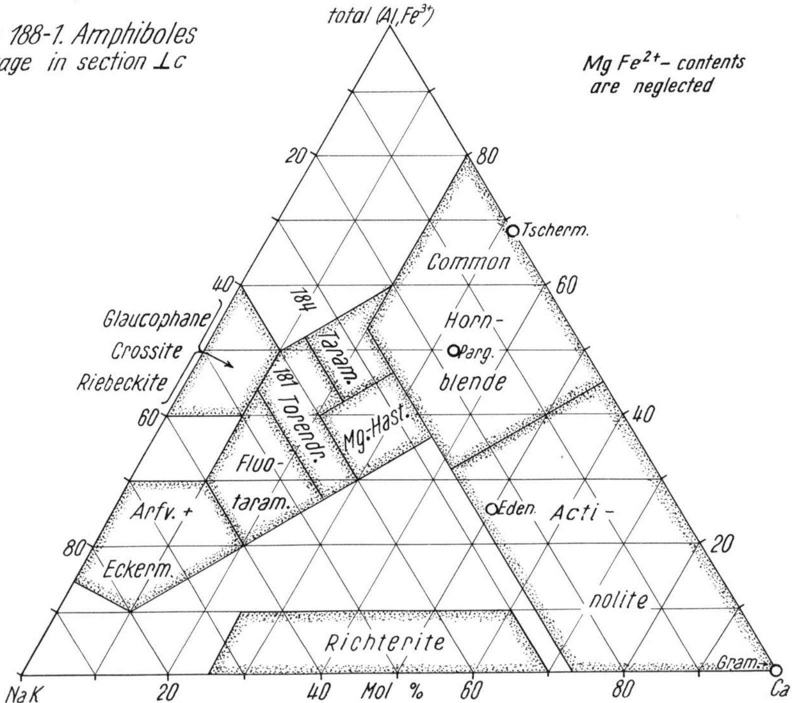


179-187. Chemical variation of the calcic and sodic amphiboles

TRÖGER after SUNDIUS (Sver. geolog. Unders. Arsb. 40₄, 34)

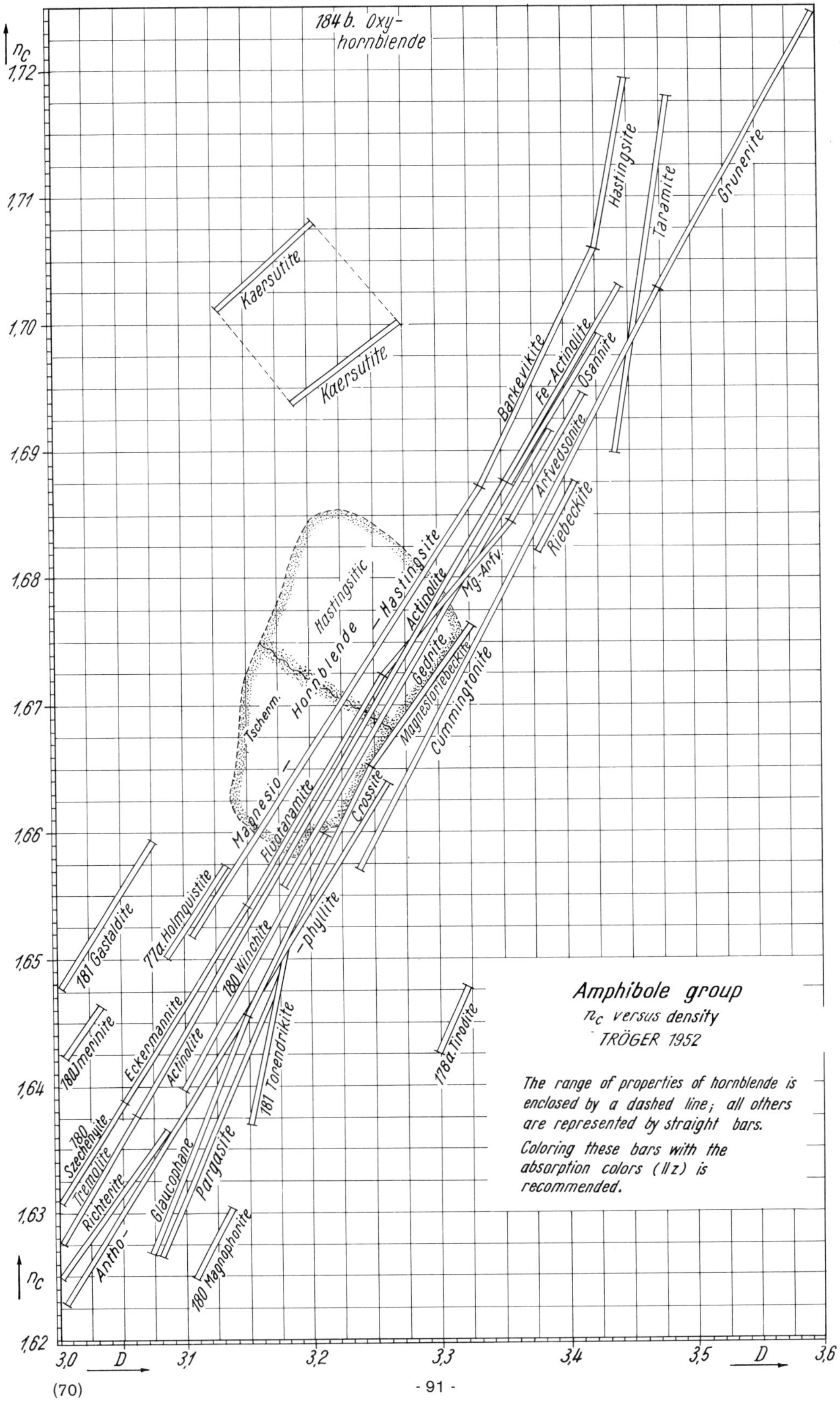

Amphibole group
$n_c$ versus density
TRÖGER 1952
The range of properties of hornblende is enclosed by a dashed line; all others are represented by straight bars.
Coloring these bars with the absorption colors (‖z) is recommended.
184 b. Oxy-hornblende
$n_c$
1,72
1,71
1,70
1,69
1,68
1,67
1,66
1,65
1,64
1,63
1,62
3,0
3,1
3,2
3,3
3,4
3,5
3,6
D
Kaersutite
Kaersutite
Hastingsite
Taramite
Grunerite
Barkevikite
Fe-Actinolite
Osannite
Arfvedsonite
Riebeckite
Mg-Arfv.
Hastingsitic
Hornblende
Hastingsite
Actinolite
Tschenm.
Magnesio-
Gedrite
Magnesioriebeckite
Cummingtonite
Crossite
Fluotaramite
-phyllite
181 Gastaldite
177a. Holmquistite
180 Winchite
180 Imerinite
Eckermannite
Actinolite
181 Torendrikite
178a. Tirodite
180 Szechenyiite
Tremolite
Richterite
Glaucophane
Pargasite
Antho-
180 Magnophorite

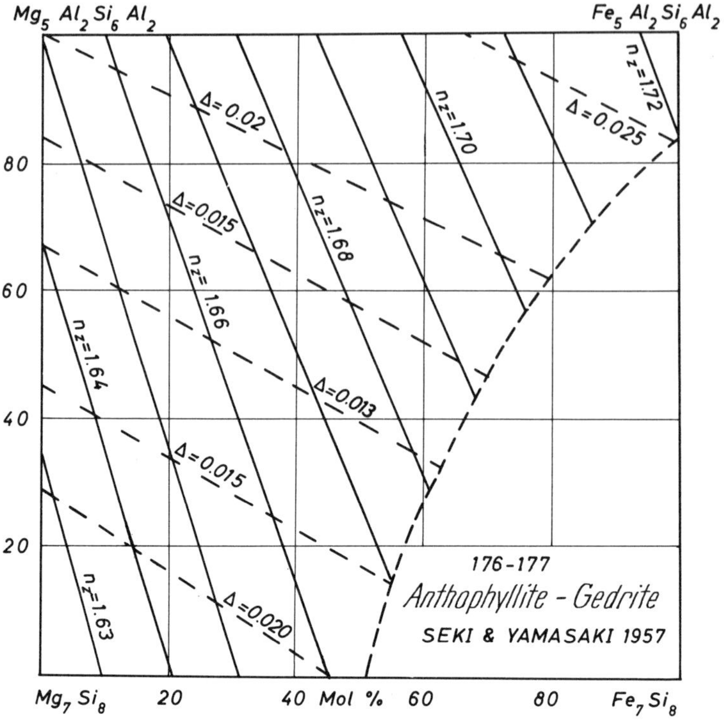


176-177
Anthophyllite - Gedrite
SEKI & YAMASAKI 1957

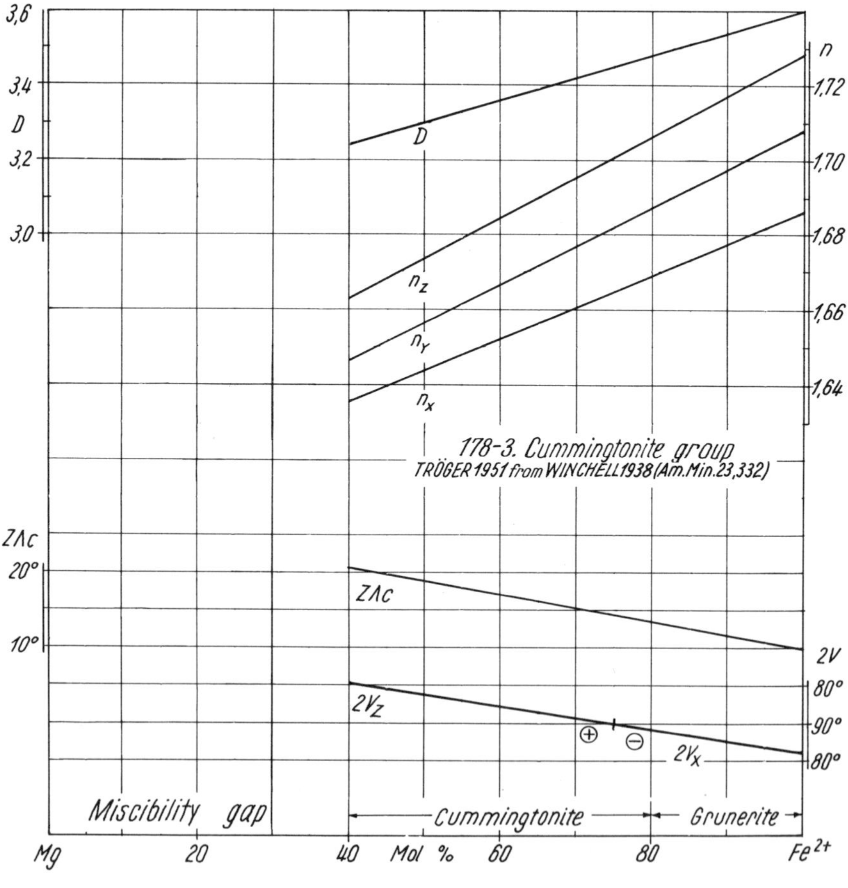


178-3. Cummingtonite group
TRÖGER 1951 from WINCHELL 1938 (Am. Min. 23, 332)

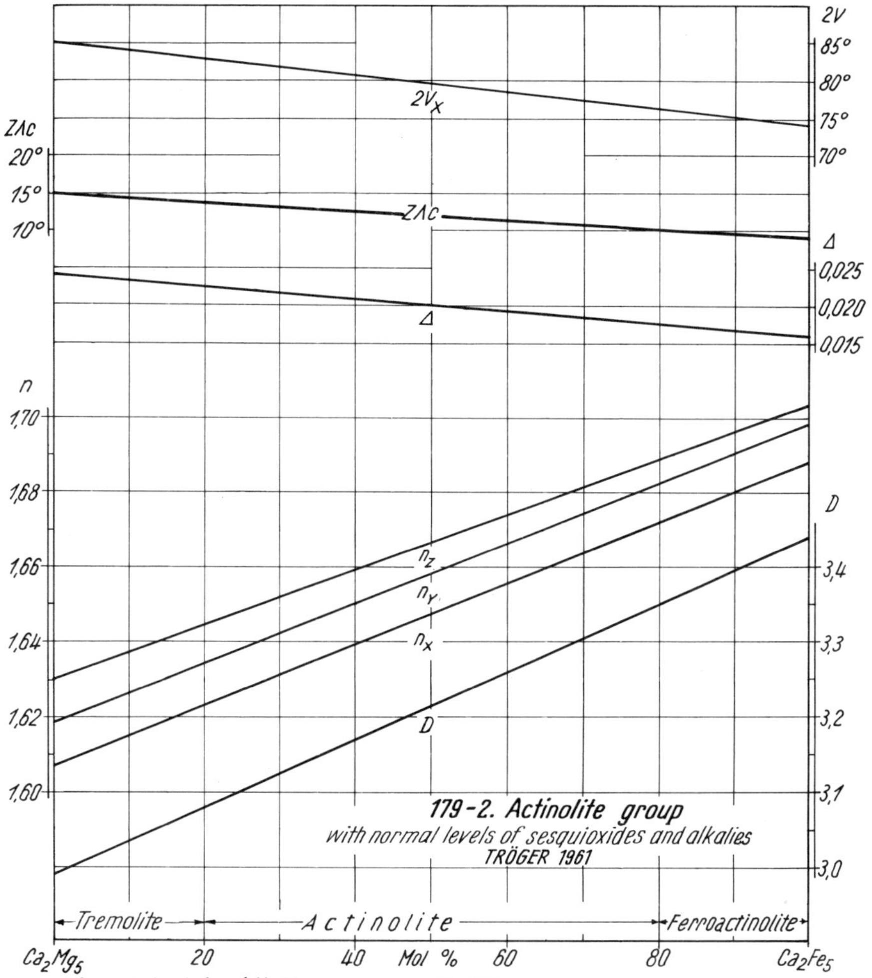


**179-2. Actinolite group**
*with normal levels of sesquioxides and alkalies*
TRÖGER 1961

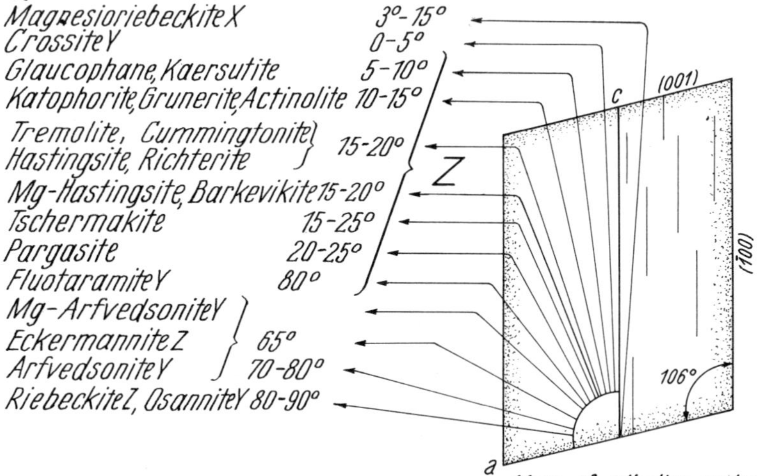


178 to 188-2. Clinoamphiboles

Mean of extinction angles
X, Y or Z∧c
in section ∥ (010)

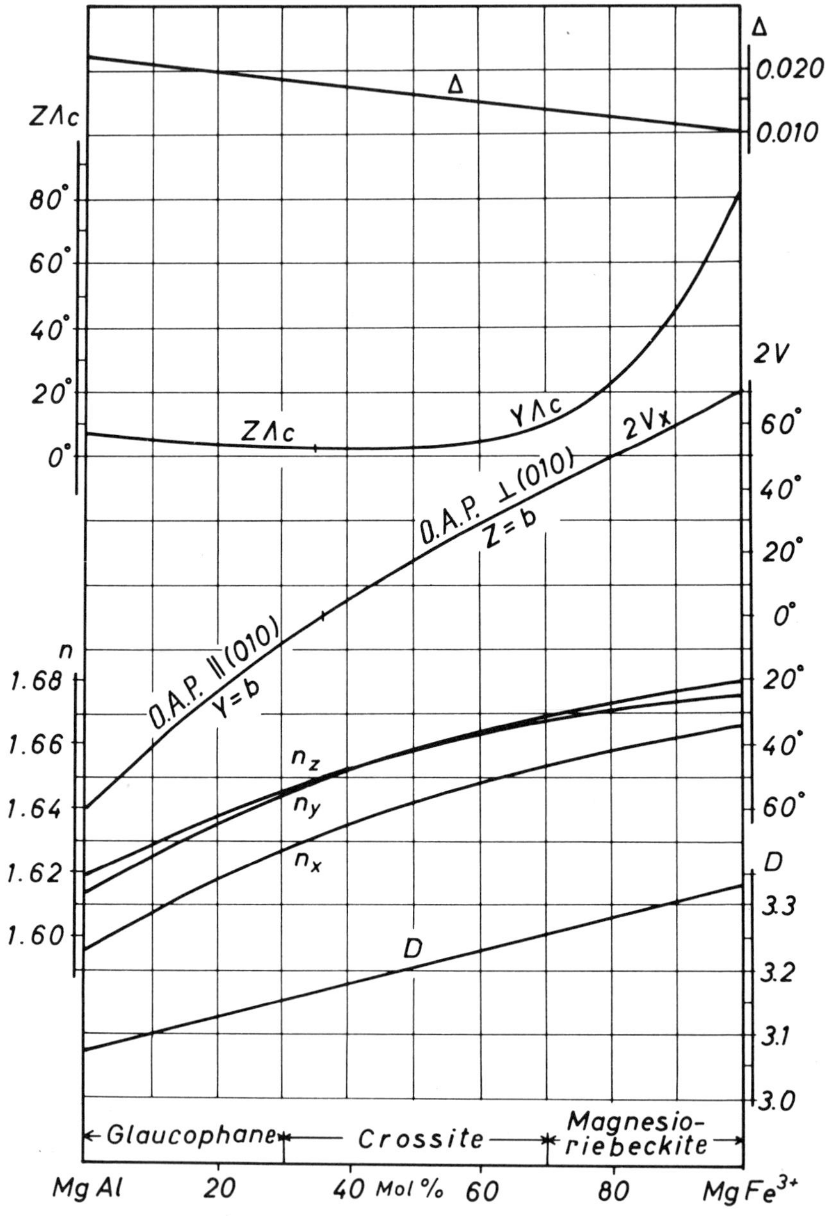


*181–4. Glaucophane group*
TRÖGER 1961

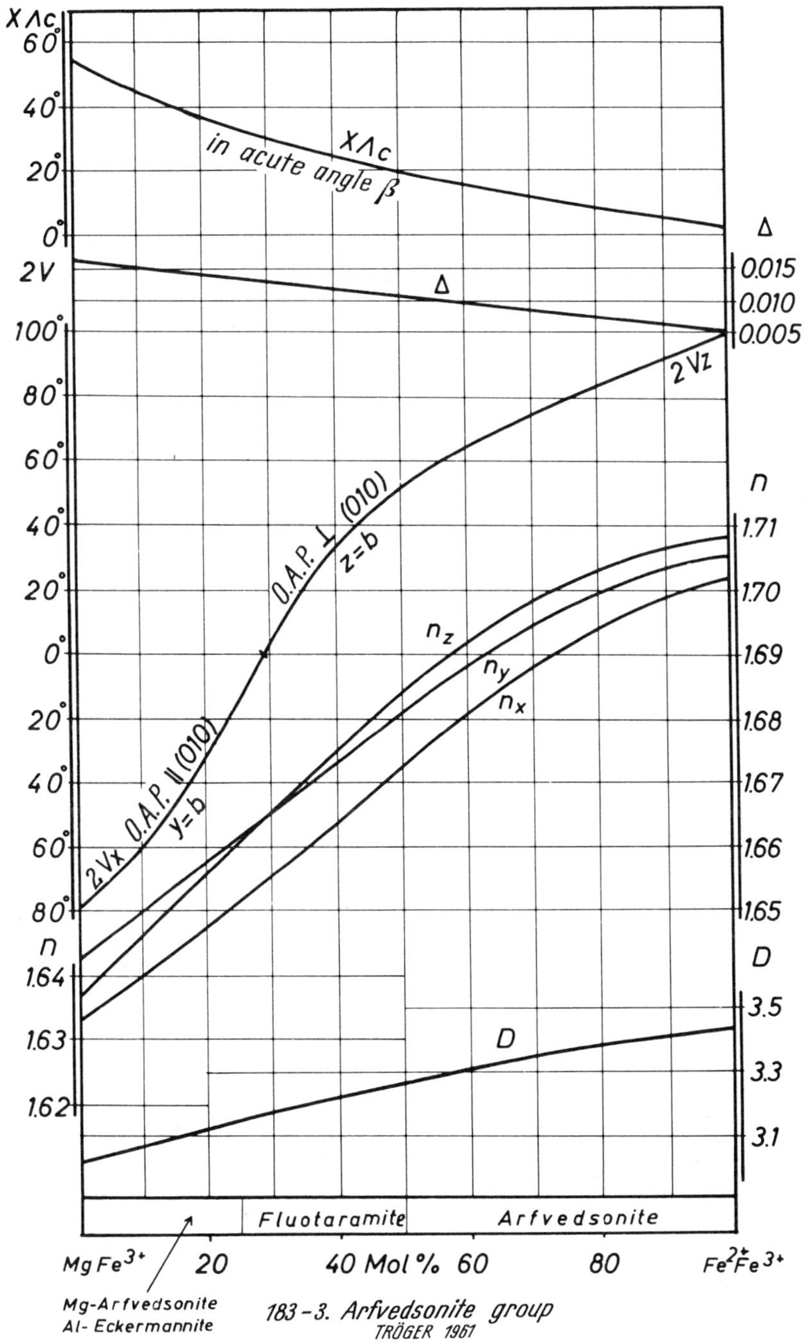


183 – 3. Arfvedsonite group
TRÖGER 1961

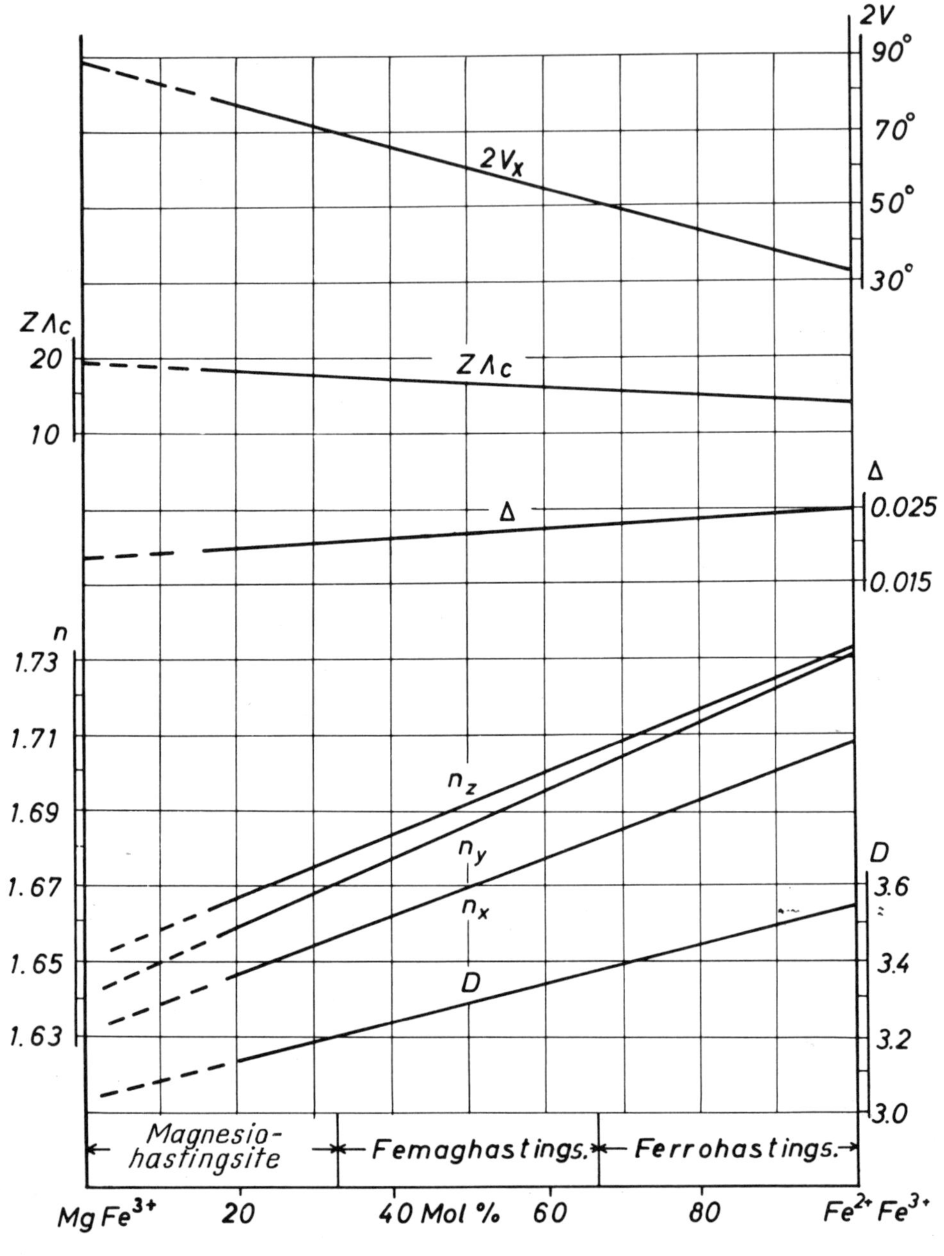


184 – 3. Hastingsite group
(partly covering common hornblende)
TRÖGER 1961

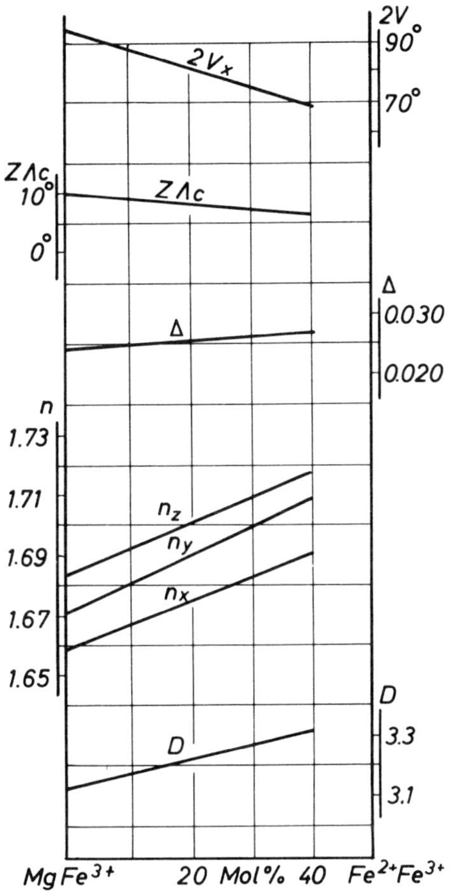


*185 - 2. Kaersutite*
TRÖGER 1961

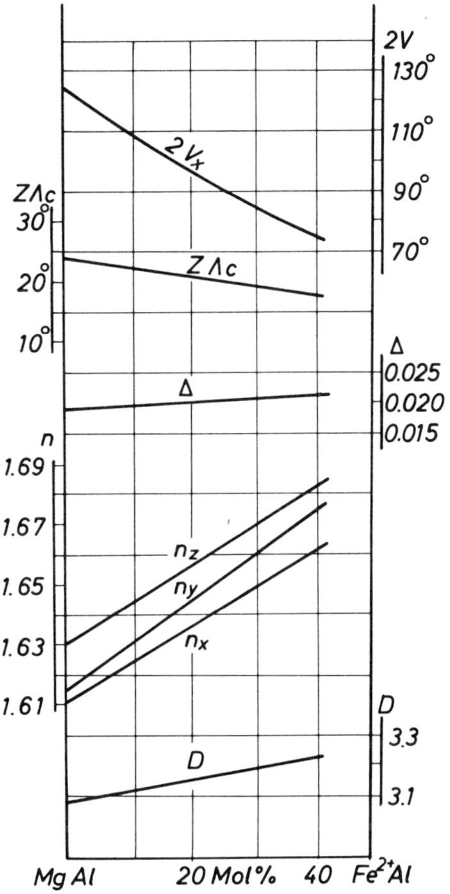


*186 - 2. Pargasite group*
TRÖGER 1961

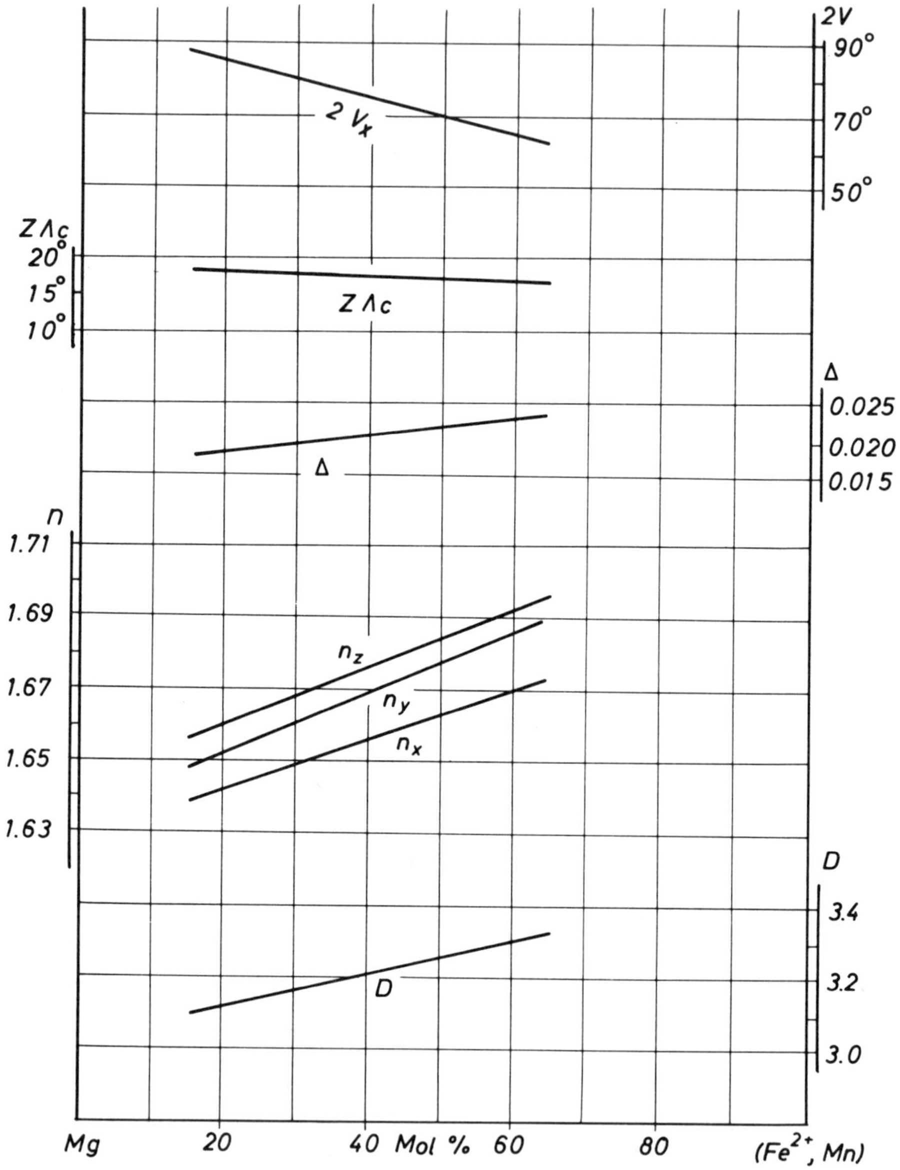


**187–2. Tschermakite group**
(partly covering common hornblende)
TRÖGER 1961

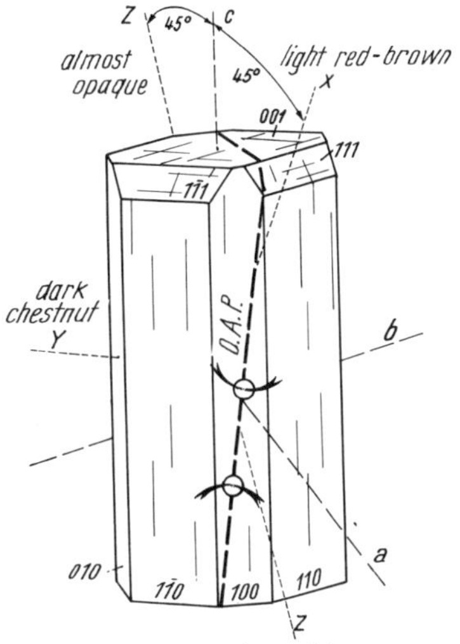


189-1. Aenigmatite

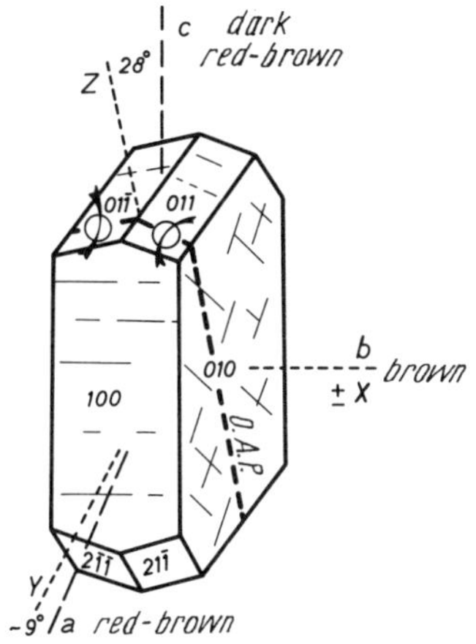


190-1. Rhoenite
WALENTA,1968

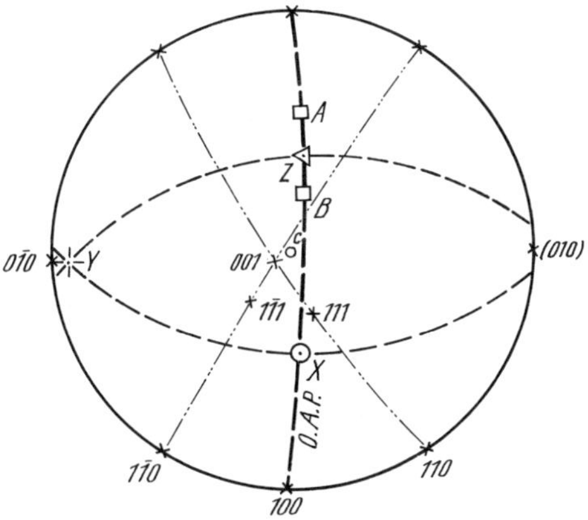


189-2. Aenigmatite

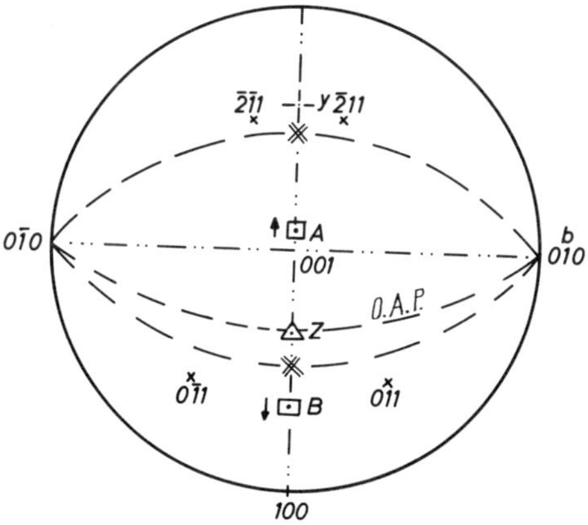


190-2. Rhoenite

| | Name<br>Symmetry<br>Chem. Composition | Common Crystal Forms, Twinning<br>Crystal Habit<br>Aggregates | Cleavage<br>Translation Gliding<br>Deform. Twinning<br>Hardness, Density | Optic<br>Orientation | Refractive Indices<br>Birefringence |
|---|---|---|---|---|---|
| 191 | Sillimanite orh<br>$Al^{[6]}Al^{[4]}[O/SiO_4]$ | {010},{110}, 88°;<br>acicular-fibrous ∥ c,<br>stellate-reticulated | # (010) perf.,<br>T = (010), t = [001];<br>flexible<br>H = 6 – 7 D = 3.24 ± | X ∥ a<br>Y ∥ b, O.A.P. (010)<br>Z ∥ c, bi;<br>l = (+)<br>fig. p. 104 | $n_x$ = 1.657 – 1.660<br>$n_y$ = 1.658 – 1.661<br>$n_z$ = 1.677 – 1.682<br>⊕Δ = 0.020 – 0.022 |
| 192 | Mullite orh<br>$Al^{[6]}Al^{[6]}[O_3(O_{0.5},OH,F)/Si_3AlO_{16}]$ | {110}, 89°;<br>acicular-prismatic ∥ c | # (010) good;<br>H = 6 – 7 D = 3.19 ± | X ∥ a<br>Y ∥ b, O.A.P. (010)<br>Z ∥ c, bi;<br>l = (+)<br>fig. p. 104 | pure Fe-Ti-cont.<br>$n_x$ = 1.639 – 1.661<br>$n_y$ = 1.641 – 1.665<br>$n_z$ = 1.653 – 1.683<br>⊕Δ = 0.014 – 0.021 |
| 193 | Dumortierite orh<br>$Al_7[O_3/BO_3/(SiO_4)_3]$ | {110},{120},{100},<br>{102};<br>◇ {110} lam. 60°;<br>pseudohexagonal<br>acicular, fibrous ∥ c,<br>reticulated | # (100) good.<br># {110},{210} poor,<br>(001) parting;<br>H = 7 D = 3.35 ± | X ∥ c, bi;<br>Y ∥ b, O.A.P. (010)<br>Z ∥ a<br>l = (–)<br>fig. p. 104 | $n_x$ = 1.659 – 1.686<br>$n_y$ = 1.684 – 1.72<br>$n_z$ = 1.686 – 1.723<br>⊖Δ = 0.027 – 0.037 |
| 194 | Carpholite orh<br>$MnAl_2[(OH)_4/Si_2O_6]$ | almost never<br>idiomorphic,<br>acicular-fibrous ∥ c | # (010) perf.,<br># {110} good, 68°;<br>H = 5½ D = 2.9 ± | X ∥ b, bi;<br>Y ∧ a ~ 0°–4°,<br>O.A.P. (100)<br>Z ∧ c = 0° – 4°<br>l' = (+)<br>fig. p. 104 | $n_x$ = 1.611 – 1.617<br>$n_y$ = 1.628 – 1.632<br>$n_z$ = 1.630 – 1.639<br>⊖Δ = 0.019 – 0.022 |
| 195 | Chloritoid trcl, mcl<br>$(\underline{Fe},Mg,Mn)_2Al_4$<br>$[(OH)_4/O_2/(SiO_4)_2]$ | {001},{hhl}, {h0l};<br>β ~ 102°; pseudohex.;<br>◇ [110],[130],[100]<br>(triplets); polysynthetic,<br>all intergrown in (001);<br>tabular (001),<br>lenticular, clusters | # (001) perf.,<br># {hhl} poor ~120° > ~120°;<br>{h0l} parting<br>brittle;<br>H = 6½ – 7 D = 3.51 ± | trcl.:<br>X near a<br>Y ∧ b ~ 20°, O.A.P. ~(010)<br>Z ∧ ⊥(001) = 16° ± 5°, bi;<br>l' = (–) fig. p. 104<br>mcl:<br>X ∥ b, O.A.P. ⊥(010)<br>Y ∧ a = –15° ± 10°<br>Z ∧ ⊥(001) = +15° ± 10°, bi;<br>l' = (–) fig. p. 104 | ⁝ Mg<br>$n_x$ = 1.713–1.728 ⁝ 1.682<br>$n_y$ = 1.719–1.734 ⁝ 1.688<br>$n_z$ = 1.723–1.740 ⁝ 1.694<br>⊕Δ = 0.010–0.012 ⁝ 0.012 |
| 196 | Pyrophyllite mcl<br>$Al_2[(OH)_2/Si_4O_{10}]$ | tabular (001), elongated;<br>β = 100°;<br>flaky, reticulated,<br>divergent, spherulitic,<br>massive | # (001) perf.;<br>no distinct percussion figure; flexible, but not elastic;<br>H = 1½ D = 2.79 ± | X ~ ⊥ (001), bi;<br>Y ~ ∥ a<br>Z ∥ b, O.A.P. ⊥(010)<br>l = (+)<br>fig. p. 104 | $n_x$ = 1.552 – 1.556<br>$n_y$ = 1.588 – 1.589<br>$n_z$ = 1.600 – 1.601<br>⊖Δ = 0.048 – 0.045 |
| 197 | Talc mcl<br>$Mg_3[(OH)_2/Si_4O_{10}]$ | β = 100¼;<br>◇ rare;<br>tabular ∥ (001),<br>pseudohex., elongated,<br>flaky, scaly, massive | # (001) perf.;<br>no distinct percussion figure; flexible but hardly elastic;<br>H = 1 D = 2.78 ± | X ~ ⊥ (001), bi;<br>Y ~ ∥ a<br>Z ∥ b, O.A.P. (010)<br>l = (+)<br>fig. p. 104 | $n_x$ = 1.539 – 1.550<br>$n_y$ = 1.589 – 1.594<br>$n_z$ = 1.589 – 1.596<br>⊖Δ = 0.050 – 0.046 |
| 197a | Minnesotaite mcl<br>$(Fe^{2+}_{2.25}Mg_{0.75})$<br>$[(OH)_2/Si_4O_{10}]$ | β = ?<br>tabular ∥ (001),<br>pseudohexagonal,<br>elongated; acicular,<br>fibrous, divergent,<br>massive | # (001) perf.;<br>H = 2 – 2½<br>D = 2.99 – 3.02 | X ~ ⊥ (001), bi;<br>Y ~ ∥ a<br>Z ∥ b, O.A.P. ⊥(010)<br>l = (+)<br>fig. like talc, Nr. 197 | $n_x$ = 1.580<br>$n_y$ = 1.612<br>$n_z$ = 1.615<br>⊖Δ = 0.035 |
| 198 | Paragonite mcl<br>$NaAl_2$<br>$[(OH)_2/AlSi_3O_{10}]$ | rarely idiomorphic;<br>β = 94½;<br>tabular (001),<br>flaky, scaly | # (001) perf.;<br>percussion figure:<br>(010),{110}; elastic;<br>H = 2½ D = 2.89 ± | X ~ ⊥ (001), bi;<br>Y ~ ∥ a<br>Z ∥ b, O.A.P. ⊥(010)<br>l = (+)<br>fig. p. 104 | $n_x$ = 1.564–1.577–1.580<br>$n_y$ = 1.594–1.599–1.609<br>$n_z$ = 1.600–1.605–1.609<br>⊖Δ = 0.036–0.028–0.029<br>diagr. p. 117 |
| 199 | Muscovite mcl<br>$KAl_2$<br>$[(OH)_2/AlSi_3O_{10}]$ | {001},{110},{010},<br>{$\bar{1}$11};<br>β = 95½°, pseudohex.;<br>◇ [310], comp. plane<br>(001); scaly, flaky in<br>metamorphic rocks,<br>tabular (001) to columnar ∥ c in pegmatites | # (001) perf.;<br>percussion figure:<br>(010), {110};<br>very elastic<br>H = 2 – 2½ D = 2.83 ± | X ∧ a = +90° to +88°, bi;<br>Y ∧ a = +0.5° to +2°<br>Z ∥ b, O.A.P. ⊥(010)<br>l = (+)<br>fig. p. 104 | Fe-free Fe-rich<br>$n_x$ = 1.552 – 1.570<br>$n_y$ = 1.582 – 1.619<br>$n_z$ = 1.588 – 1.624<br>⊖Δ = 0.036 – 0.054<br>diagr. p. 106 |

| Angle and Dispersion of Optic Axes | Color in Thin Section Pleochroism | Chemical Reactions Special Remarks | Distinguishing Features of Similar Minerals | Occurrences | |
|---|---|---|---|---|---|
| $2V_z = 21^o - 30^o$<br>r > v strong | usually colorless, rarely brownish, spotted, then X<Y<Z, weak | not decomposed by HF; not granular, not massive; subparallel intergrowth with quartz | kyanite: gr. 2V, gr. n;<br>andalusite: sm. Δ, sm. n;<br>zoisite: sm. Δ, gr. n;<br>carpholite: diff. opt. orientation | with cordierite, garnet, biotite and kyanite in medium to high grade metam. pelites; rarely in eclogitic and amphibolitic rocks | 191 |
| $2V_z = 45^o - 50^o$<br>r > v distinct | colorless, if Fe,Ti-content then:<br>X = pale pink,<br>Y = colorless,<br>Z = colorless | insoluble in HF | sillimanite: gr. n, gr. Δ | in high grade contact-metam. pelitic and siliceous rocks; synthetic in china, porcelain | 192 |
| $2V_x = 13^o - 63^o$<br>r ≪ v strong<br>rarely r ≫ v (?) | X = cobalt-blue to violet to greenish<br>Y, Z colorless to light yellowish, pink to light blue;<br>X≫Y>Z | not decomposed by HF; partly radioactive, causes pleochroic haloes; alteration to sericite; often abnormal interference colors | tourmaline: absorp. ⊥ l ≫ ∥ l;<br>grandidierite: O.A.P. ⊥ l;<br>riebeckite, glaucophane: sm. Δ, # | high grade metam. in gneisses, kinzigites, anatexites; rarely in aplites; pegmatitic-pneumatol. with B-, Zr-, Ce-, Y-minerals | 193 |
| $2V_x = 50^o - 70^o$<br>r > v | X = greenish yellow to yellow,<br>Y = greenish yellow to yellow,<br>Z = colorless<br>X = Y > Z | slightly decomposed by HCl; turbid upon heating | sillimanite: ⊕, sm. 2V;<br>epidote: O.A.P. ⊥ l;<br>actinolite: gr. Z∧c;<br>tourmaline: uniaxial | with quartz in low grade metam. schists; pneumatolytic with quartz in fissures | 194 |
| $2V_z = (36^o) - 58^o - (70^o)$,<br>rarely $2V_z \sim 120^o$: ⊖!<br>r > v very strong ($\sim 20^o$) | X = grey-green to olive-green,<br>Y = grey-blue,<br>Z = colorless to yellowish green to light blue;<br>X<Y>Z | soluble in $H_2SO_4$, magnetic after heating to red heat; strong dispersion of extinction; zonal structure, hour-glass structure<br><u>ottrelite</u>: Mn-chloritoid | xanthophyllite: ⊖, sm. 2V; stilpnomelane, biotite: gr. Δ; chlorite: sm. n; sapphirine: ⊖, sm. Δ; Na-amphiboles: diff. ◊, diff. # | low grade metam. with almandine, rutile in phyllites, chlorite and mica schists; with margarite in emery rocks; in glaucophane rocks; rarely contact-metam. and hydrotherm.; never together with stilpnomelane | 195 |
| $2V_x = 53^o - 62^o$<br>r > v weak | usually colorless;<br>X<Y~Z | slightly soluble in $H_2SO_4$, more soluble in HF;<br>straight extinction in sections ∥ (010) | talc, paragonite, muscovite: sm. 2V, sm. $n_z$-$n_y$;<br>kaolinite: sm. Δ | in low grade metam. pelitic sediments from kaolinite + quartz; hydrotherm. in veins and lenses of metam. schists; diaphtoritic from kyanite, feldspar and Al-minerals | 196 |
| $2V_x = 0^o - 30^o$<br>r > v | colorless | insoluble in acids; straight extinction in sections ∥ (010); resistant to weathering | pyrophyllite: gr. 2V;<br>brucite: ⊕ uniaxial;<br>muscovite: gr. 2V;<br>gibbsite: ⊕, sm. Δ, oblique extinction in (010) | with chlorite and tremolite e.g. metasomatic and/or low grade metam. in dolomites, talc schists, serpentinites; pseudomorph after olivine, etc. | 197 |
| $2V_x$ very small | X = colorless to pale yellow,<br>Y = pale green,<br>Z = pale green;<br>X<Y~Z | straight extinction in sections ∥ (010); weathers to siderite, limonite and chalcedony | greenalite: X–ray | low grade metam. in sedim. siliceous iron ores with quartz, siderite, stilpnomelane, magnetite, greenalite | 197a |
| $2V_x = 40^o - 50^o$<br>r > v<br>(3T: $2V = 0^o - < 5^o$!)<br>diagr. p. 106 | usually colorless;<br>X<Y~Z | insoluble in HCl and $H_2SO_4$;<br>approx. straight extinction in all sections ⊥ # (001) | muscovite, pyrophyllite, talc: X-ray;<br>margarite: gr. n | metam. with kyanite, staurolite, muscovite in mica and paragonite schists; in glaucophane rocks; diaphtoritic from plagioclase, andalusite, chloritoid, corundum | 198 |
| $2V_x = 35^o - 50^o$<br>r > v<br>diagr. p. 106 | usually colorless; rarely pale yellow, pale green, pale red-brown;<br>X<Y~Z | insoluble in HCl, $H_2SO_4$; epitactic relations to other minerals in (001); rarely pleochroic haloes; <u>sericite:</u> fine scaly filling of feldspar, foids, andalusite and others | phengite: sm. 2V;<br>lepidolite: Li-color in flame; paragonite: X-ray;<br>talc: sm. 2V; <u>fuchsite:</u> Y = yellowish green, Z = bluish green | magmatic only in acid plutonites; in pegmatites; detritic in sands; metam. in phyllites, mica schists, gneisses; contact-metam. in hornfelses; autohydroth. to diaphtoritic in feldspar, etc. as sericite | 199 |

| No. | Name<br>Symmetry<br>Chem. Composition | Common Crystal Forms, Twinning<br>Crystal Habit<br>Aggregates | Cleavage<br>Translation Gliding<br>Deform. Twinning<br>Hardness, Density | Optic Orientation | Refractive Indices<br>Birefringence |
|---|---|---|---|---|---|
| **199a** | Phengite mcl<br>$KAl(\underline{Al},\underline{Mg},Fe)$<br>$[(OH)_2/(\underline{Al},Si)Si_3O_{10}]$ | rarely idiomorphic;<br>tabular ‖ (001)<br>flaky, scaly | # (001) perf.;<br>percussion figure:<br>(010), {110};<br>very elastic;<br>H = 2 – 2½ D = 2.81± | X ~ ⊥ (001), bi;<br>Y ~ ‖ a<br>Z ‖ b, O.A.P. ⊥ (010)<br>l = (+)<br>fig. p. 104 | Fe-free Fe-rich<br>$n_x$ = 1.547 – 1.571<br>$n_y$ = 1.584 – 1.610<br>$n_z$ = 1.587 – 1.612<br>⊖Δ = 0.040 – 0.041<br>diagr. p. 106 |
| **199b** | Hydromuscovite mcl<br>$(K,H_3O)Al_2$<br>$[(OH)_2/AlSi_3O_{10}]$ | rarely idiomorphic;<br>β = 94¾°;<br>tabular ‖ (001),<br>scaly, reticulated,<br>massive | # (001) perf.;<br>elastic;<br>H = 2 D = 2.75± | X ~ ⊥ (001), bi;<br>Y ~ ‖ a<br>Z ‖ b, O.A.P. ⊥ (010)<br>l = (+)<br>fig. like muscovite<br>p. 104 | $n_x$ = 1.555 – 1.575<br>$n_y$ ~ 1.577 – 1.606<br>$n_z$ = 1.580 – 1.610<br>⊖Δ = 0.025 – 0.035 |
| **200** | Margarite mcl<br>$(Ca,Na)Al_2$<br>$[(OH)_2/Al_2Si_2O_{10}]$ | {001},{110},{010},<br>β = 95½°, pseudohex.;<br>◊ [310], comp. plane<br>(001);<br>thin flakes ‖ (001),<br>thin tabular | # (001) perf.;<br>percussion figure:<br>(010), {110};<br>rather brittle;<br>H = 3½ – 4½<br>D = 3.02± | X∧a = +84° to + 82°, bi;<br>Y∧a = –6° to –8°<br>Z‖b, O.A.P. ⊥ (010)<br>l' = (+)<br>fig. p. 105 | $n_x$ = 1.630 – 1.638<br>$n_y$ = 1.643 – 1.648<br>$n_z$ = 1.644 – 1.650<br>⊖Δ = 0.014 – 0.012<br>diagr. p. 117 |
| **201** | Phlogopite mcl<br>$KMg_3$<br>$[(OH),F)_2/AlSi_3O_{10}]$ | {001},{110},{010},<br>{hhl};<br>β = 100¼°, pseudohex.;<br>◊ [310], composition<br>plane (001);<br>tabular ‖ (001) | # (001) perf.;<br>percussion figure:<br>(010), {110};<br>very elastic;<br>H = 2 – 2½ D = 2.83± | X∧ā = +88° to +86°, bi;<br>Y ‖ b, O.A.P. (010)<br>Z∧a = +2° to +4°<br>l' = (+)<br>fig. p. 105 | Fe-free Fe-rich<br>$n_x$ = 1.522 – 1.568<br>$n_y$ = 1.548 – 1.609<br>$n_z$ = 1.549 – 1.613<br>⊖Δ = 0.027 – 0.045<br>diagr. p. 107 |
| **202** | Biotite group mcl<br>Meroxene-Lepido-<br>melane-Siderophyllite<br>$K(Mg,Fe^{2+})_{3-1.5}$<br>$(Al,Fe^{3+})_{0-1}$<br>$[(OH,F)_2/Al_{1-1.5}$<br>$Si_{3-2.5}O_{10}]$ | {001},{110},{010},<br>{$\bar{1}$11},{hhl};<br>β = 99¼°, pseudohex.;<br>◊ [310] comp. pl. (001);<br>tabular ‖ (001),<br>flaky, scaly,<br>elongated ‖ [100] | # (001) perf.;<br>percussion figure:<br>(010), {110};<br>elastic;<br>H = 2½ – 3<br>D = {2.90 mer. / 3.30 sid.} | X∧ā = +90° to (+81°), bi;<br>Y ‖ b, O.A.P. (010)<br>Z∧a = 0° (to +9°)<br>l = (+)<br>fig. p. 105 | mer. lep. sid.<br>$n_x$ = 1.571 – 1.598 – 1.616<br>$n_y$ = 1.609 – 1.651 – 1.696<br>$n_z$ = 1.610 – 1.652 – 1.697<br>⊖Δ = 0.039 – 0.054 – 0.081<br>diagr. p. 107 |
| **202a** | Hydrobiotite mcl<br>$K_{<1}(\underline{Mg},\underline{Fe}^{2+},Fe^{3+},$<br>$Al)_{<3}[(OH)_2/$<br>$AlSi_3O_{10}]\cdot(H_2O)_{\sim2}$ | {001},{110},{010},<br>{hhl};<br>β = 100°, pseudohex.;<br>tabular ‖ (001),<br>flaky, scaly | # (001) perf.;<br>H = 2 – 2½<br>D ~ 2.80 – 3.20 | X∧a ~ +90°, bi;<br>Y ‖ b, O.A.P. (010)<br>Z∧a ~ 0°<br>l = (+)<br>fig. like biotite p. 105 | $n_x$ =, $n_y$ =, $n_z$ =, ⊖Δ = } somewhat smaller than in biotite |
| **203** | Titanbiotite mcl<br>$K(\underline{Mg},Fe^{2+},Ti^{4+})_3$<br>$[(O,OH,F)_2/$<br>$AlSi_3O_{10}]$ | {001},{110},{010};<br>β = 99¼°, pseudohex.;<br>◊ [310], comp. pl. (001);<br>tabular ‖ (001),<br>flaky | # (001) perf.;<br>percussion figure:<br>(010), {110};<br>elastic;<br>H = 2½ – 3 D = 3.10± | X∧a = +87° to + 81°, bi;<br>Y ‖ b, O.A.P. (010)<br>Z∧a = +3° to +9°<br>l' = (+)<br>fig. p. 105 | $n_x$ = 1.599 – 1.655<br>$n_y$ = 1.642 – 1.692<br>$n_z$ = 1.643 – 1.695<br>⊖Δ = 0.044 – 0.040 |
| **203a** | Oxybiotite mcl<br>$KMgFe_2^{3+}$<br>$[O_2/AlSi_3O_{10}]$ | {001},{110},{010},<br>{$\bar{1}$11},{hhl};<br>β = 99¼°, pseudohex.;<br>◊ [310], comp. pl. (001);<br>tabular (001), flaky | # (001) perf.;<br>percussion figure:<br>(010), {110};<br>elastic;<br>H = 2½ – 3 D = 3.12± | X∧ā = +90° (to +83°), bi;<br>Y ‖ b, O.A.P. (010)<br>Z∧a = 0° (to +7°)<br>l = (+) | $n_x$ = 1.600 – 1.610<br>$n_y$ = 1.677 – 1.722<br>$n_z$ = 1.680 – 1.730<br>⊖Δ = 0.080 – 0.120 |
| **204** | Protolithionite group mcl<br>Protolithionite-<br>Li-Siderophyllite<br>$KLi_{0.6}Fe_{0.6}^{2+}Al_{1.4}$<br>$[(OH)_{1.3}F_{0.7}/$<br>$AlSi_3O_{10}]$ | {001},{110},{010};<br>β = 99¾°, pseudohex.;<br>◊ [310], rare;<br>tabular ‖ (001),<br>flaky | # (001) perf.;<br>percussion figure:<br>(010), {110};<br>elastic;<br>H = 2¼ – 4 D = 2.98± | X ~ ⊥ a, bi;<br>Y∧a ~ 0°<br>Z ‖ b, O.A.P. (010)<br>l = (+)<br>fig. p. 105 | protol. Li-sid.<br>$n_x$ = 1.541 – 1.605 – 1.619<br>$n_y$ = 1.570 – 1.645 – 1.678<br>$n_z$ = 1.570 – 1.645 – 1.678<br>⊖Δ = 0.029 – 0.040 – 0.059<br>diagr. p. 107 |
| **205** | Lepidolite group mcl<br>Polylithionite-<br>Lepidolite<br>$KLi_{1.4}Al_{1.4}[(OH)_{0.6}$<br>$F_{1.4}/Al_{0.6}Si_{3.4}O_{10}]$ | {001},{110},{010};<br>β = 100¾°, pseudohex.;<br>◊ [310], rare;<br>tabular ‖ (001),<br>scaly-massive,<br>divergent, spherulitic | # (001) perf.;<br>percussion figure:<br>(010), {110};<br>very elastic;<br>H = 2½ – 4 D = 2.84± | X∧ā = +90° to +87°, bi;<br>Y ‖ b, O.A.P. (010)<br>Z∧a = 0° to +3°<br>l' = (+)<br>fig. p. 105 | $n_x$ = 1.524 – 1.537<br>$n_y$ = 1.543 – 1.563<br>$n_z$ = 1.545 – 1.566<br>⊖Δ = 0.021 – 0.029<br>diagr. p. 107 |

| Angle and Dispersion of Optic Axes | Color in Thin Section Pleochroism | Chemical Reactions Special Remarks | Distinguishing Features of Similar Minerals | Occurrences | |
|---|---|---|---|---|---|
| $2V_x = 24^o - 36^o$ (rarely $\sim 0^o$) $r > v$ diagr. p. 106 | usually colorless; $X < Y \sim Z$ | slightly soluble in acids; pinite: pseudomorph. after cordierite; sericite: fine scales in metam. schists and as filling of feldspar etc. | kaolinite: adsorbs aniline colors; muscovite: gr. 2V; paragonite: gr. 2V; talc: usually sm. 2V; zinnwaldite: Li-color in flame | almost only as sericite; low grade metamorphic in pelitic schists and phyllites; high pressure metam. with glaucophane, diaphtoritic in schists, as fillings in feldspar or with chlorite from muscovite and biotite | 199a |
| $2V_x = 5^o - 10^o - (40^o)$ $r > v$ | usually colorless; guembelite: pale green, pale brown; $X \sim Y < Z$ | slightly soluble in acids; often pseudomorph after muscovite; illite: hydromusc. of clay particle size ($< 2\,\mu$); guembelite: hydrophengite | sericite: gr. Δ; colorless chlorite: sm. Δ | hydrothermal from muscovite in granites, pegmatites; alteration product of muscovite; detritic and authigenic in pelitic sediments | 199b |
| $2V_x = (40^o) - 45^o - (67^o)$ $r < v$ distinct diagr. p. 106 | colorless | not completely soluble in $H_2SO_4$; distinct oblique extinction in sections ‖ (010) | muscovite, paragonite, talc: sm. n, gr. Δ; chloritoid: gr. n, l' = (−); diaspore: gr. n, O.A.P. ‖ # | metam. with chloritoid, diaspore, etc. in emery rocks, in phyllites, in chlorite and mica schists; hydroth. with carbonates in fissures; diagenetic in bauxites | 200 |
| $2V_x = 0^o - 20^o$ $r < v$ diagr. p. 106 | colorless; with Fe colored: X = colorless, Y = Z = yellowish or brownish; $X \ll Y < Z$ | bleached by conc. $H_2SO_4$; Fluor-reaction; weathers to hydrophlogopite, then to vermiculite | muscovite: gr. 2V, O.A.P. ⊥ (010); biotite: gr. n, deeper colors; protolithionite: O.A.P. ⊥ (010), Li-color in flame; vermiculite: sm. Δ | magmatic in basic to ultrabas. plutonites and dike rocks: kimberlite, peridotite, lamproite, serpentinite; pneum. to hydroth. in anatectic rocks with diopside, calcite; metam. in marbles and cordierite gneisses; in sanidinites | 201 |
| $2V_x = 0^o - 33^o$ $r \gtrsim v$ (Mg-rich) to $r < v$ (Fe-rich) diagr. p. 106 | X = grey-yellow to brown-green to orange-brown, Y = dark brown to dark green to dark red-brown, Z = dark brown to dark green to dark red-brown; $X \ll Y \sim Z$ diagr. p. 105 | bleached by $H_2SO_4$; O.A.P. ⊥ (010): anomite: (fig. p. 105); ⊥ (001) most intensive absorption ‖ l; pleochroic haloes frequent; in effusive rocks often with resorption-rim; near extinct.-position: "birdseye maple" mottling | xanthophyllite: sm. Δ; stilpnomelane: absorption color = golden yellow, parting (010); hydrobiotite: bleached; vermiculite: sm. n, Δ; tourmaline: most intensive absorption ⊥ l; | magmatic especially in acid to intermediate plutonites; phenocryst in volcanic rocks; pegmatitic; metamorphic in medium and high grade schists, gneisses etc. | 202 |
| $2V_x$ = always small, usually $\sim 0^o$ $r \gtrsim v$ | due to bleaching usually: X = colorless, Y = yellowish, Z = yellowish; $X \lessdot Y \sim Z$ | strong acids leach out metal-cations: = bauerite; hard to distinguish from biotites | biotite: X-ray, chem. anal. (alkalies, $H_2O$); vermiculite: sm. n, sm. Δ | weathering of biotite, in all appropriate rocks; hydroth. in fissures; detrital in soils | 202a |
| $2V_x = 5^o - 38^o$ $r \ll v$ diagr. p. 106 | X = light orange to yellowish, Y = orange to orange-brown, Z = brown-red to dark brown-red; $X \ll Y < Z$ | bleached by $H_2SO_4$; weathers to hydrobiotite, then to vermiculite | biotite: no reddish absorption colors; oxybiotite: gr. n, gr. Δ; phlogopite: sm, n; manganophyllite: different paragen. | magmatic in alkaline rocks: shonkinites, K-lamproites and other such volcanics; contact-metasom. with pleonaste in biotite olivine rocks | 203 |
| $2V_x = 20^o - 30^o$ $r < v$ distinct diagr. p. 106 | X = yellow-brown to reddish brown, Y = brown-red to dark brown-red, Z = brown-red to dark brown-red; $X \lessdot Y \sim Z$ | bleached by $H_2SO_4$; often rimmed by opacite (augite + rhoenite + sanidine and others) | biotite: sm. n, Δ; titanbiotite: sm. Δ (see Nr. 202) | magmatic as phenocryst, often with oxyhornblende, in intermediate volcanics: andesite, latite, quartz latite and others | 203a |
| $2V_x = (5^o) - 18^o - (28^o)$ also $2V_x \sim 0^o$ (3T – Li-sid.) | protol. Li-sid. X = pale brown to yellowish brown, Y = brownish to dark brown, Z = brownish to dark brown; $X \ll Y \sim Z$ | bleached by acids; pleochroic haloes; intergrown with zinnwaldite | biotite, oxybiotite: O.A.P. (010), no Li-absorp. spectr.; zinnwaldite: sm. n, paler absorption | with spodumene or topaz in granite-pegmatites; pneumatolytic or metasom. in wolframite quartz veins, in greisen and tourmaline granites | 204 |
| $2V_x = (23^o) - 41^o - (63^o)$ rarely $2V_x \sim 0^o$ $r > v$ diagr. p. 106 | colorless; lepidolite also: X = alm. colorless, Y = Z = pink, pale violet; $X < Y \sim Z$ | almost insoluble in acids; pleochroic haloes | muscovite: gr. n, gr. Δ; zinnwaldite: gr. n, gr. Δ, sm. 2V | with amblygonite, spodumene or topaz, tourmaline, beryl e.g. in granite-pegmatites, in miarolitic cavities esp. in granite, in greisen | 205 |

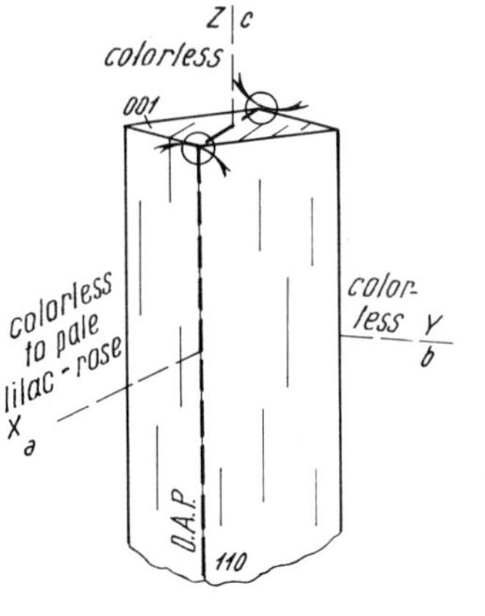


192. Mullite

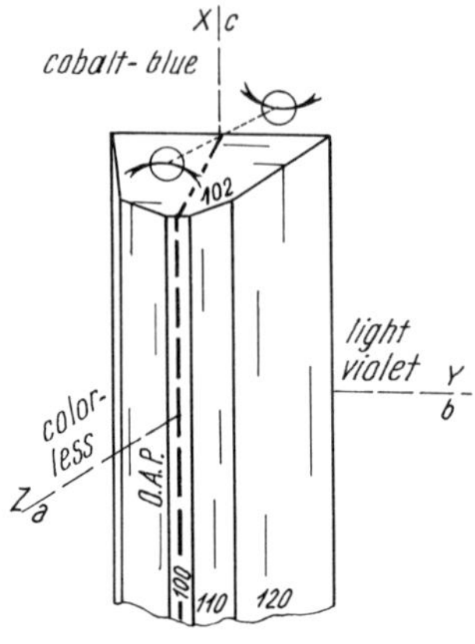


193. Dumortierite

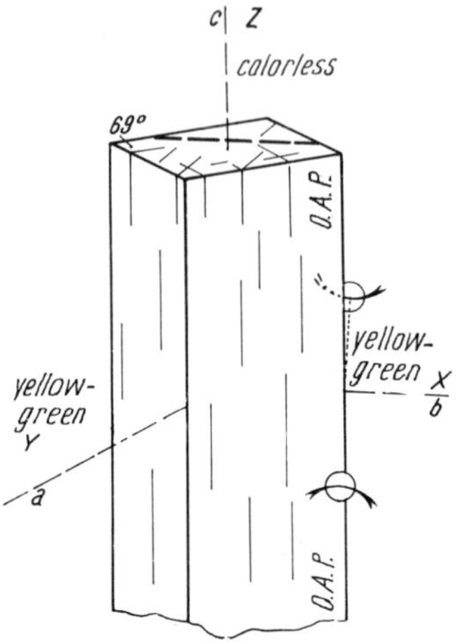


194. Carpholite

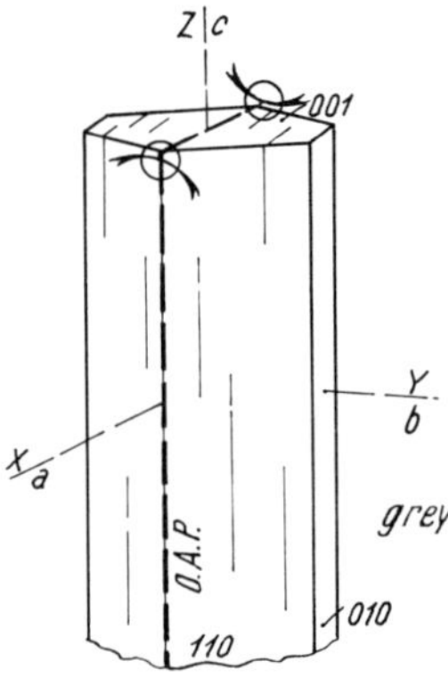


191. Sillimanite

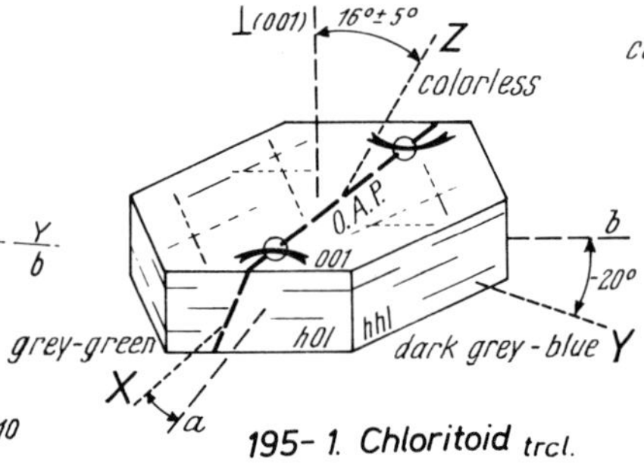


195- 1. Chloritoid trcl.

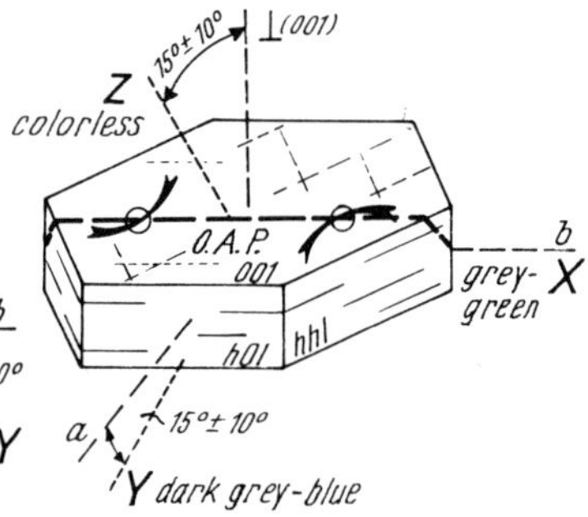


195-2. Chloritoid mcl.

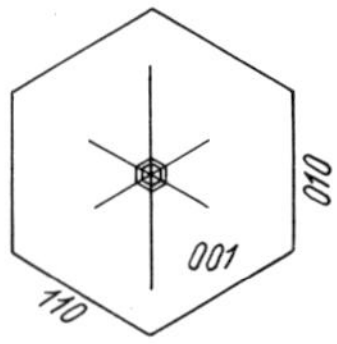


Percussion figure of phyllosilicates

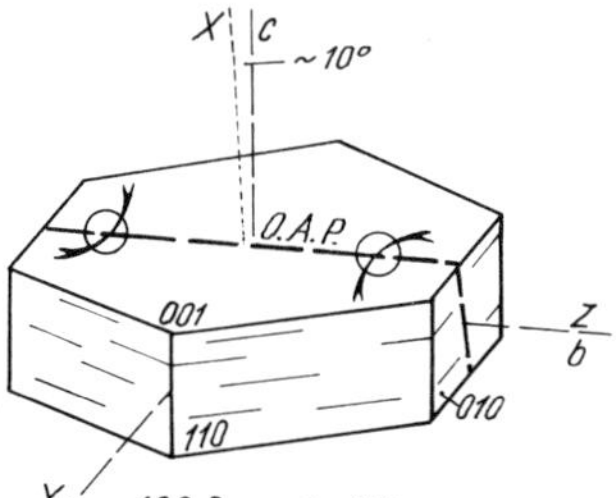


196. Pyrophyllite

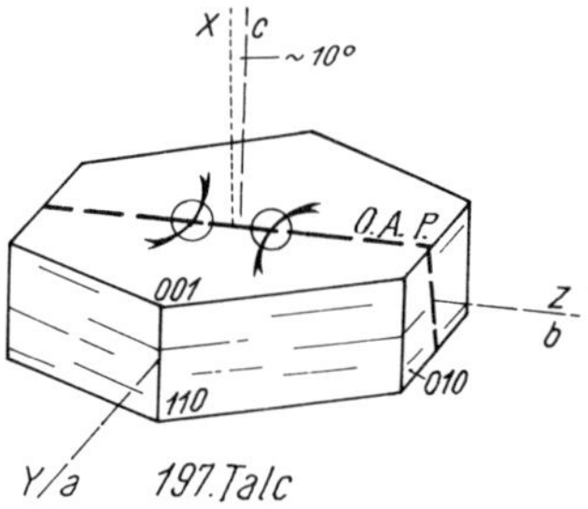


197. Talc

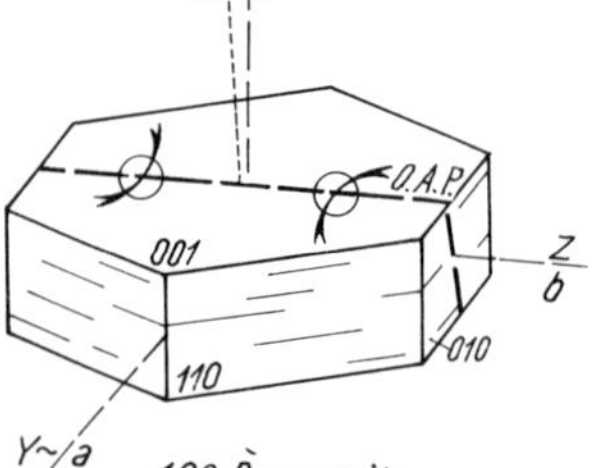


198. Paragonite

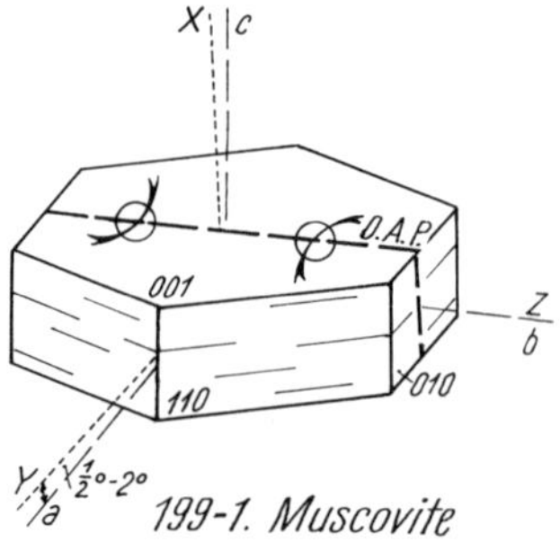


199-1. Muscovite

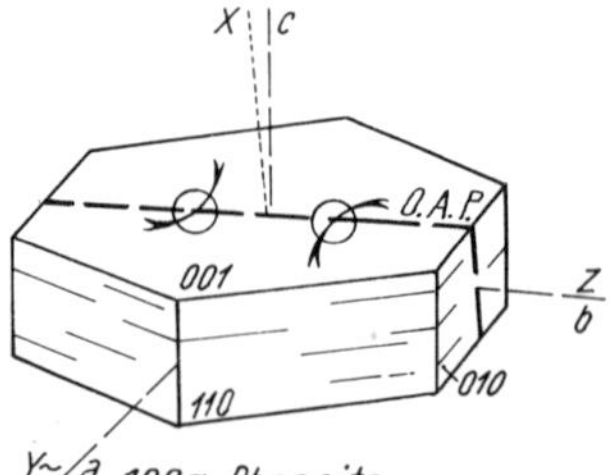


199a. Phengite

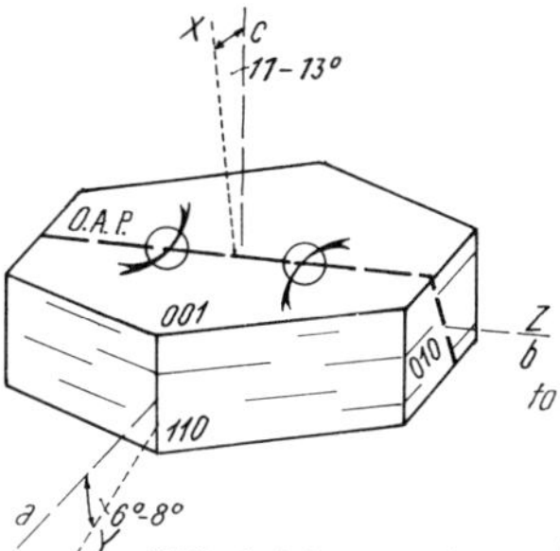


200-1. Margarite

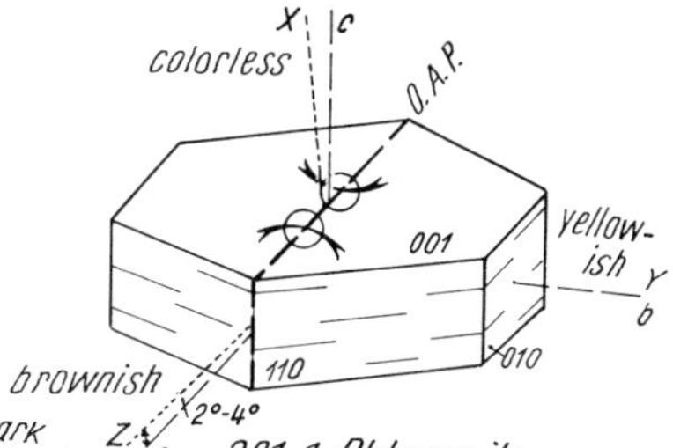


201-1. Phlogopite

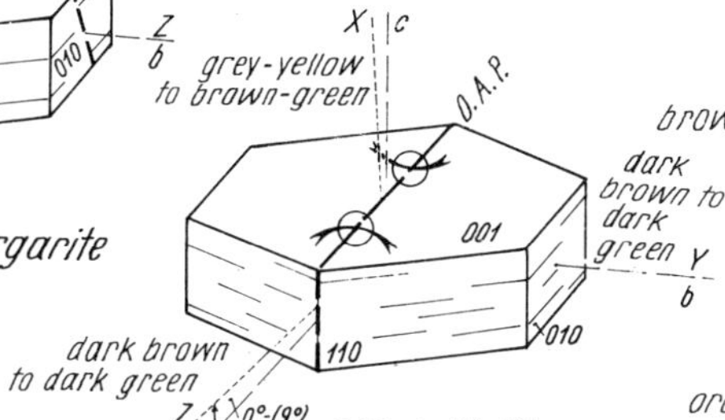


202-1. Biotite

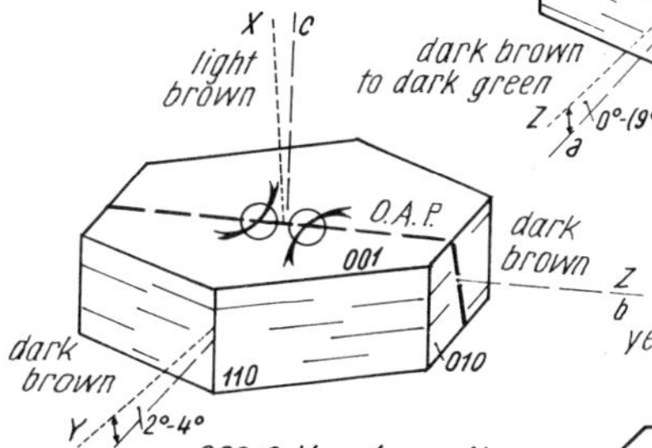


202-2. Var. Anomite

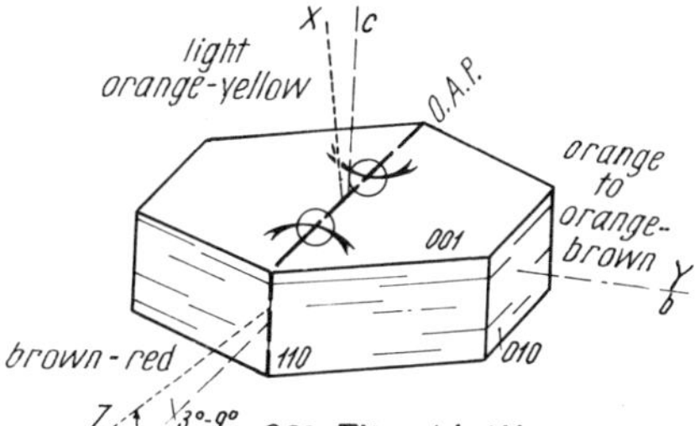


203. Titanbiotite

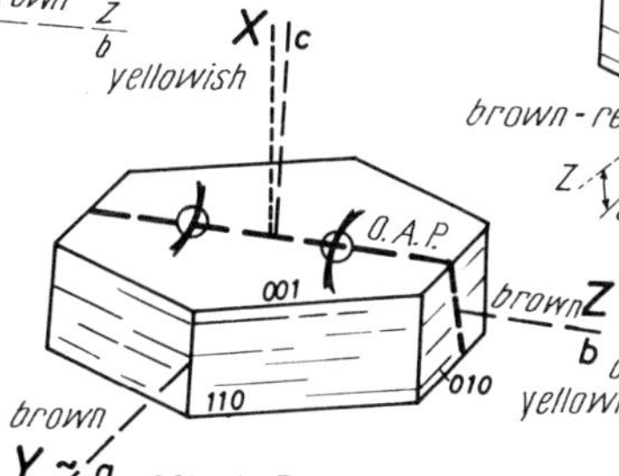


204-1. Protolithionite

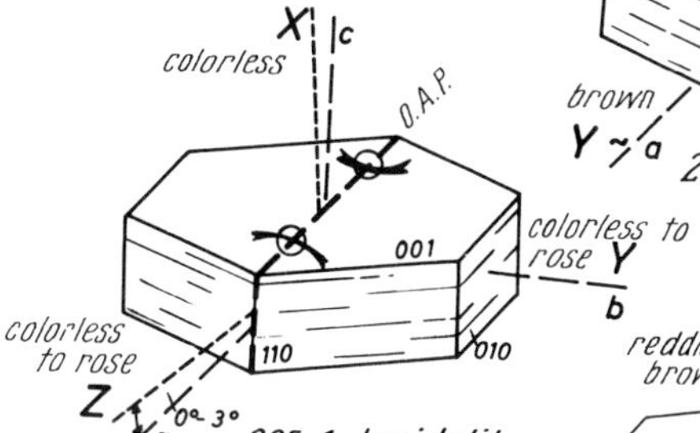


205-1. Lepidolite

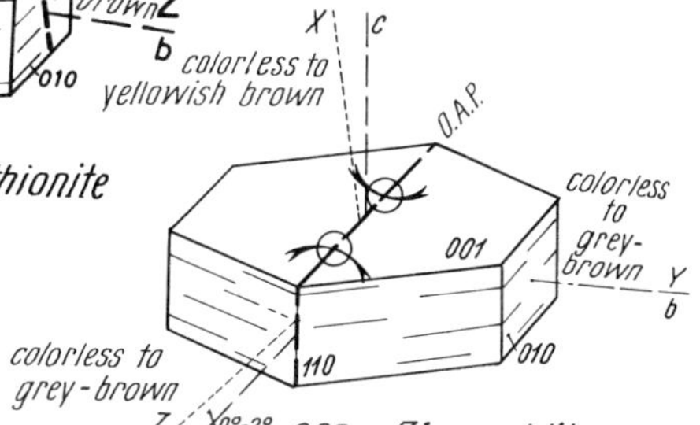


205a Zinnwaldite

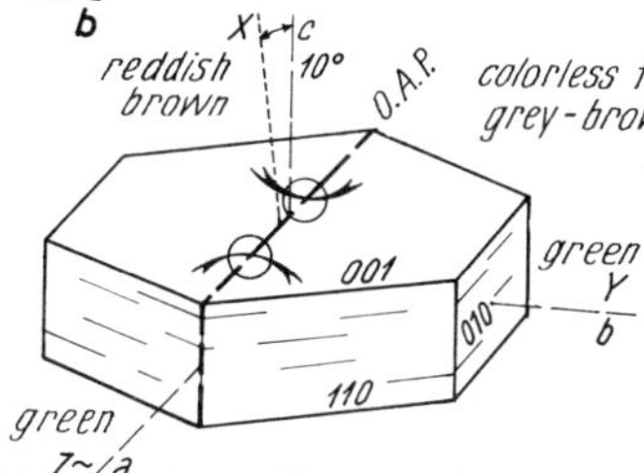


206-1. Xanthophyllite

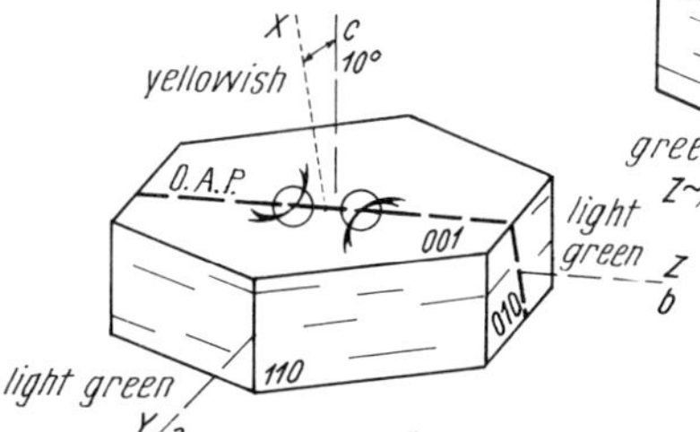


206-2. Var. Clintonite

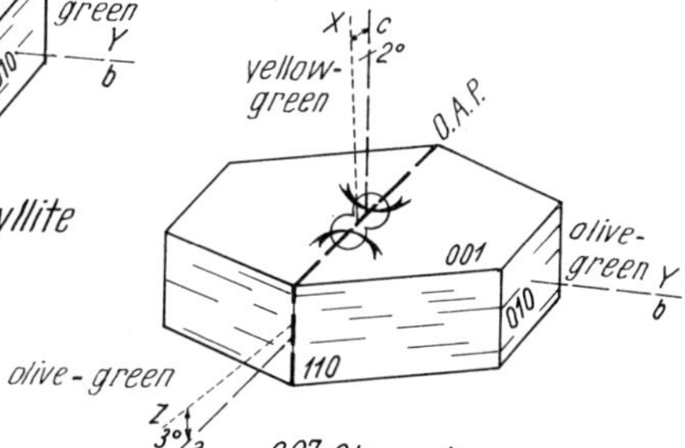


207. Glauconite

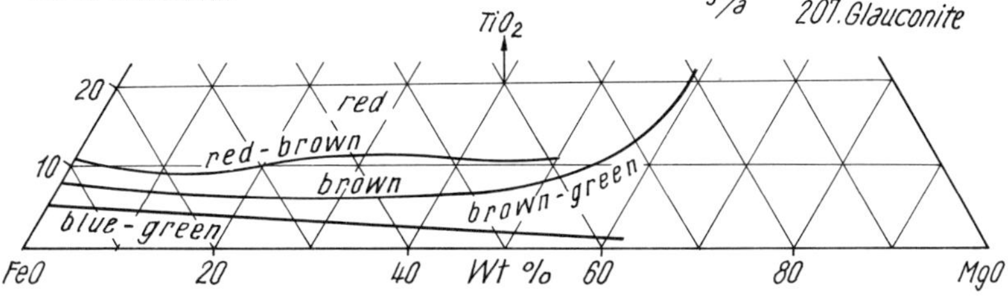


**202-3. Biotite group: color of Z**

HALL 1941 (Am. Min. 26, 30)

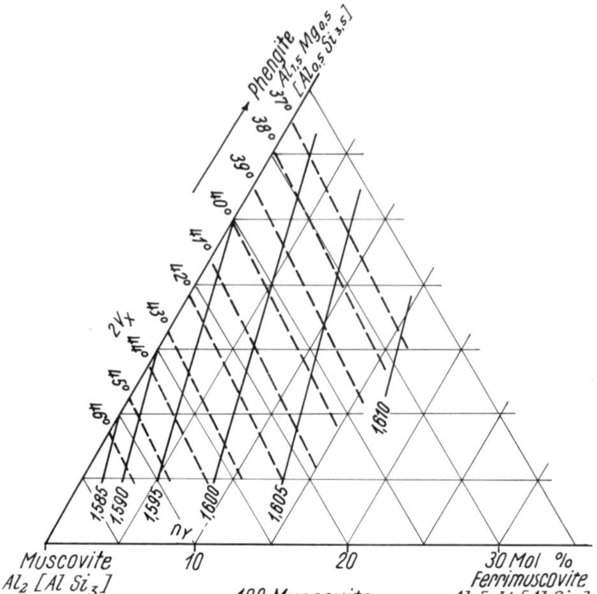


199. Muscovite
VOLK 1939 Am.Min. 24,255

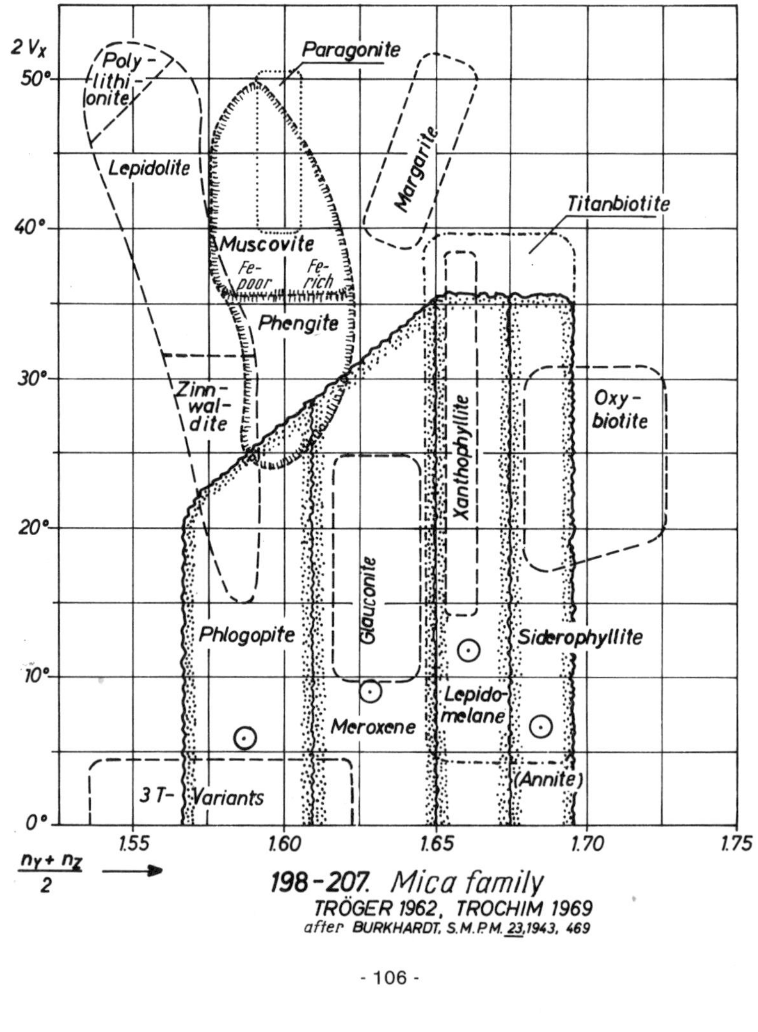


**198-207.** *Mica family*
TRÖGER 1962, TROCHIM 1969
after BURKHARDT, S.M.P.M. 23,1943, 469

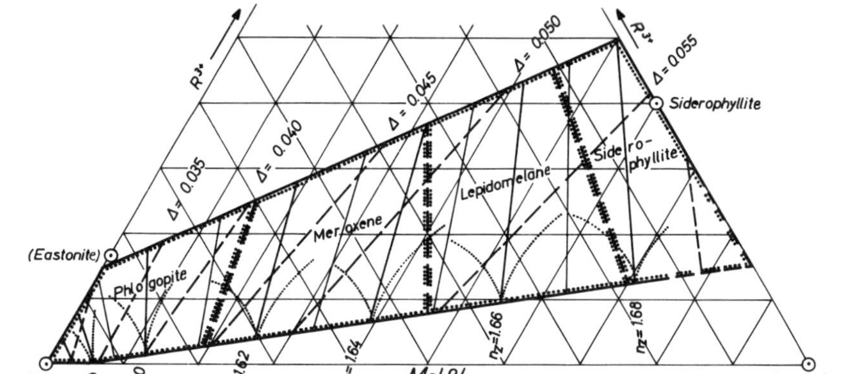


Mol %
**201-202. Biotite** *group*
TROCHIM 1966 *after* FOSTER 1960
(U.S.Geol.Surv.Prof.Pap. 354 B)

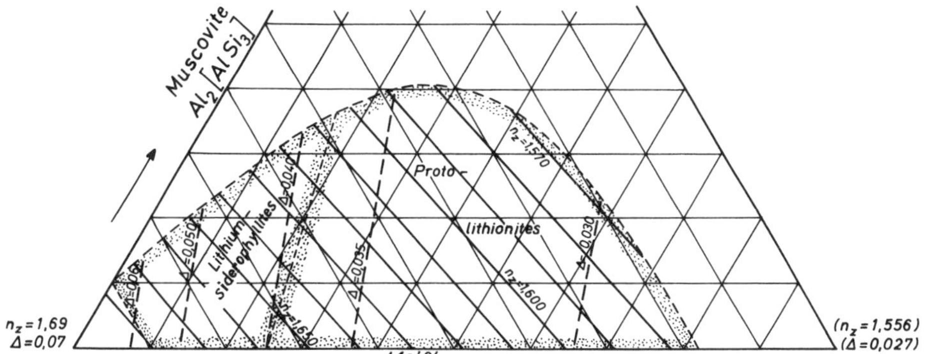


Mol %
**204-2. Protolithionite** *group*
TRÖGER 1962 (Beitr. Min. Petr. 8, 430)

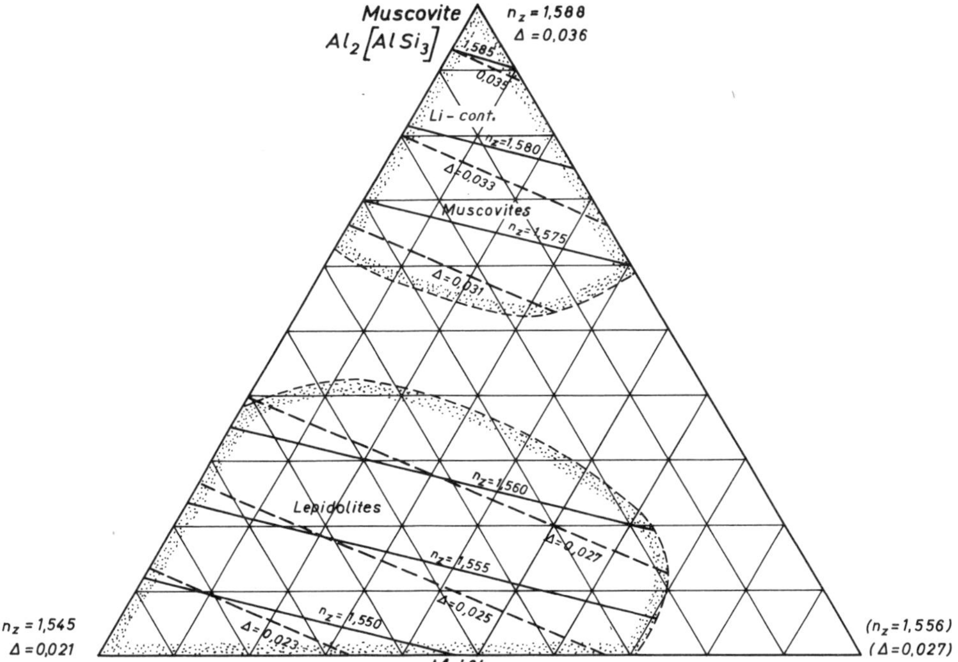


Mol %
**205-2. Lepidolite** *group*
TRÖGER 1962
(Beitr. Min. Petr. 8, 428)

| No. | Name<br>Symmetry<br>Chem. Composition | Common Crystal Forms, Twinning<br>Crystal Habit<br>Aggregates | Cleavage<br>Translation Gliding<br>Deform. Twinning<br>Hardness, Density | Optic<br>Orientation | Refractive Indices<br>Birefringence |
|---|---|---|---|---|---|
| 205a | Zinnwaldite mcl<br>KLi $Fe^{2+}_{0.65}$ Al<br>[$F_{1.6}(OH)_{0.4}/Al_{0.75}$<br>$Si_{3.25}O_{10}$] | {001},{110},{010},<br>{$\bar{1}$11},{100};<br>β = 100°, pseudohex.;<br>◈ [310], comp.pl.(001);<br>tabular ∥ (001),<br>divergent | # (001) perf.;<br>percussion figure:<br>(010), {110};<br>very elastic;<br>H = 2¼–4 D = 2.97± | X∧ā = +90° to 88°, bi;<br>Y ∥ b, O.A.P. (010)<br>Z∧a = 0° to +2°<br>l' = (+)<br>fig. p. 105 | $n_x$ = 1.541 – 1.551<br>$n_y$ = 1.571 – 1.578<br>$n_z$ = 1.573 – 1.581<br>⊖Δ = 0.032 – 0.030 |
| 206 | Xanthophyllite mcl<br>$CaMg_{2.25}Al_{0.75}$<br>[$(OH)_2/Al_{2.75}Si_{1.25}O_{10}$] | {001},{110},{010};<br>β = 100°, pseudohex.;<br>◈ [310], polysynth.,<br>compl.pl. (001);<br>tabular ∥ (001),<br>pseudorhomboedr. | # (001) perf.;<br>percussion figure:<br>(100); very brittle;<br>H = 3½ – 6 D = 3.10± | xanth.<br>X ∧ ā = +89½°, bi;<br>Y ∥ b, O.A.P. (010)<br>Z ∧ a = +½° l = (+)<br>clintonite:<br>X ∧ a ~ 90°, bi;<br>Y ∧ a ~ 0°<br>Z ∥ b, O.A.P. ⊥ (010)<br>l = (+) fig. p. 105 | $n_x$ = 1.643 – 1.649<br>$n_y$ = 1.655 – 1.660<br>$n_z$ = 1.655 – 1.661<br>⊖Δ = 0.012 – 0.012 |
| 207 | Glauconite mcl<br>$(K,Ca,Na)_{<1}$ ($Fe^{3+}$,<br>$Al,Mg,Fe^{2+})_2$<br>[$(OH)_2/Al_{0.5-0}$<br>$Si_{3.5-4}O_{10}$] | {001}; β = 95°;<br>seldom idiomorphic;<br>scaly, granular, globular,<br>divergent aggregates | # (001) perf.;<br>H = 2 D = 2.69± | X∧ā = +87°, bi;<br>Y ∥ b, O.A.P. (010)<br>Z∧a = +3°<br>l' = (+)<br>fig. p. 105 | $n_x$ = 1.592 – 1.612<br>$n_y$ = 1.613 – 1.643<br>$n_z$ = 1.614 – 1.644<br>⊖Δ = 0.022 – 0.032 |
| 208 | Montmorillonite mcl<br>$Al_{1.67}Mg_{0.33}$<br>[$(OH)_2/Si_4O_{10}$]<br>$\cdot Na_{0.33} \cdot (H_2O)_4$ | {001}, usually crypto-<br>crystalline ;<br>β ~ 90°; pseudohex.;<br>flaky ∥ (001), vermicular<br>or reticulated | # (001) perf.,<br># {110} distinct, 63°;<br>H = 1 – 2 D = 2.0± | X∧a ~ 90°, bi;<br>Y ∥ b, O.A.P. (010)<br>Z∧a ~ 0°<br>l = (+)<br>fig. p. 114 | $n_x$ = 1.475 – 1.503<br>$n_y$ = 1.499 – 1.533<br>$n_z$ = 1.500 – 1.534<br>⊖Δ = 0.025 – 0.031<br>fig. p. 116 |
| 209 | Nontronite mcl<br>$Fe^{3+}_2$ [$(OH)_2/Al_{0.33}$<br>$Si_{3.67}O_{10}$] $\cdot Na_{0.33}$<br>$\cdot (H_2O)_4$ | {001}, usually crypto-<br>crystalline;<br>β ~ 90°, pseudohex.;<br>flaky ∥ (001), bladed,<br>radial or reticulated,<br>spherulitic | # (001) perf.,<br># {110} distinct, 63°;<br>H = 1 – 2 D = 2.3± | X∧a ~ 90°, bi;<br>Y ∥b, O.A.P. (010)<br>Z∧a ~ 0°<br>l = (+)<br>fig. p. 114 | $n_x$ = 1.567 – 1.600<br>$n_y$ = 1.604 – 1.632<br>$n_z$ = 1.605 – 1.640<br>⊖Δ = 0.038 – 0.040<br>diagr. p. 116 |
| 210 | Vermiculite mcl<br>($\underline{Mg}$,$Fe^{3+}$, $Al)_3$<br>[$(OH)_2/Al_{>1}Si_{<3}O_{10}$]<br>$\cdot Mg_{0.33} \cdot (H_2O)_4$ | {001},{110},{010};<br>β = 97°, pseudohex.;<br>tabular ∥ (001),<br>scaly, vermicular<br>sinuously bent, occurs<br>in thick stacks | # (001) perf.;<br>not elastic;<br>H = 1½ – 2 D = 2.30± | X∧ā = +88° to +89°, bi;<br>Y ∥ b, O.A.P. (010)<br>Z∧a = +1° to +2°<br>l' = (+)<br>fig. p. 114 | $n_x$ = 1.525 – 1.561<br>$n_y$ = 1.545 – 1.581<br>$n_z$ = 1.545 – 1.581<br>⊖Δ = 0.020 – 0.020 |
| 210a | Saponite group mcl<br>Saponite-Griffithite<br>($\underline{Mg}$,$Fe^{2+}Fe^{3+}$)<br>[$(OH)_2/Al_{0.33}$<br>$Si_{3.67}O_{10}$] $\cdot Na_{0.33}$<br>$\cdot (H_2O)_4$ | {001},{110},{010};<br>β ~ 97°, pseudohex.;<br>tabular ∥ (001),<br>bladed ∥ a, flaky or<br>subparallel, spherulites | # (001) perf.;<br>H = 1½ – 2 D = 2.35± | oblique extinction ~ 0°;<br>X = bi, O.A.P.?<br>l' = (+) | Fe-poor Fe-rich<br>$n_x$ ~ 1.48 – 1.54<br>$n_y$ ~ 1.50 – 1.58<br>$n_z$ ~ 1.50 – 1.58<br>⊖Δ ~ 0.02 – 0.04 |
| 211 | Stilpnomelane group mcl<br>Ferrostilpnomelane-<br>(Ferri)-Stilpnomelane<br>$(K,Na,Ca)_{0-0.4}$<br>($\underline{Fe}^{2+}$, Mg, $\underline{Fe}^{3+}$, $Al)_{3.4-3.9}$<br>[$(OH)_2/Si_4O_{10}$]<br>$\cdot (H_2O)_2$ | {001},{110},{010};<br>β = 97°, pseudohex.;<br>tabular, flaky,<br>columnar, radiated,<br>stellate, massive | # (001) good,<br>(010) parting;<br>brittle;<br>H = 3–4 D = {2.59 $Fe^{2+}$ / 2.90 $Fe^{3+}$} | X∧a = 90°, bi;<br>Y ∥ b, O.A.P. (010)<br>Z∧a = 0°<br>l = (+)<br>fig. p. 114 | ferro (ferri-)<br>$n_x$ = 1.543 – 1.634<br>$n_y$ = 1.576 – 1.745<br>$n_z$ = 1.576 – 1.745<br>⊖Δ = 0.033 – 0.111<br>diagr. p. 116 |

| Angle and Dispersion of Optic Axes | Color in Thin Section Pleochroism | Chemical Reactions Special Remarks | Distinguishing Features of Similar Minerals | Occurrences | |
|---|---|---|---|---|---|
| $2V_X = (15^o)–24^o–(32^o)$<br>$r > v$ weak<br>diagr. p. 106 | colorless to colored:<br>X=alm. colorless,<br>Y=Z=brownish grey;<br>$X<Y<Z$ | bleached by acids, but slightly soluble; pleochroic haloes | lepidolite: sm. n, sm. Δ, gr. 2V; muscovite: usually colorless, gr. n, gr. Δ | pneumatolytic with topaz, cassiterite, wolframite in greisen; in granite-pegmatites with muscovite; in contraction fissures of tinbearing granites | **205a** |
| $2V_X = (2^o)–17^o–(40^o)$<br>$r < v$ weak<br>diagr. p. 106 | X=colorless to pale orange to red-brown,<br>Y=pale brownish yellow to pale green,<br>Z=pale brownish yellow to pale green;<br>$X<Y\sim Z$ | insoluble in cold HF; hydrothermal alteration to chlorite | stilpnomelane: gr. Δ; biotite: gr. Δ; chloritoid: ⊕, gr. $2V_Z$, l' = (−); chlorites: sm. Δ, gr. dispersion; muscovite: sm. n, gr. Δ, colorless | contact-metam. with grossularite, vesuvianite, monticellite e.g. in marble, calc-silicate rocks; metam. with magnetite, spinel in talc and chlorite schists | **206** |
| $2V_X = 10^o – 24^o$<br>$r > v$<br>diagr. p. 106 | X=dark blue to green to greenish yellow,<br>Y=brownish yellow to dark olive-green to blue-green,<br>Z=brownish yellow to dark olive-green;<br>$X<Y\sim Z$<br>rarely $X>Y\sim Z$ | decomposed by hot HCl, $SiO_2$-skeleton remains; celadonite = similar, hydroth., ⊕Δ? | chlorites: sm. Δ; thuringites, chamosite: gr. n, sm. Δ; biotite: gr. Δ | glauconite: synsedim. to diagenetic in coastal marine sediments; celadonite: hydrotherm. pseudom. after olivine (and hypersthene); in basalt vesicles; with zeolites in tuffacious rocks | **207** |
| $2V_X = 7^o – 25^o$ | colorless | decomposed by acids: $SiO_2$-skeleton; adsorbs aniline colors; n becomes larger with heating, n changes due to immersion-liquid; ($n_y \sim n_z = 1.50 – 1.64$) | kaolinite, dickite, nacrite: gr. n, sm. Δ; illite: gr. n, gr. Δ; vermiculite: X-ray; chlorite: sm. Δ | autometasom.-secondary in vesicles, fissures of basalt; alteration product of volc. glasses, tuffs; synsedim. in bentonites, soils, alluvial and lacustrine clays; diagenetic in salt-cont. pelites, in limestones, in gypsum deposits | **208** |
| $2V_X = 25^o – 68^o$ | X=yellowish,<br>Y=yellow-green (to dark brown),<br>Z=olive-green (to light brown);<br>$X<Y<Z$ or $X<Z<Y$ | soluble in NaOH-sol.; HCl leaches out cations: $SiO_2$-skeleton remains; adsorbs aniline colors; n becomes larger with heating;<br>$Y \sim Z$ = red-brown after heating | vermiculite: sm. n, sm. Δ; chlorite: sm. Δ; epidote: gr. n, no # (001) | hydrothermal as alteration product of basalts; coating in joints; pseudom. after hornblende and pyroxene; impregnation in porous rocks near olivine-basalt; in graphite and sulfide deposits | **209** |
| $2V_X = 0^o – 8^o$<br>$r < v$ weak | X=colorless to light brown,<br>Y=pale brownish to pale yellow-green<br>Z=pale brownish to pale yellowish;<br>$X<Y\sim Z$ | soluble in hot dilute $H_2SO_4$: $SiO_2$-skeleton remains; swells upon heating due to opening up of layers; interchangeable stacking with chlorite (=corrensite) and biotite: X-ray | hydrobiotite: gr. n, X-ray; biotite, phlogopite: gr. n, gr. Δ; chlorite, antigorite: sm. Δ; talc: gr. Δ; montmorillonite: X-ray | low grade metam. in reaction rims around serpentinite lenses; hydroth. to metasom. in contact-zones between acid and basic rocks; with chlorite in veinlets of serpentine; pseudomorph after biotite etc.; authigenic in pelites | **210** |
| $2V_X = 0^o – 40^o$ | sap. griff.<br>X=colorless to light yellow to green-brown,<br>Y=colorless to brown-green to dark brown,<br>Z=colorless to brown-green to dark brown-green;<br>$X<Y\sim Z$ | gelatinizes with HCl; oriented pseudomorphs after olivine; resistant against weathering | montmorillonite, nontronite: X-ray; vermiculite, antigorite, chrysotile: somewhat gr. $n_y$; chlorite, biotite, celadonite: distinctly gr. $n_y$ | autohydroth. with zeolites, calcite in vesicles of basalt; in basaltic glass; late hydroth. in fissures of calc-silicates and Fe-skarns, amphibolites, serpentinites | **210a** |
| $2V_X \sim 0^o$<br>(rarely up to $40^o$) | ferro- (ferri-)<br>X=light yellow to golden-yellow,<br>Y=dark green to dark olive-brown,<br>Z=dark green to dark olive-brown;<br>$X \ll Y \sim Z$ | slowly decomposable in strong acids: $SiO_2$-skeleton remains; easily confused with biotite, but stilpnomelane has lower relief; parsettensite: (Mn-rich) Y=Z=greenish yellow | biotite: better # (001), no parting (010), X=brown or green, near extinct.posit.: "birdseye maple" mottling; chloritoid: gr. n, sm. Δ, gr. $2V_Z$; xanthophyllite: sm. Δ | low grade metam. with chlorite, muscovite, albite in greenschists, phyllites, calc schists etc.; in fissures in low grade Fe-ores with magnetite, quartz; with glaucophane, garnet in blue schists | **211** |

(80, 86)

| | Name<br>Symmetry<br>Chem. Composition | Common Crystal Forms, Twinning<br>Crystal Habit<br>Aggregates | Cleavage<br>Translation Gliding<br>Deform. Twinning<br>Hardness, Density | Optic Orientation | Refractive Indices<br>Birefringence |
|---|---|---|---|---|---|
| 212 | Mg-Chlorites mcl<br>Pennine-Clinochlore-Grochauite<br>$(Mg,Fe^{2+},Al)_6$<br>$[(OH)_8/Al_{>0.5}Si_{<3.5}$<br>$O_{10}]$ | {001},{1̄01},{010},<br>{112},{1̄11},{132};<br>β = 97°; pseudohex.;<br>pseudorhombohedr.;<br>◇ (001); ◇[310], comp. plane (001);<br>tabular ‖ (001), scaly to fibrous, massive | # (001) perf.;<br>percussion figure: (010);<br>flexible, inelastic;<br>H = 2–2½ D = {2.64 pe. / 2.74 gr.} | pennine ⊖<br>XΛa ~ 90°, bi;<br>Y ‖ b, A.O.P. (010)<br>ZΛa ~ 0°<br>l = (+); fig. p. 114<br>- - - - - - - -<br>penn. to groch. ⊕<br>XΛa = 0° to +2°<br>Y ‖ b, O.A.P. (010)<br>ZΛā = 90° to +88°, bi;<br>l' = (–); fig. p. 114 | pe. – clin. – groch.<br>$n_x$ = 1.562 – 1.581 – 1.594<br>$n_y$ = 1.565 – 1.581 – 1.594<br>$n_z$ = 1.565 – 1.586 – 1.606<br>Δ = ⊖0.003 – ⊕0.005 – ⊕0.012<br>diagr. p. 117 – 118 |
| 212a | Cr-Chlorites mcl<br>Kaemmererite = Cr-pennine<br>> 2 weight % $Cr_2O_3$<br>Kotschubeite = Cr-clinochlore<br>> 2 weight % $Cr_2O_3$ | {001},{1̄01},{010},<br>{112};<br>β = 97°; pseudohex.,<br>pseudorhombohedr.;<br>◇ (001); [310], comp. plane (001);<br>tabular ‖ (001),<br>scaly, massive | # (001) perf.;<br>percussion figure: (010);<br>flexible, inelastic;<br>H = 2½ – 3 D = 2.66± | < ~ 6 weight % $Cr_2O_3$:<br>XΛa = 0° to +2°<br>Y ‖ b, O.A.P. (010)<br>ZΛā = 90° to +88°, bi;<br>l' = (–);<br>- - - - - - - -<br>> ~ 6 weight % $Cr_2O_3$:<br>XΛā = 90° to +88°, bi;<br>Y ‖ b, O.A.P. (010)<br>ZΛa = 0° to +2°<br>l' = (+); fig. p. 114 | < – 6 weight % – > $Cr_2O_3$<br>$n_x$ = 1.576 – 1.584 – 1.585<br>$n_y$ = 1.576 – 1.584 – 1.590<br>$n_z$ = 1.580 – 1.584 – 1.591<br>Δ = ⊕0.004 – ⊕/⊖0.000 – ⊖0.006 |
| 213 | $Mg\text{-}Fe^{3+}$-Chlorites mcl<br>Diabantite-Pycnochlorite-Rhipidolite<br>$(Mg,Fe^{2+},Al)_6$<br>$[(OH)_8/Al_{>0.5}$<br>$Si_{<3.5}O_{10}]$ | {001},{110},{010},<br>{hkl}, {h0l};<br>β = 97°, pseudohex.;<br>◇ (001); ◇ [310],<br>comp. pl. (001);<br>tabular ‖ (001),<br>scaly to fibrous,<br>vermicular, massive | # (001) perf.;<br>percussion figure: (010);<br>flexible, inelastic;<br>H = 2 D = {2.82± diab. / 2.95± rhip.} | diabantite: ⊖<br>XΛa ~ 90°, bi;<br>Y ‖ b, O.A.P. (010)<br>ZΛa ~ 0°<br>l = (+)<br>- - - - - - - -<br>rhipidolite: ⊕<br>XΛa ~ 0°<br>Y ‖ b, O.A.P. (010)<br>ZΛa ~ 90°, bi;<br>l = (–); fig. p. 114 | Mg-diab. – Fe-rhip. – corundoph.<br>$n_x$ = 1.589 – 1.627 – 1.605<br>$n_y$ = 1.595 – 1.627 – 1.608<br>$n_z$ = 1.595 – 1.629 – 1.615<br>Δ = ⊖0.006 – ⊕0.002 – ⊕0.010<br>diagr. p. 117 – 118 |
| 214 | $Fe^{2+}$-Chlorites mcl<br>Brunsvigite-Aphrosiderite-Bavalite<br>$(Fe^{2+},Mg,Al,Fe^{3+})_6$<br>$[(OH)_8/Al_{>0.7}Si_{<3.3}$<br>$O_{10}]$ | {001},{110},{hkl},<br>{h0l};<br>β = 97°, pseudohex.;<br>◇ (001) and ◇ [310] rare;<br>rarely tabular ‖ (001),<br>scaly, rounded-concentric masses, massive | # (001) perf.;<br>flexible, inelastic;<br>H = 1–2 D = {3.10 aphr. / 3.21 bav.} | XΛā = 90° to +88°<br>Y ‖ b, O.A.P. (010)<br>ZΛa = 0° to +2°<br>l' = (+)<br>fig. p. 114 | brunsv. – aphr. – bav.<br>$n_x$ = 1.632 – 1.647 – 1.665<br>$n_y$ = 1.638 – 1.651 – 1.675<br>$n_z$ = 1.638 – 1.651 – 1.675<br>⊖Δ = 0.006 – 0.004 – 0.010<br>diagr. p. 117 – 118 |
| 215 | Leptochlorite mcl<br>Delessite-14 Å-Chamosite-Thuringite<br>$(Fe^{2+},Fe^{3+},Mg,Al)_6$<br>$[(OH)_8/Al_{>0.5}Si_{<3.5}$<br>$O_{10}]$ | {001},{110},{010},<br>{hkl}, {h0l};<br>β = 97°, pseudohex.;<br>◇ (001); ◇ [310] rare;<br>tabular ‖ (001),<br>scaly to fibrous,<br>rounded concentric masses, massive | # (001) perf.;<br>flexible, inelastic;<br>H = 1–2 D = {2.82 del. / 3.22 thur.} | XΛā = +90° to +88°, bi;<br>Y ‖ b, O.A.P. (010)<br>ZΛa = 0° to +2°<br>l' = (+); fig. p. 115<br>- - - - - - - -<br>rarely:<br>XΛa ~ 0°<br>Y ‖ b, O.A.P. (010)<br>ZΛa ~ 90°, bi;<br>l = (–) | del. – thur.<br>$n_x$ = 1.595 – 1.671<br>$n_y$ = 1.599 – 1.684<br>$n_z$ = 1.599 – 1.685<br>⊖Δ = 0.004 – 0.014 |
| 216 | Sudoite mcl<br>$(Al,Mg)_{>4}[(OH)_8/$<br>$Al_{0.6-1.3}Si_{3.4-2.7}O_{10}]$ | {001}; only cryptocrystalline;<br>β = 97°;<br>◇ unknown;<br>finest scales ‖ (001),<br>reticul.-massive | # (001)?<br>H = 1 – 2? D = 2.66± | XΛa ~ 90°, bi;<br>Y ‖ b, O.A.P. (010)} uncertain<br>ZΛa ~ 0°<br>l = (+)<br>fig. p. 115 | $n_x$ = 1.568 – 1.574<br>$n_y$ = 1.572 – 1.580<br>$n_z$ = 1.572 – 1.580<br>⊖Δ = 0.004 – 0.006 |
| 216a | Cookeite mcl<br>$LiAl_4$<br>$[(OH)_8/AlSi_3O_{10}]$ | {001},{110},{010};<br>β ~ 98°; pseudohex.;<br>◇ (001); ◇ [310],<br>comp. pl. (001),<br>triplets?; tabular ‖ (001),<br>flaky, subrounded,<br>barrel-shaped, fibrous,<br>felty | # (001) perf.;<br>flexible, inelastic;<br>H = 2–2½ D = 2.68± | XΛa = 0° to +3°<br>Y ‖ b, O.A.P. (010)<br>ZΛā = 90° to +87°, bi;<br>l' = (–)<br>fig. p. 115 | $n_x$ = 1.572 – 1.565<br>$n_y$ = 1.578 – 1.572<br>$n_z$ = 1.589 – 1.595<br>⊕Δ = 0.017 – 0.030 |
| 217 | Antigorite mcl<br>$Mg_6[(OH)_8/Si_4O_{10}]$ | {001},{110},{010},<br>{hk0}, {h0l};<br>β = 91½°;<br>◇ (001);<br>tabular ‖ (001),<br>bladed, scaly, massive | # (001) perf.;<br>percussion figure: (010);<br>flexible, inelastic;<br>H = 3–4 D = 2.56± | XΛa ~ 90°, bi;<br>Y ‖ b, O.A.P. (010)<br>ZΛa = 0°<br>l = (+)<br>fig. p. 115 | Fe-free – Fe-rich<br>$n_x$ = 1.546 – 1.595<br>$n_y$ = 1.551 – 1.603<br>$n_z$ = 1.552 – 1.604<br>⊖Δ = 0.006 – 0.009<br>diagr. p. 118 |

| Angle and Dispersion of Optic Axes | Color in Thin Section Pleochroism | Chemical Reactions Special Remarks | Distinguishing Features of Similar Minerals | Occurrences | |
|---|---|---|---|---|---|
| $2V_X = 0^o - 5^o$ r > v (pennine); $2V_Z = 0^o–30^o–(70^o)$ r < v (clin-groch.) | Fe-free: colorless; otherwise weak pleochroism: X=light yellow-green to light bluish gr., Y=light greenish yellow to light bluish green X=light green-yellow to light blue-green; X<Y∼Z: penn. ⊖, X∼Y>Z: penn.-groch. ⊕ | soluble in hot $H_2SO_4$; if Δ small: abnormal interf. colors; pleochr. haloes conspicuous; often intergrown with biotite or phlogopite; leuchtenbergite = extremely Fe-poor pennine or clinochl.; sheridanite = extremely Fe-poor grochauite | antigorite: ⊖, gr. Δ, normal interf. colors; amesite: ⊕, gr. Δ, normal interf. colors; Mg-Fe-chlorites: gr. n, stronger pleochroism; micas, talc, kaolinite: gr. Δ, normal interf. colors | prim. magmatic in peridotite and lherzolite; sec. hydroth. in serpent. ultramafic rocks; pseudomorph after biotite and others; metam. in marble, skarns and calc-silicate rocks; low grade metam. in greenschists, amphibolites; in corundum rocks; detrital in sedimentary rocks | 212 |
| $2V = (30^o)–0^o–(30^o)$ r ≶ v | deeper colors with increasing Cr-content: pale yellowish pink to violet-pink to purple; ⊕: X∼Y>Z, ⊖: X<Y∼Z | slightly soluble in HCl, easily soluble in $H_2SO_4$; often zonal structure due to variable Cr-content; intergrowth in (001) with Mg-chlorite, abnormal interf. colors up to Δ=0.003 | quite distinctive: characteristic absorption color, small Δ and abnormal interf. colors | primary magmatic in dunite, peridotite, serpentinite; autometasomatic to hydroth. in chromitite and chromite serpentinite | 212a |
| $2V_X = 0^o–(10^o)$; diab.-pycn.: $2V_Z = 0^o–(20^o)$; rhip.-corundoph.: $2V_Z = 10^o–40^o$; r<v | X=light blue-green to yellow-green, Y=grey-green to brown-green, Z=grey-green to brown-green; X<Y∼Z: diab.-rhip. ⊖; X∼Y>Z: pycn.-rhip. ⊕ | soluble in HCl: $SiO_2$-skeleton remains; intense brown staining upon heating; distinct pleochroism; if Δ small: abnormal interf. colors; intergrowth with biotite ∥(001); pseudomorph after garnet, biotite etc.; corundophilite = extremely Al-rich rhipidol. | micas, nontronite: gr. Δ, norm. interf. colors; antigorite, Mg-chlorite: weaker pleochroism; alumoserpentine: no abnormal interference colors, often isotropic aggregates, X-ray | diab.-pycn.: autohydroth. in basic magmatites; in vesicles, in fissures of basalt, diabase and spilite; pycn.-rhip.: metam. in phyllites, mica schists, greenschists, chlorite schists; hydroth. in fissures; authigenic in sedim. rocks; corundoph.: in corundum rocks | 213 |
| $2V_X = 0^o–(30^o)$ r>v | X=pale yellowish to brownish, Y=dark olive-green to dark brown-green, Z=dark olive-green to dark brown-green; X≪Y∼Z | easily soluble in HCl; strong pleochroism; rarely abnormal interference colors (blue-grey) | green biotite: gr. Δ; glauconite, nontronite: sm. n; leptochlorite: chem. anal.; alumoserpentine: X-ray! | postmagmatic in pegmat. to hydroth. veins; in sedim. to weakly metam. iron oolites, iron ores, weakly metam. diabases; rarely in para-metamorphic rocks | 214 |
| $2V_X = 0^o–(30^o)$ r>v | del. to thur. X=yellow-green to light brown-green, Y=green to dark green, Z=green to dark green; X<Y∼Z (rarely X∼Y>Z) | soluble in HCl: $SiO_2$-skeleton remains; thuringite gelatinizes with HCl; pleochroism increases with Fe-content; if Δ small then abnorm. interf. colors; klementite: leptochlorite with high Al-, low Fe-content | green micas: gr. Δ; nontronite: gr. Δ; orthochlorite: chem. anal.; alumoserpentine: X-ray! | del.: autohydroth. in groundmass, fissures, vesicles of basic volcanics; cham., thur: hydroth. in fissures of granites, pegmat.; in sedim. and weakly metam. iron oolites, in iron ores, weakly metam. diabases | 215 |
| $2V_X = 0^o$ to very small | colorless to pale greenish | insoluble in diluted acids; usually clay-particle-size < 2 μ; no abnormal interf. colors; isotropic aggregates; intergrown (stacking) with hydromuscovite | chlorites: chem. anal.; micas: gr. Δ; kaolinite and other clay minerals: X-ray, DTA | in hydrotherm. altered rhyolitic tuffs; sedim. in marls; diagenetic with sericite in sandstones; in (paleo-) soils; in bauxites | 216 |
| $2V_Z = 0^o - 80^o$ very variable r<v | X=pale green to pink, Y=pale green to pink, Z=colorless to pale yellow; X∼Y>Z | slightly soluble in diluted acids; no abnormal interf. colors, but sixfold division with diff. opt. orientation | chlorites: sm. Δ, abnorm. interf. colors; Li-micas: ⊖; other micas: gr. Δ | primary in hydroth. gold-quartz veins; sec. as hydroth. alteration prod. of lepidolite, spodumene, tourmaline, petalite, e.g. in Li-pegmatites | 216a |
| $2V_X = 27^o - 60^o$ r>v | almost colorless; Fe-rich: X=greenish yellow, Y=Z=greenish; X∼Y∼Z | decomposed by HCl, forms $SiO_2$-gel; jenkinsite: Fe-rich variety (30–40 atom % $Fe^{2+}$ instead of Mg) | crysotile: finely fibrous; alumoserpentine: n > 1.61; Mg-chlorites: usually ⊕ or sm. Δ and abnormal interf. colors; micas: gr. Δ | low grade metam. from olivine, pyroxene, and other Mg-silicates during serpentinization of ultrabasic to basic magmatites (stress) | 217 |

| | Name<br>Symmetry<br>Chem. Composition | Common Crystal Forms, Twinning<br>Crystal Habit<br>Aggregates | Cleavage<br>Translation Gliding<br>Deform. Twinning<br>Hardness, Density | Optic Orientation | Refractive Indices<br>Birefringence |
|---|---|---|---|---|---|
| **217a** | Greenalite mcl<br>$Fe^{2+}_{4.5}Fe^{3+}$<br>$[(OH)_8/Si_4O_{10}]$ | {001}; β = 90°;<br>cryptocrystalline;<br>reticulated and spherulitic aggregates, rarely flakes ‖ (001) | # (001)?<br>H = ? D = 3.3 ± | unknown | $n_y$ = 1.660 – 1.675<br>disp.: $n_F$-$n_c$ = 0.016<br>Δ = small<br>diagr. p. 118 |
| **218** | Chrysotile mcl<br>$Mg_6[(OH)_8/Si_4O_{10}]$ | rolled-up tubes ‖ [100], rarely ‖ [010] of antigorite-struct.; β = 93¼°;<br>fibrous ‖ a, asbestiform, reticulated, irregular | pseudocleavage poor, 59°;<br>H = 3–4 D = 2.51 ± | chrys.-Z: Z‖a, bi; O.A.P. ‖a; l = (+) ⁞ chrys.-X: X‖a, bi; O.A.P. ‖a; l = (−)<br>fig. p. 115 | chrys.-Z ⁞ chrys.-X<br>$n_x$ = 1.532–52 ⁞ 1.538–60<br>$n_y$ = 1.545–61 ⁞ 1.546–67<br>Δ = ⊕0.013–09 ⁞ ⊖0.008–07 |
| **218a** | Lizardite (orh. hex)<br>$Mg_6[(OH)_8/Si_4O_{10}]$ | {001}, never idiomorph.; β = 90°;<br>finest scales ‖ (001), cap-like bent individua, fibrous, felty | # (001) perf.;<br>H = 3–4 D = 2.53 ± | XΛa = 90°, bi;<br>Y ‖ b, O.A.P. (010)<br>ZΛa = 0°<br>l = (+)<br>fig. p. 115 | $n_x$ = 1.538 – 1.550<br>$n_y$ = 1.546 – 1.560<br>$n_z$ = 1.546 – 1.560<br>⊖Δ = 0.008 – 0.010 |
| **219** | Alumoserpentine mcl<br>7 Å-Chamosite-Berthierine<br>$(\underline{Fe}^{2+},Mg,\underline{Al},Fe^{3+})_6$<br>$[(OH)_8/Al_{0-2}Si_{4-2}O_{10}]$ | {001}, rarely idiom.;<br>β = 104½°<br>fine scales ‖ (001), rounded concentric aggregates, irregular scaly, massive | # (001) perf.;<br>H = 1–2 D = 2.8–3.2 | XΛa ~ 90°, bi;<br>Y ‖ b, O.A.P. (010)<br>ZΛa ~ 0°<br>l = (+)<br>fig. p. 115 | $n_x$ = 1.616 – 1.665<br>$n_y$ = 1.620 – 1.672<br>$n_z$ = 1.620 – 1.672<br>⊖Δ = 0.004 – 0.007<br>diagr. p. 118 |
| **219a** | Amesite mcl<br>$(\underline{Mg},Fe^{2+})_4Al_2$<br>$[(OH)_8/Al_2Si_2O_{10}]$ | {001},{110},{010};<br>β = 90°; pseudohex.;<br>◇ (001), ◇ [310];<br>sixfold division;<br>tabular ‖ (001); flaky, irregular scaly, massive | # (001) perf.;<br>H = 3–4 D = 2.79 ± | XΛa = 0° to +20°<br>Y ‖ b, O.A.P. (010)<br>Z ~ ⊥(001), bi;<br>l' = (−)<br>fig. p. 115 | $n_x$ = 1.588 – 1.597<br>$n_y$ = 1.588 – 1.597<br>$n_z$ = 1.611 – 1.612<br>⊕Δ = 0.023 – 0.015<br>diagr. p. 118 |
| **219b** | Cronstedtite mcl, hex, trig<br>$Fe^{2+}_4Fe^{3+}_2$<br>$[(OH)_8/Fe^{3+}_2Si_2O_{10}]$ | {001} (mcl),<br>{10$\bar{1}$0} (hex),<br>{10$\bar{1}$9} (trig);<br>β = 104½° (mcl)<br>◇ ?<br>8 types of polymorphs;<br>hex: tabular<br>trig: pyramids, reniform, flaky, fibrous | # (001), (0001) } perf.;<br>somewhat elastic;<br>H = 3½ D = 3.45 ± | X⊥(001), (0001)<br>O.A.P. (010)?<br>ZΛa ~ 0°<br>l = (+) | $n_y$ ~ 1.80<br>(with 20 atom % Mg: $n_y$ = $n_z$ = 1.721)<br>⊖Δ = large |
| **220** | Kaolinite trcl<br>$Al_4[(OH)_8/Si_4O_{10}]$ | {001},{110},{010};<br>β = 104½°, pseudohex.;<br>◇ rare;<br>tabular, scaly, massive, vermicular | # (001) perf.;<br>flexible, but inelastic<br>H = 2–2½ D = 2.63 ± | XΛa = +88° to 86°, bi;<br>YΛa = +2° to +4°, O.A.P. ⊥(010)<br>Z ‖ b<br>l' = (+)<br>fig. p. 119 | $n_x$ = 1.553 – 1.563<br>$n_y$ = 1.559 – 1.569<br>$n_z$ = 1.560 – 1.570<br>⊖Δ = 0.007 – 0.007 |
| **221** | Dickite mcl<br>$Al_4[(OH)_8/Si_4O_{10}]$ | {001},{110},{010}, {0kl};<br>β = 103½°, pseudohex.;<br>tabular ‖ (001);<br>elongated ‖ [100], columnar, vermicular | # (001) perf.;<br>somewhat elastic;<br>H = 2½–3 D = 2.62 ± | XΛā = −78° to −70°, r < v<br>YΛa = +12° to +20°, r > v, O.A.P. ⊥(010)<br>Z ‖ b, bi;<br>l' = (+)<br>fig. p. 119 | $n_x$ = 1.560 – 1.562<br>$n_y$ = 1.562 – 1.564<br>$n_z$ = 1.566 – 1.569<br>⊕Δ = 0.006 – 0.007 |
| **222** | Nacrite mcl<br>$Al_4[(OH)_8/Si_4O_{10}]$ | {001},{110},{010};<br>β = 113¾°, pseudohex.;<br>triplets, comp.pl. (110);<br>tabular ‖ (001)<br>scaly, radiate | # (001) perf.,<br># (010) good,<br># {110} good;<br>inelastic;<br>H = 2½–3 D = 2.58 ± | XΛā = +80° to +78°, bi;<br>YΛa = +10° to +12° O.A.P. (010)<br>Z ‖ b<br>l' = (+)<br>fig. p. 119 | $n_x$ = 1.557 – 1.560<br>$n_y$ = 1.562 – 1.566<br>$n_z$ = 1.563 – 1.568<br>⊖Δ = 0.006 – 0.008 |
| **223** | Halloysite mcl<br>$Al_4[(OH)_8/Si_4O_{10}]$<br>$\cdot 4H_2O$ | xenomorphic;<br>β = 100¼°;<br>cylindric tubes of ~<1μ diameter, porcelain-like, earthy | not recognizable;<br>H = 1–2 D = 2.1 ± | l' = (−)<br>fig. p. 119 | $n_y$ = 1.528 – 1.542<br>Δ = 0.000 – 0.004<br>meta-halloysite:<br>$n_y$ = 1.541 – 1.552 |

| Angle and Dispersion of Optic Axes | Color in Thin Section Pleochroism | Chemical Reactions Special Remarks | Distinguishing Features of Similar Minerals | Occurrences | |
|---|---|---|---|---|---|
| unknown | light green, almost colorless; pleochroism barely discernable | soluble in acids; brown staining after heating and due to weathering; often isotropic aggregates | X-ray | diagen. in sedim. siliceous Fe-deposits; hydroth. to metasom. from siderite in magnetite ores | **217a** |
| $2V_z = 10^\circ - 90^\circ$ (chrys. - Z),<br>$2V_x = 30^\circ - 35^\circ$ (chrys.-X) | colorless; with Fe-cont. very pale:<br>X=yellowish to greenish yellow,<br>Z=yellow-green to orange-yellow;<br>X > Z (chr.-Z)<br>X ⩽ Z (chr.-X) | decomposed by HCl: $SiO_2$-gel; aggregates often pseudom. after olivine | antigorite: flaky; alumoserpentine: n > 1.61; chlorite: sm. Δ, abnormal interf. colors; fibrous amphib.: gr. n, gr. Δ, partly oblique extinction | from olivine under low and medium grade metam. conditions; peridotite → serpentine rocks; asbestos = veinfilling; metasomatic from dolomite | **218** |
| $2V_x$ = very small, often ⊖ uniaxial | colorless, rarely pale greenish | decomposed by HCl: $SiO_2$-gel; bastite: oriented lizardite pseudomorph after orthopyroxene | hard to distinguish from chrysotile and antigorite in massive aggregates: X-ray | only in massive serpentinites, often intimately mixed with chrysotile-Z and antigorite | **218a** |
| $2V_x = 0^\circ - (7^\circ)$ | X=light brownish green,<br>Y ~ Z = light green;<br>X < Y ~ Z | easily soluble in HCl; no abnormal interference colors; often isotropic aggregates | X-ray; Fe-chlorites: deeper absorption color; Mg-chlorites: abnormal interf. colors; antigorite: sm. n, colorless; amesite: sm. n, gr. Δ | diagen. with siderite in marine-sed. Fe-oolites; with kaolinite, hematite in slates, laterites, bauxites; hydroth. in ore lodes; pseudom. after cordierite | **219** |
| $2V_z = 0^\circ - 20^\circ$<br>r < v | colorless | slightly soluble in HCl; no abnormal interf. colors; chromium-amesite (1.5 wt % $Cr_2O_3$): pink-violet | antigorite: sm. Δ, ⊖; Mg-chlorites: sm. Δ, abnormal interf. colors; diaspore: gr. n, gr. Δ, gr. 2V; mica, talc, pyrophyllite: gr. Δ, ⊖ | diagenetic to low grade metam. with diaspore, magnetite, corundophilite in $SiO_2$-poor, Al-rich rocks | **219a** |
| $2V_x \sim 0^\circ$<br>usually uniaxial | X=emerald-green to reddish brown,<br>Y=Z=olive-green, almost opaque;<br>X < Y = Z | decomposed by HCl: $SiO_2$-gel; nearly opaque in thin sections | strongest absorption colors of all phyllosilicates with perf. cleavage; n ~ 1.8! | hydroth. with siderite, pyrite, sphalerite in quartz veins | **219b** |
| $2V_x = (24^\circ) - 42^\circ - (50^\circ)$<br>r > v weak | colorless | slightly soluble in HCl or $H_2SO_4$, easily soluble in HF or NaOH-sol.; adsorbs aniline colors; often vermicular aggregates | dickite, gibbsite: ⊕; nacrite: gr. Y Λ a; montmorillonite: sm. n, gr. Δ; sericite, illite, talc, pyrophyllite: gr. Δ | hydroth. from feldspar, foids and Al-silicates in granites, gneisses etc.; intensive weathering to kaolinization at pH ~ 5 – 6; in soils; detritic in marine clays; diagen. in lacustr. carbonate-poor sediments; diagen. in psammites and psephites | **220** |
| $2V_z = (52^\circ) - 66^\circ - (80^\circ)$<br>r < v | colorless | slightly soluble in HCl and $H_2SO_4$; hardly adsorbs aniline colors; only opt. ⊕ clay-mineral | kaolinite: ⊖, Y Λ a ~ 3°; nacrite: ⊖, Y Λ a ~ 11°; sericite, illite, talc, pyrophyllite, montmorillonite: ⊖, gr. Δ, about parallel extinction | hydroth. in veins with sulph. ores and quartz; coating on quartz, calcite etc.; not in pelites, clays; authigen. as cement in sandstone | **221** |
| $2V_x = 40^\circ$ or<br>$2V_x \sim 90^\circ$ | colorless | slightly soluble in HCl and $H_2SO_4$: aggregates open up on heating with blowpipe: hardly adsorbs aniline colors; rare | dickite: ⊕, Y Λ a = 12° – 20°; kaolinite: Y Λ a ~ 3°; sericite, illite, talc, pyrophyllite, montmorillonite: gr. Δ, about parallel extinction | pneumatol. to hydroth. with quartz, calcite, barite, sulfides in veins, fissures, cavities of pegmatites, porphyries, marbles etc.; not in young sediments | **222** |
| unknown | colorless | adsorbs methylene-blue; n changes with immersion-medium; Δ barely visible; meta-halloysite: irreversibly dehydrated halloysite | X-ray; kaolinite, dickite, nacrite: gr. n, gr. Δ; montmorillonite: much gr. Δ | diagen.-concretionary with kaolinite or diaspore or gibbsite or montmorillonite in (paleo-) soils, kaolin deposits, bauxites | **223** |

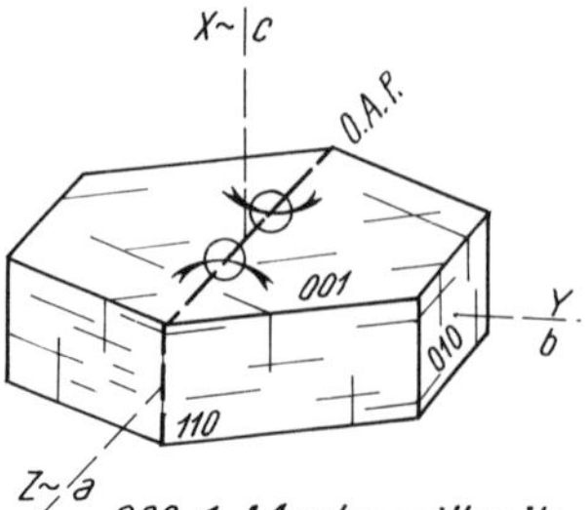

208-1. Montmorillonite

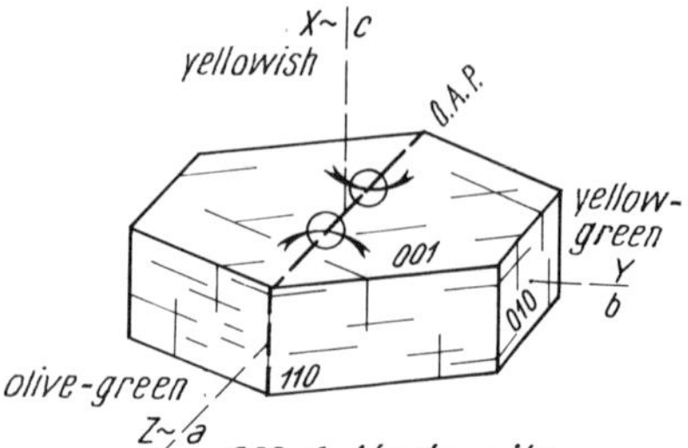

209-1. Nontronite

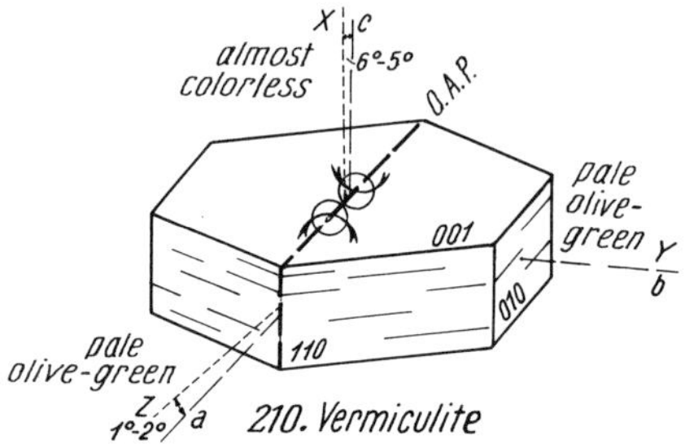

210. Vermiculite

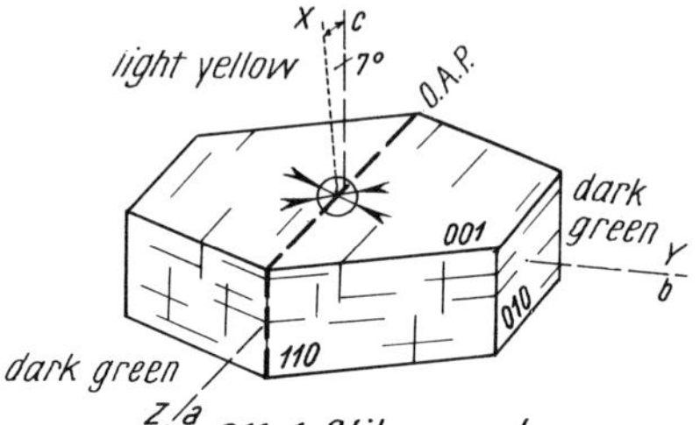

211-1. Stilpnomelane

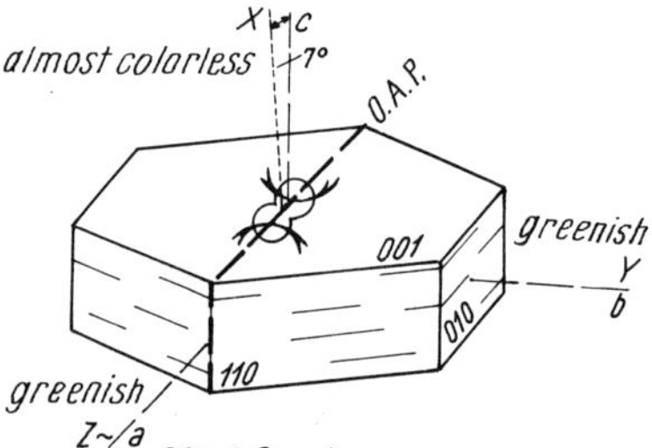

212-1. Pennine ⊖

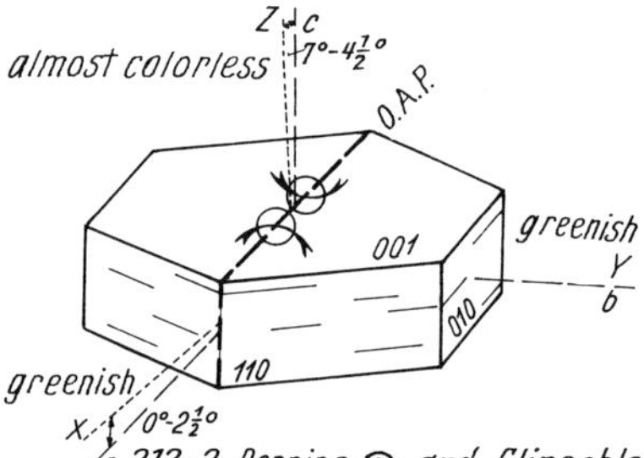

212-2. Pennine ⊕ and Clinochlore

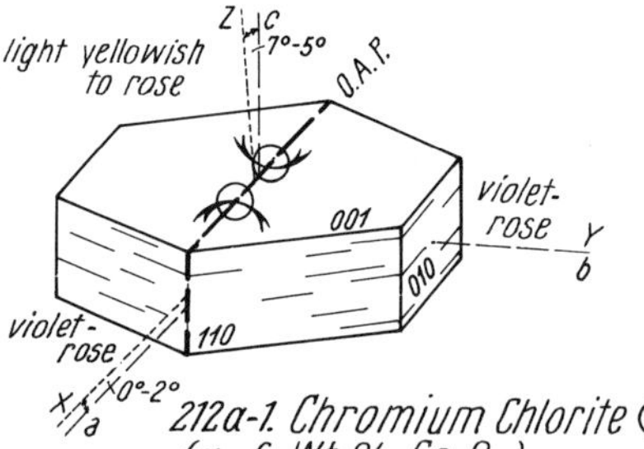

212a-1. Chromium Chlorite ⊕
(<~6 Wt % $Cr_2O_3$)

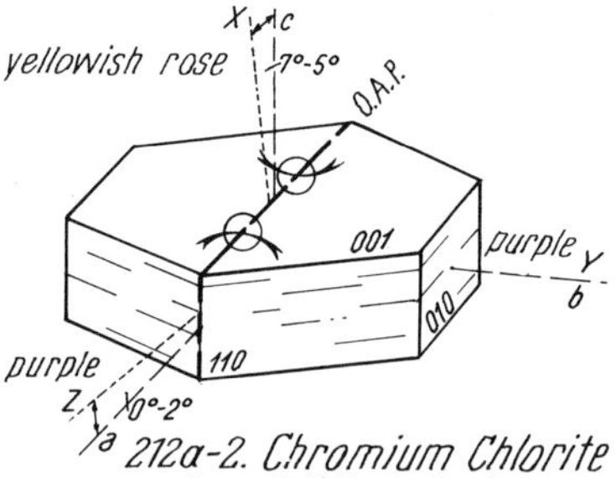

212a-2. Chromium Chlorite ⊖
(>~6 Wt % $Cr_2O_3$)

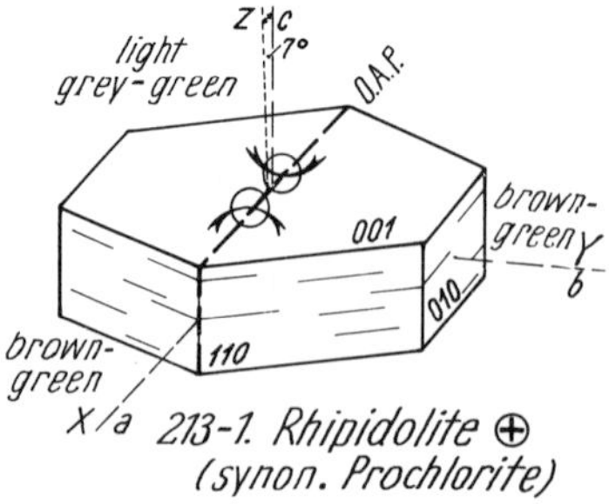

213-1. Rhipidolite ⊕
(synon. Prochlorite)

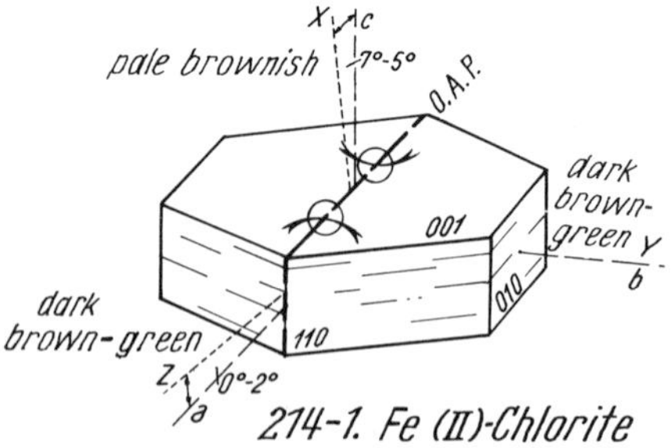

214-1. Fe (II)-Chlorite

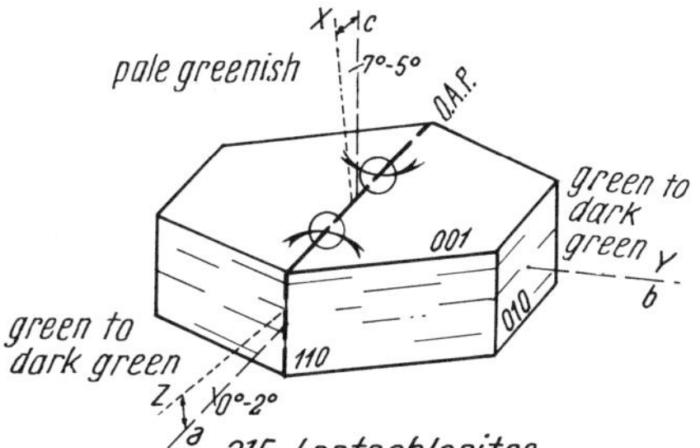


215. Leptochlorites
(Delessite-Thuringite)

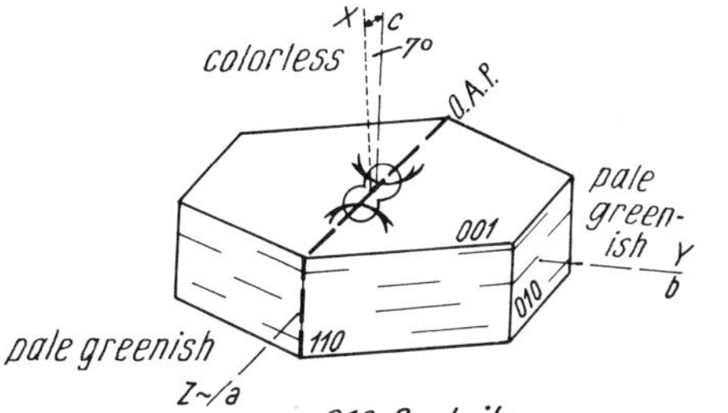


216. Sudoite
(opt. orient. still uncertain)

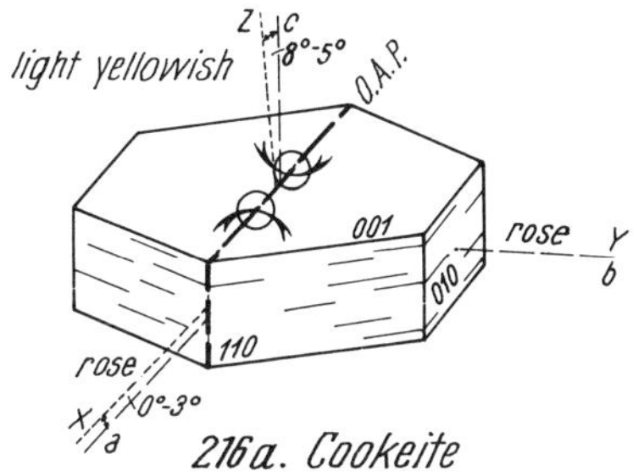


216a. Cookeite

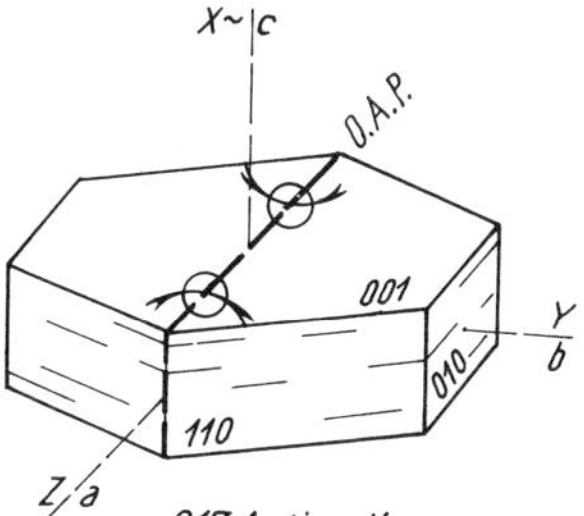


217. Antigorite

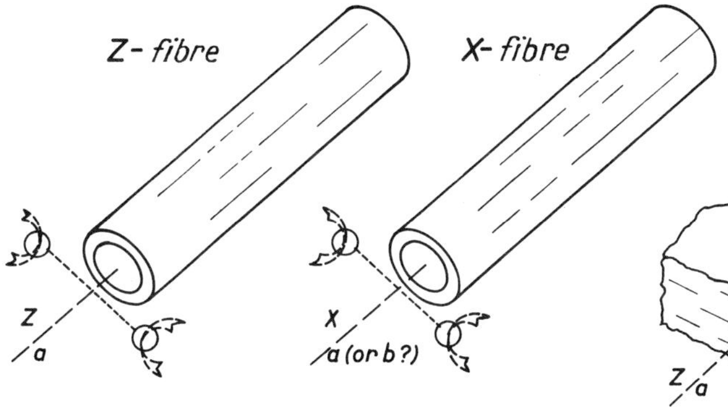


218. Chrysotile

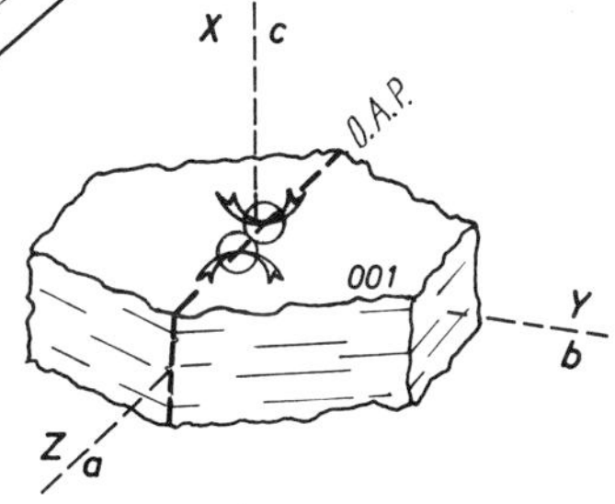


218 a. Lizardite

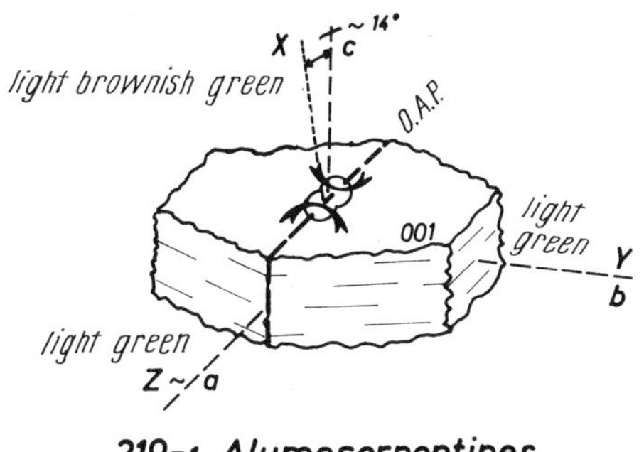


219-1. Alumoserpentines

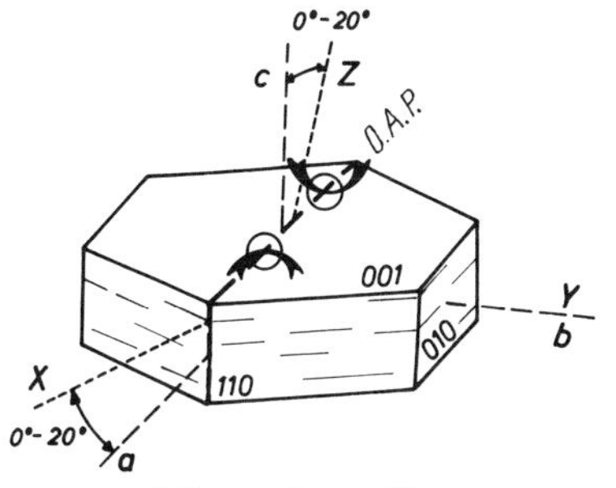


219a. Amesite

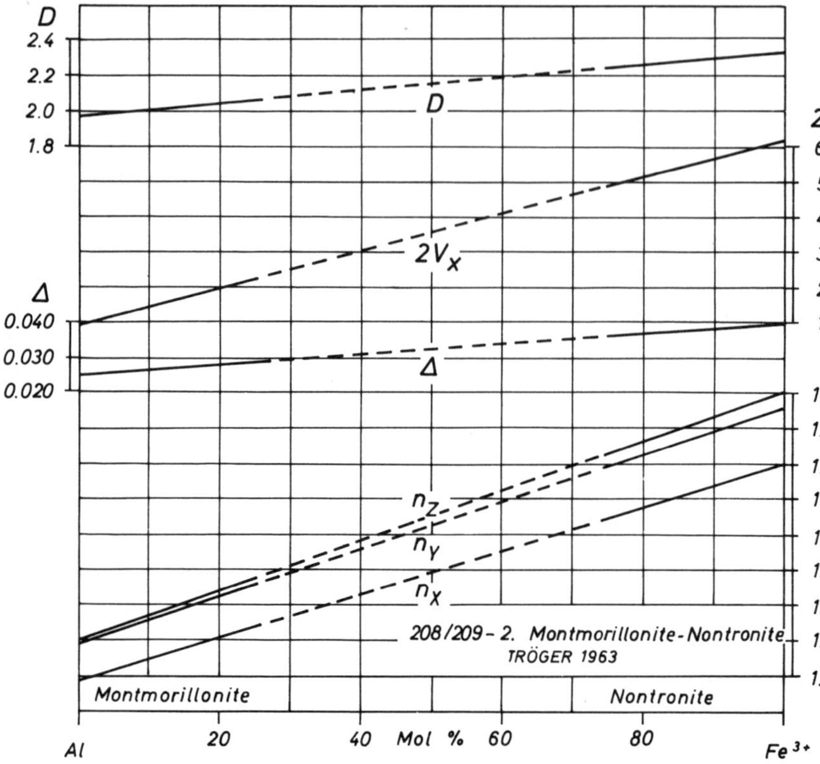


208/209 - 2. Montmorillonite-Nontronite
TRÖGER 1963

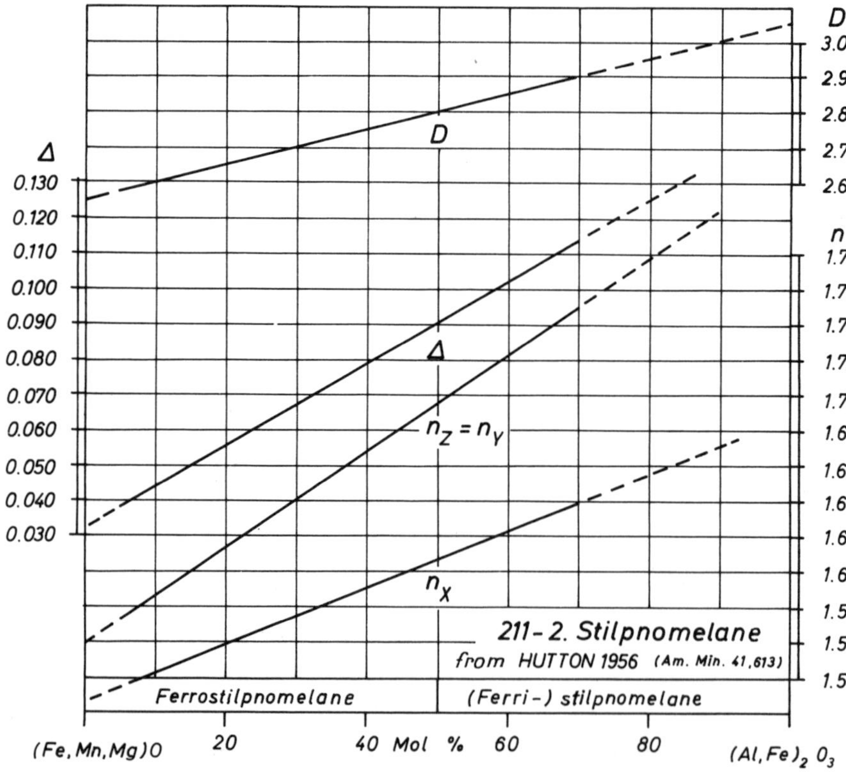


211 - 2. Stilpnomelane
from HUTTON 1956 (Am. Min. 41,613)

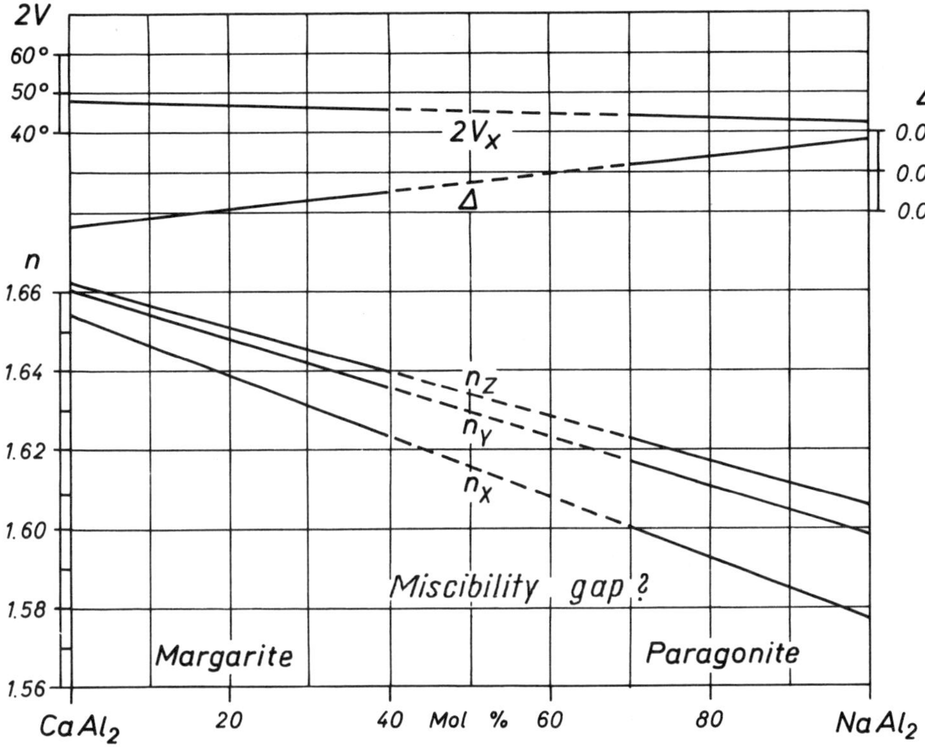


## 200-2. Margarite
TRÖGER 1963

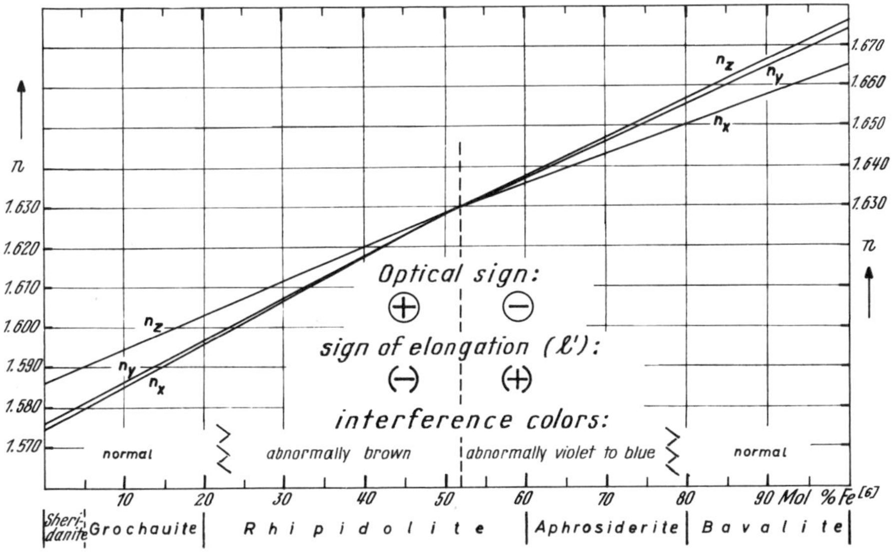


## 212-214-3. Chlorites containing 60-75 mol % Al[4]
TROCHIM 1966 from ALBEE 1962 (Am. Min. 47, 865)

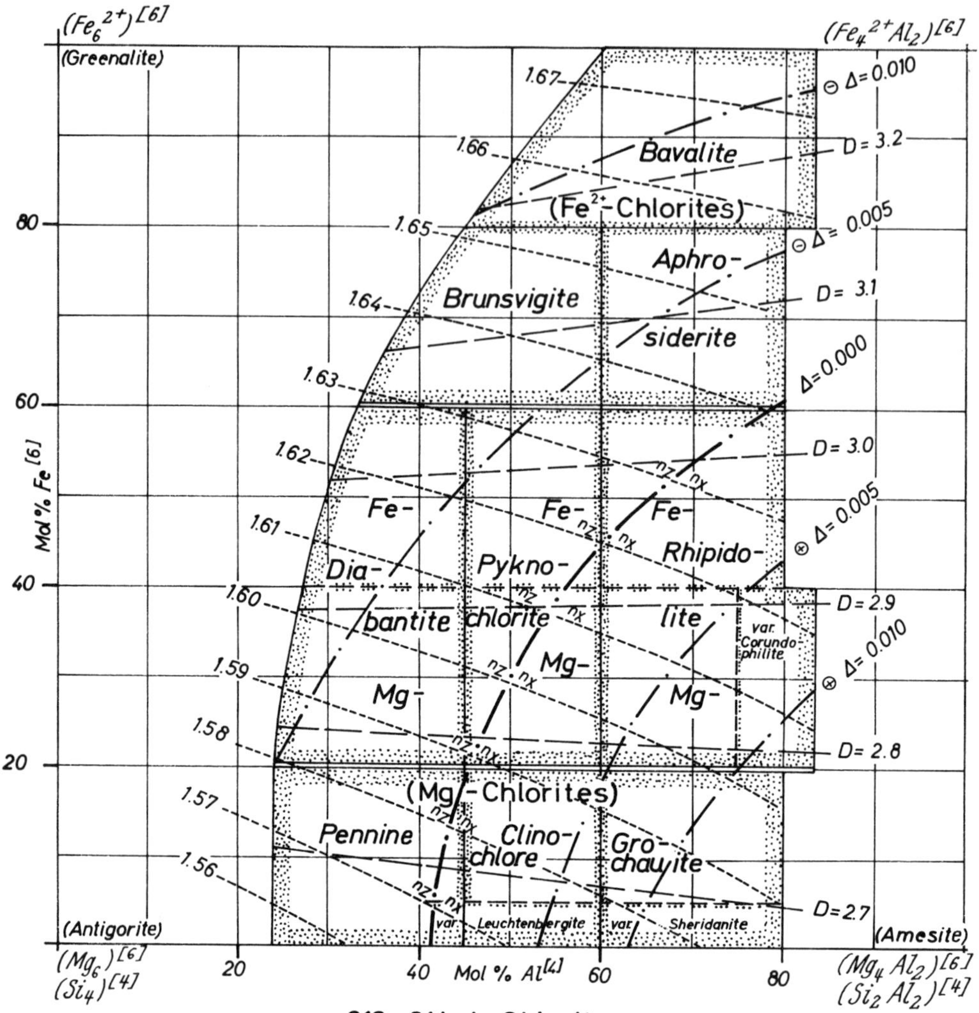


**212-214-4. Chlorite group**

TRÖGER 1963, TROCHIM 1966

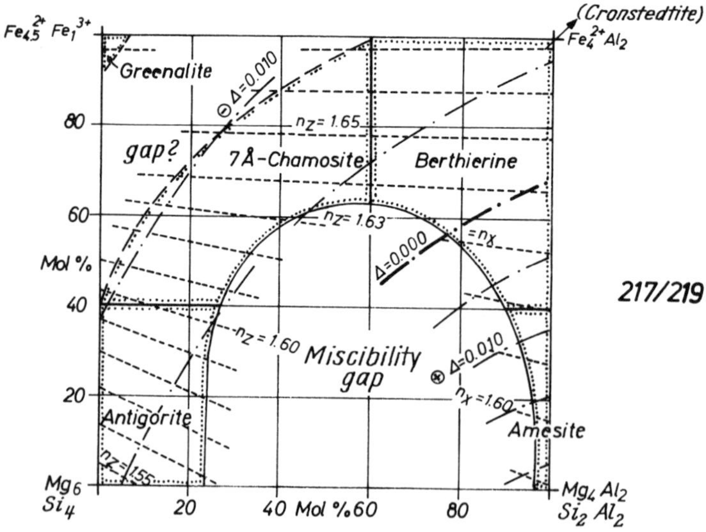


**217/219-2. Antigorite/Alumoserpenti**

TROCHIM 1966

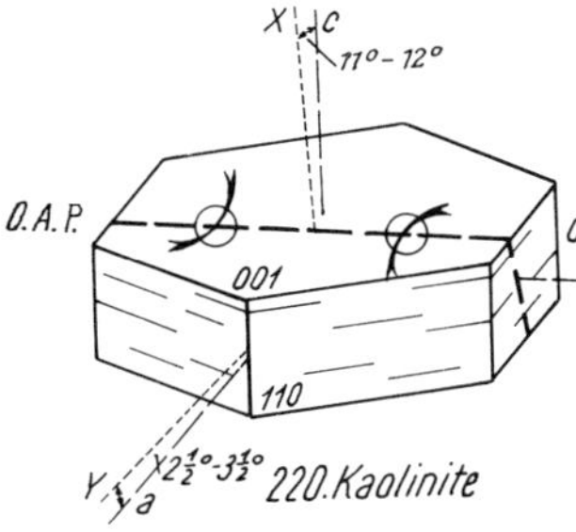

220. Kaolinite

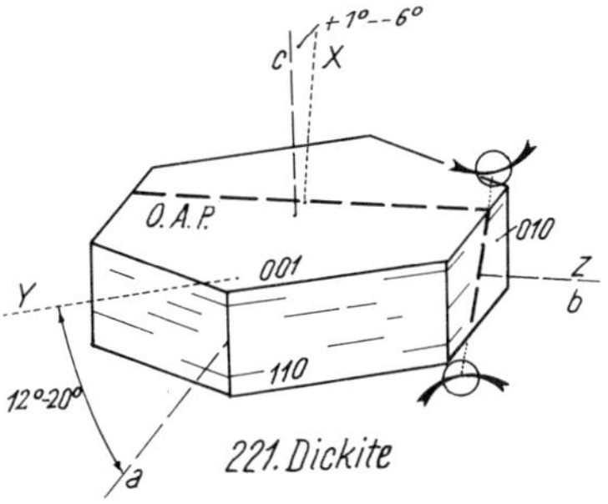

221. Dickite

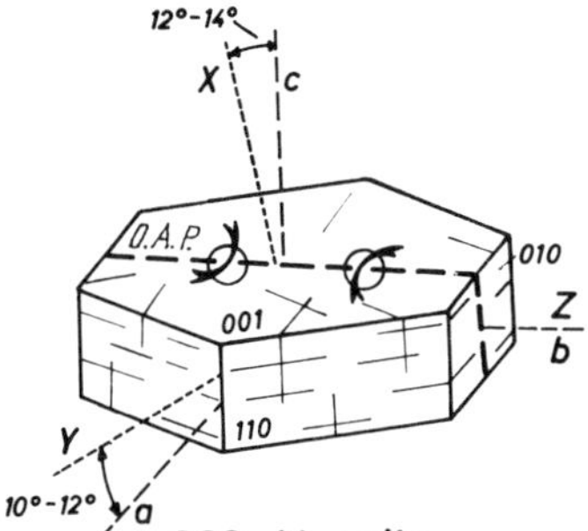

222. Nacrite

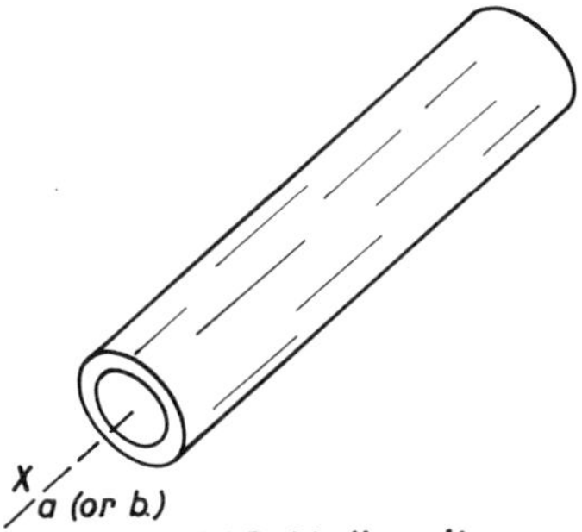

223. Halloysite

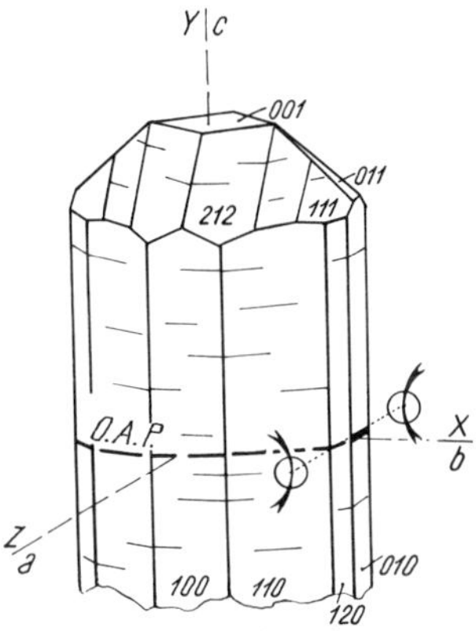

224. Danburite
for Na-light: 2V~90°
(KOENIGSBERGER)

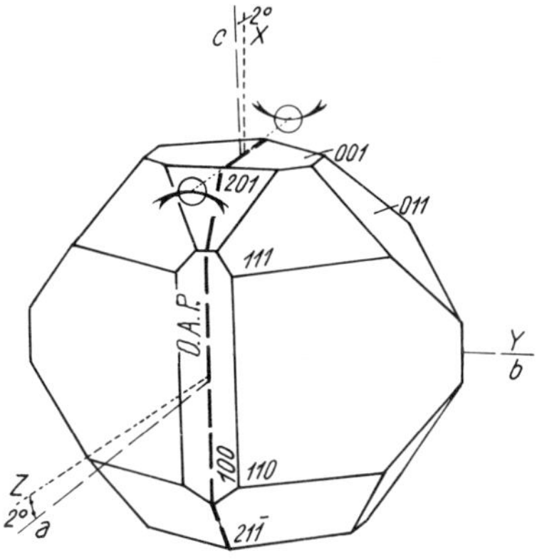

225-1. Datolite

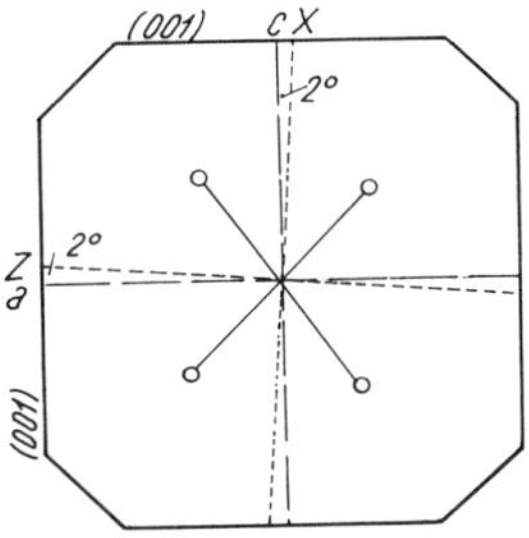

225-2. Datolite
section ‖(010)

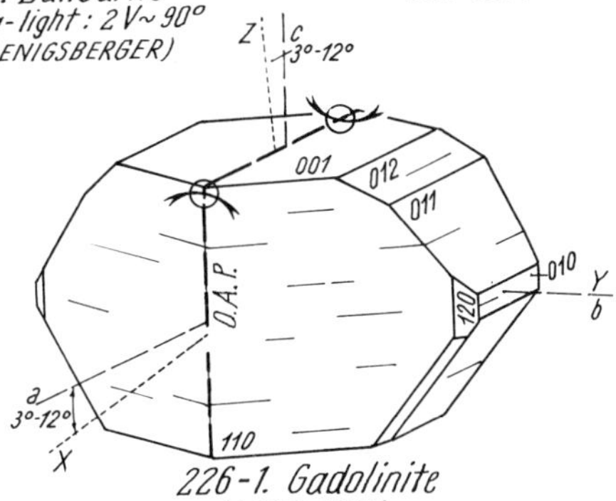

226-1. Gadolinite
(LANG 1864)

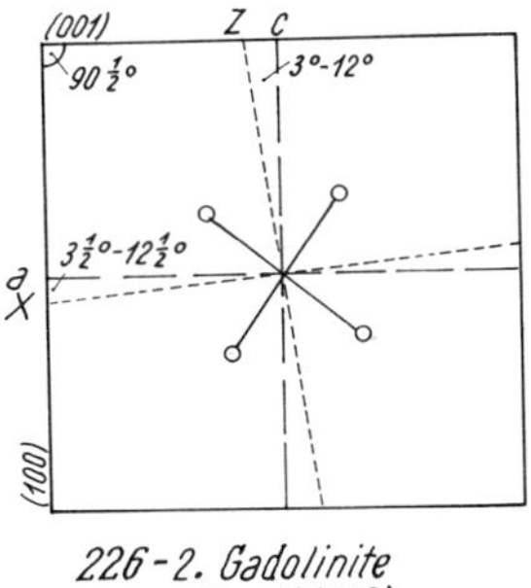

226-2. Gadolinite
section ‖(010)

| No. | Name<br>Symmetry<br>Chem. Composition | Common Crystal Forms, Twinning<br>Crystal Habit<br>Aggregates | Cleavage<br>Translation Gliding<br>Deform. Twinning<br>Hardness, Density | Optic Orientation | Refractive Indices<br>Birefringence |
|---|---|---|---|---|---|
| 224 | Danburite orh<br>$Ca[B_2Si_2O_8]$ | {100},{010},{001},{hk0},{hkl}; cryst. setting: GOLDSCHMIDT;<br>columnar ∥ c, granular | # (001) poor;<br>H = 7 D = 2.99 | X∧a, bi;<br>Y ∥ c, O.A.P. (001)<br>Z ∥ b (for λ=$Na_D$)<br>I = (+), (−)<br>fig. p. 119 | λ = 6710 5830 4050 Å<br>$n_x$ = 1.628 – 1.630 – 1.646<br>$n_y$ = 1.630 – 1.633 – 1.648<br>$n_z$ = 1.633 – 1.636 – 1.652<br>Δ = ⊖0.005–⊖0.006–⊕0.006 |
| 225 | Datolite mcl<br>$CaB[OH/SiO_4]$ | {110},{001},{111},{011};<br>β = 90½°; cryst. setting: DAUBER;<br>short columnar ∥ c or ∥ a, tabular ∥ (100),<br>granular, fibrous-stellate | H = 5 – 5½<br>D = 2.99 | X∧c = ~ −2°, bi;<br>Y ∥ b, O.A.P. (010)<br>Z∧a = ~ +2°<br>pseudo-orthorhomb.<br>fig. p. 119 | $n_x$ = 1.622–1.626–1.631<br>$n_y$ = 1.648–1.652–1.658<br>$n_z$ = 1.665–1.675–1.700<br>⊖Δ = 0.043–0.049–0.069 |
| 226 | Gadolinite mcl<br>$Y_2FeB_2[O/SiO_4]_2$ | {110},{001},{0kl};<br>β = 90½°;<br>tabular, columnar ∥ c,<br>massive | # (001) (absent after isotropization)<br>H = 6½ – 7<br>D = 4.6–4.1 (metamict) | X∧a = −3½ to 12½°<br>Y ∥ b, O.A.P. (010)<br>Z∧c = +3° to +12°, bi;<br>fig. p. 119 | $n_x$ = 1.801, $n_y$ = 1.812, $n_z$ = 1.824 } metamict down to n ~ 1.71<br>⊕Δ = 0.023 Δ ~ 0 |
| 227 | Petalite mcl<br>$Li[AlSi_4O_{10}]$ | {001},{010},{201},{110};<br>β = 112½°;<br>◇ (001) lam. (rare);<br>columnar ∥ a, tabular ∥ (010), spathic | # (001) perf.<br># (201) good } > 38½°<br>H = 6½ D = 2.41 | X∧a = −2° to −9° (r < v)<br>Y∧c = +24° to +30°, O.A.P. ⊥ (010)<br>Z ∥ c, bi;<br>fig. p. 122 | $n_x$ = 1.504 – 1.507<br>$n_y$ = 1.510 – 1.512<br>$n_z$ = 1.516 – 1.518<br>⊕Δ = 0.012 – 0.011 |
| 228 | (K,Na)-Sanidine mcl<br>$(K,Na)[AlSi_3O_8]$<br>K:Na ≳ 1:2<br>low: Al,Si-disordered<br>high: Al,Si-ordered | {010},{001},{201},{110}; β = 116°;<br>◇ [001], rarely:<br>◇ (001),(021);<br>tabular ∥ (010)<br>columnar ∥ a<br>fig. 228–1, 229–1 p. 122 | # (001) good<br># (010) distinct } > 90°<br># {110} poor, 61°;<br>H = 6<br>D = 2.56 Or, 2.60 $Ab_2Or$ ⇅<br>diagr. p. 123 | X∧a ~ 0° to +9°<br>Y ∥ b, O.A.P. (010) rare<br>Z ∥ b, O.A.P. ⊥ (010) freq.<br>fig. p. 122<br>Optics depending on (K, Na) and degree of Al, Si-order | Or Ab-rich<br>$n_x$ = 1.518 – 1.525<br>$n_y$ = 1.522 – 1.530<br>$n_z$ = 1.523 – 1.532<br>⊖Δ = 0.005 – 0.007 |
| 229 | Orthoclase<br>paramorph of microcline-perthite after sanidine<br>Or > Ab<br>Or: $K[AlSi_3O_8]$<br>Ab: $Na[AlSi_3O_8]$ | mcl morphology, see sanidine; habit frequently columnar ∥ a;<br>microcline-component is crypto twinned;<br>(see Vol. 2, p. 681 ff.) | # (001) perf.<br># (010) good } 90°<br># {110} poor, 61°;<br>(100), ($\bar{7}$01) parting;<br>H = 6<br>D = { 2.56 Or, 2.59 OrAb<br>diagr. p. 123 | "average" optics usually mcl, may also be trcl;<br>homogeneous to undulatory;<br>X∧a = +3° to +12°<br>r < v, bi;<br>Z ∥ b, O.A.P. ⊥ (010)<br>fig. p. 124 | Or ~OrAb<br>$n_x$ = 1.518 – 1.523<br>$n_y$ = 1.522 – 1.528<br>$n_z$ = 1.523 – 1.530<br>⊖Δ = 0.005 – 0.007 |
| 230 | Microcline trcl<br>$(\underline{K},Na)[AlSi_3O_8]$ | rarely idiomorphic;<br>β = 116°; usually as ◇ domains in alkali feldspar paramorphs;<br>◇ (010), [010]: lam.<br>"cross hatching",<br>rhombic section: ~ +97° | # (001) perf.<br># (010) good } ~ 90°<br># (110), (1$\bar{1}$0) poor ≃ 61°;<br>H = 6 D = 2.56 | X'∧a in (010) = +5° to +6°<br>X'∧a in (001) = (~0° to −10°) to max. −18°, O.A.P. ~⊥ (010)<br>fig. and stereogr. p. 124 | $n_x$ = 1.518 – 1.520<br>$n_y$ = 1.522 – 1.524<br>$n_z$ = 1.524 – 1.526<br>⊖Δ = 0.006 – 0.006<br>see Nr. 228 (sanidine) |
| 231 | (K,Ca)-Analbite trcl<br>$(\underline{Na},K,Ca)$<br>$[Al(\underline{Si},Al)Si_2O_8]$<br>(formerly: "high albite") | finely lamellar domains in certain anorthoclases<br>◇ (010), [010] lam.<br>"cross hatching",<br>rhombic section: ~ −5° to ~ 0° | # (001) perf.<br># (010) good } ~ 90°<br>H ~ 6<br>D = { 2.62 (Ab-rich), 2.60 (Or-rich) | X'∧a in (001) = ~ 0° to +13° (usually 0° to +3°)<br>X'∧a in (010) = +8° to 9°, O.A.P. near ⊥ (010)<br>stereogr., fig. p. 124 | Ab-rich Or-rich<br>$n_x$ = 1.527 – 1.524<br>$n_y$ = 1.532 – 1.529<br>$n_z$ = 1.535 – 1.531<br>⊖Δ = 0.008 – 0.007 |
| 232 | Anorthoclase<br>paramorph of perth. to antiperthitic (K,Ca)-analbite or albite after monoclinic alkali feldspar (Ab,Or,An),<br>Or < Ab, Ab > An | mcl morphology:<br>{010},{001},{110},{$\bar{2}$01}; β ~ 116°;<br>◇ [001], (001), (021);<br>columnar ∥ c, rarely column. ∥ a, tabular ∥ (010), partly "rhombic", cross sections;<br>fig. p. 124 | # (001) perf.<br># (010) good } > 90°<br># {110} poor;<br>H = 6<br>D = 2.63 (Ab), 2.59 (Or,Ab) | ± homogeneous, "average" optics:<br>X'∧a in (001) ~ 0 – 3°<br>X'∧a in (010) ~ 8 – 9°<br>O.A.P. near ⊥ (010);<br>fig. p. 124<br>inhomogeneous:<br>see also analbite;<br>fig. p. 124 | Ab-rich Or-rich<br>$n_x$ = 1.527 – 1.522<br>$n_y$ = 1.533 – 1.528<br>$n_z$ = 1.535 – 1.529<br>⊖Δ = 0.008 – 0.007 |
| 233 | Plagioclase group trcl<br>(Na,Ca)<br>$[Al(Si,Al)Si_2O_8]$<br>Ab = $Na[AlSi_3O_8]$<br>An = $Ca[Al_2Si_2O_8]$ | {010},{001},{110},{$\bar{1}$01}; β = 116°;<br>◇ (010), [010], [001], etc., see Vol. 2, p. 667;<br>tabular ∥ (010),<br>columnar ∥ a,<br>rarely column. ∥ c | # (001) perf.<br># (010) good } > 86°<br># (110) poor<br># (1$\bar{1}$0) poor } > 59°<br>$K_1$ = (010), $\sigma_1$ = irrat. proj. of [010] on (010);<br>H = 6–6½ D = { 2.62 Ab, 2.76 An ⇅ | triclinic optics depending on An-content and Al,Si ordering;<br>see diagrams and figures p. 125 – 137 | low Ab An<br>$n_x$ = 1.529–1.554–1.575<br>$n_y$ = 1.533–1.558–1.583<br>$n_z$ = 1.539–1.561–1.588<br>\|Δ\| = 0.010–0.007–0.013<br>(low: Al,Si-ordered)<br>(high: Al,Si-disordered) |
| 234 | Hyalophane group mcl<br>(K,Ba)<br>$[Al(Si,Al)Si_2O_8]$<br>Cn ≳ 5 mol %<br>Cn = $Ba[Al_2Si_2O_8]$<br>(Celsiane) | {110},{10$\bar{1}$},{001},{010}; β ≈ 115½°;<br>◇ [001], (021), (001);<br>columnar ∥ c,<br>morphology similar to adularia, fig. p. 124 | # (001) perf.<br># (010) good } 90°<br># {110} poor;<br>H = 6 – 6½<br>D = { 2.6 (Or), 3.4 (Cn) ⇅ diagr. p. 138 | Or > Cn<br>X∧a = ±0 to −40°, bi;<br>Y∧c = +25° to +65°<br>Z ∥ b, O.A.P. ⊥ (010)<br>Cn-rich:<br>X∧c ~ −4°<br>Y ∥ b, O.A.P. (010)<br>Z∧a ~ +30°<br>fig., diagr. p. 124, 138 | Or Cn<br>$n_x$ = 1.518 – 1.585<br>$n_y$ = 1.521 – 1.589<br>$n_z$ = 1.524 – 1.595<br>⊖Δ = 0.006 – 0.010 |

| Angle and Dispersion of Optic Axes | Color in Thin Section Pleochroism | Chemical Reactions Special Remarks | Distinguishing Features of Similar Minerals | Occurrences | |
|---|---|---|---|---|---|
| $2V_X = \begin{cases} 88^\circ\ (\lambda=6710Å) \\ 88\tfrac{1}{2}^\circ\ (\lambda=5830Å) \\ 93^\circ\ (\lambda=4050Å) \end{cases}$ | colorless | gelatinizes with HCl after heating to red heat, otherwise insoluble in HCl; Boron-reaction | andalusite: gr. Δ, always l = (–); datolite: gr. Δ, smaller $2V_X$; topaz: perf. # | with B-, F-, Be-minerals pneumatolytic to hydroth; in contact-metasomatic marbles; in granite-pegmatites | 224 |
| $2V_X = 72^\circ - 74^\circ$<br>r > v weak | colorless | gelatinizes with inorganic acids, Boron-reaction | muscovite: # perf.; prehnite: $2V_Z$; topaz, danbunite: sm. Δ | hydroth. in contact-met. siliceous marbles near basic magmatites; partly also in granites; in cavities and fissures of basic magmatites with calcite, zeolites, chalcedony etc. | 225 |
| $2V_Z \sim 80^\circ$<br>r ≪ v | X = olive-green to yellowish brown, Y = grass-green to brownish yellow<br>Z = Y; pleochroism often hardly visible | gelatinizes with acids only if metamict, zonal structure | other metamict minerals: gr. D; allanite: stronger pleochroism; spinel: isotropic, {111} and diff. paragenesis; hoegbomite: uniaxial | mainly in granite-pegmatites | 226 |
| $2V_Z = 82^\circ - 84^\circ$<br>r > v weak | colorless | soluble only in HF, Li-color in flame | albite, orthoclase: gr. n, sm. Δ, diff. #; thomsonite: diff. paragenesis; spodumene: gr. n, diff. # | mainly in Li-rich pegmatites, with spodumene, lepidolite, amblygonite | 227 |
| $2V_X$ disp. O.A.P.<br>60–65° r < v ∥ (010)<br>~0° { r: ⊥ (010); v: ∥ (010) }<br>60–65° r > v ⊥ (010)<br>diagr. p. 123<br>fig. p. 122 | colorless, frequently clear | soluble only in HF; staining react. with Na-hexanitro-cobaltate III not absolutely reliable, no polysynthetic ◇; kaolinization, "hour-glass-structure" | quartz: clearer, uniaxial ⊕; orthoclase: $2V_X > 65^\circ$, gr. n; anorthoclase: gr. n, more or less inhomogeneous; nepheline: uniaxial ⊖, sm. Δ, gr. n | metastable at low temperatures; mainly in young, leucocratic volcanics and tuffs | 228 |
| $2V_X = 60^\circ – 65^\circ$ to $82^\circ$<br>r > v<br>diagr. p. 123<br>may be opt. mcl alkali feldspar with $2V_X = 80^\circ – 110^\circ$ = "iso-orthoclase",<br>see Vol. 2, p. 693 | colorless, commonly turbid, may be pale brown | soluble only in HF; staining react. with Na-hexanitrocobaltate III; often kaolinized or sericitized, often undulatory, may show beginning microcline cross hatching; micro- to cryptoperthitic; adularia: hydrothermal trachtvariety with somet. diff. optics (see Vol. 2, p. 695) | quartz: clearer, gr. n, uniaxial ⊕; nepheline: gr. n, sm. Δ; plagioclase: polysynthetic ◇ lam., usually gr. n; sanidine: sm. $2V_X$, X-ray: Vol. 2, p. 744 | magmatic in leucocratic rocks; pegmatitic with quartz; rarely hydroth.: adularia; high grade metamorphic in gneisses | 229 |
| $2V_X = 60^\circ – 65^\circ$ to $84^\circ$, (up to 115°: "isomicrocline")<br>r > v<br>diagr. p. 123 | colorless, may be turbid | soluble only in HF, often kaolinized or sericitized, however more resistant against weathering than sanidine and orthoclase; usually "cross hatching", macro- to cryptoperthitic; staining react.: see orthoclase | orthoclase: submicr. cross hatching, mcl optics; (K,Ca)-analbite: often finer cross hatching, sm. 2V; sanidine: mcl optics, no ◇ lam., sm. $2V_X$; X-ray | magmatic in leucocratic plutonites; pegmatitic with quartz, medium to high grade metam. in gneisses; authigenic in calcareous-dolomitic sedimentary rocks | 230 |
| $2V_X \sim 20^\circ – 50^\circ$<br>r > v<br>diagr. p. 123 | colorless; may be turbid | soluble in HF; usually very fine cross hatching, frequently micro-cryptoantiperthitic | microcline: usually coarser cross hatching, gr. $2V_X$, diff. rhombic section; X-ray: Vol. 2, p. 744 | component of anorthoclases in volcanics | 231 |
| $2V_X \sim 20^\circ – 110^\circ$<br>diagr. p. 123;<br>see also Nr. 229 "iso-orthoclase" | colorless; commonly turbid | soluble in HF, optically homogeneous to inhomogeneous with microscop. to submicroscop. domains; mixture of several feldspar phases; see Vol. 2, p. 680 | orthoclase: sm. n, gr. $2V_X$; X-ray: Vol. 2, p. 698 | magmatic mainly in Na-rich volcanics, rarely in plutonites | 232 |
| $2V_X = 50^\circ – 105^\circ$<br>dispersion of optic axes variable<br>diagr. p. 126 | colorless; may be turbid | soluble in HF; if < $An_{50}$ then resistant against HCl; $An_{100}$ decomp. by HCl; polysynth. ◇, zonal structure, sericitization, kaolinization, saussuritization | orthoclase: no ◇ lam., sm. n, sm. D; microcline: sm. n, sm. D, cross hatching; anorthoclase: sm. n, sm. D; analbite: fine cross hatching, sm. $2V_X$ | magmatic and metamorphic, more frequently Ab-rich in leucocratic rocks, more An-rich in melanocratic rocks; albite also pegmatitic, hydroth. and authigenic in calcareous sediments | 233 |
| $2V_X = 65^\circ – 95^\circ$<br>r > v<br>diagr. p. 138 | colorless | soluble in HF, Cn-rich members decomposed by HCl-, Ba-color in flame; no polysynthetic ◇ lam.; zonal structure | K-feldspar: sm. n, sm. D; plagioclase: polysynthetic ◇; tricl. optics; if Cn-content small: distinction from K-feldspar with chem. anal. | Ba-feldspars with Cn > 5 mol % are rare; in some volcanics, pegmatites; mainly contact-metam. with Mn-minerals; also in vesicles of dolomite | 234 |

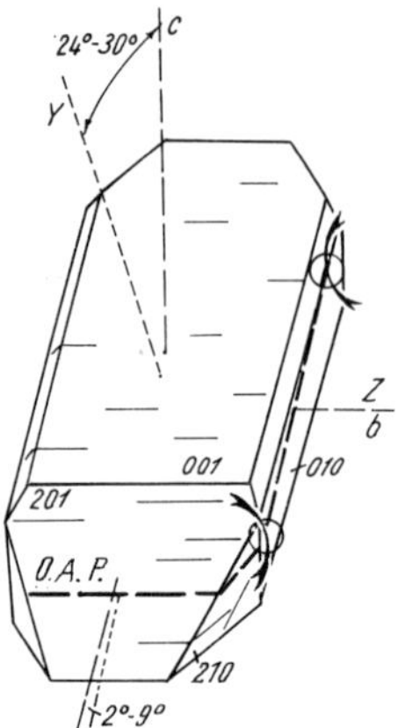

227-1. Petalite

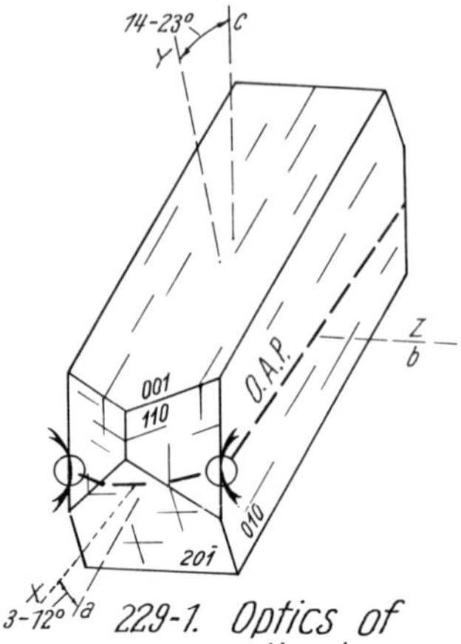

229-1. Optics of orthoclase and low sanidine

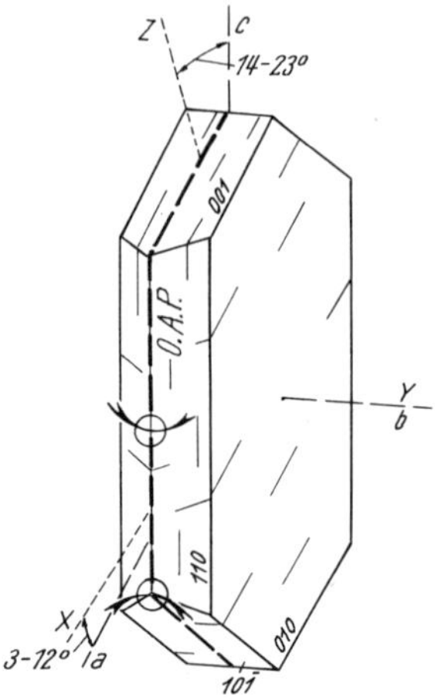

228-1. Optics of high sanidine

228-229. Morphology of alkali feldspars

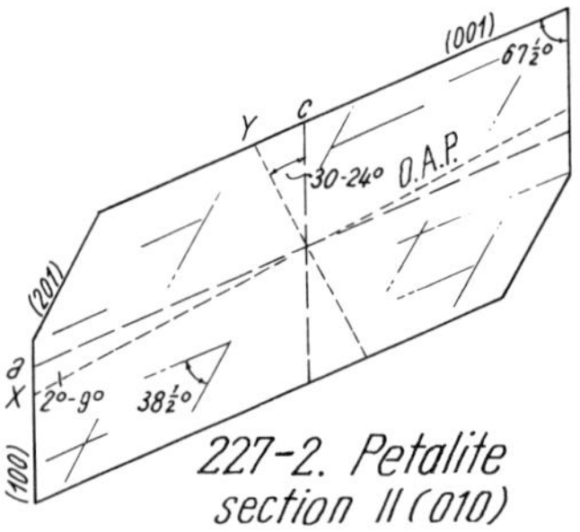

227-2. Petalite section ‖ (010)

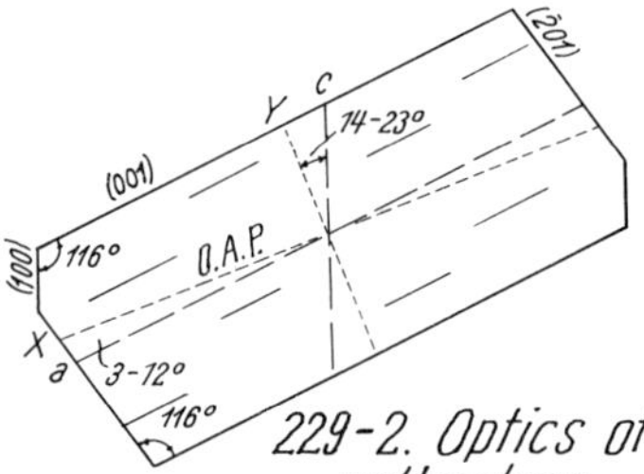

229-2. Optics of orthoclase section ‖ (010)

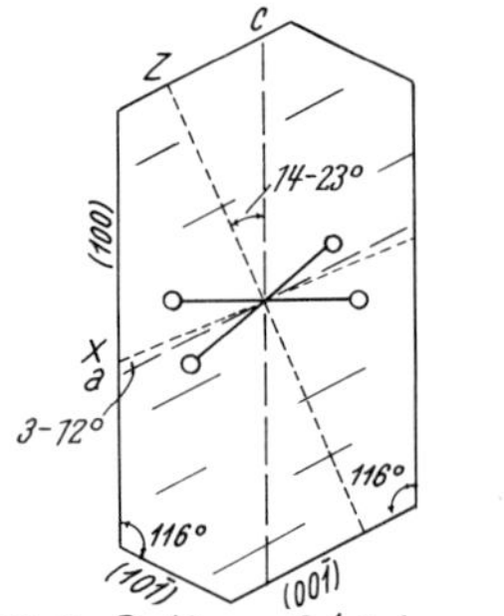

228-2. Optics of high sanidine section ‖ (010)

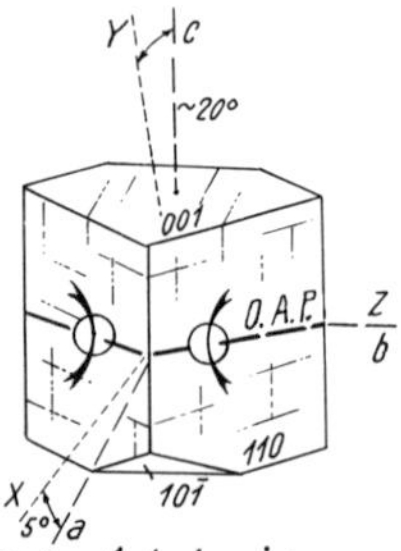

229-3. Adularia (mcl. "average" optics)

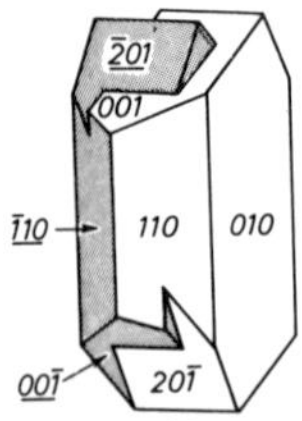

229-4. ◊ [001] Carlsbad (penetration twin)

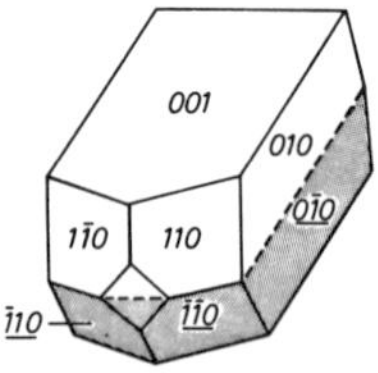

229-5. ◊ (001) Manebach

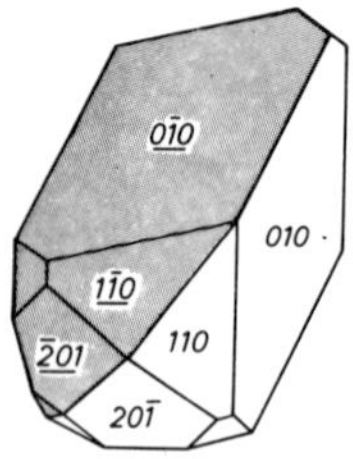

229-6. ◊ (021) Baveno

228-229. Twins of monoclinic alkali feldspars

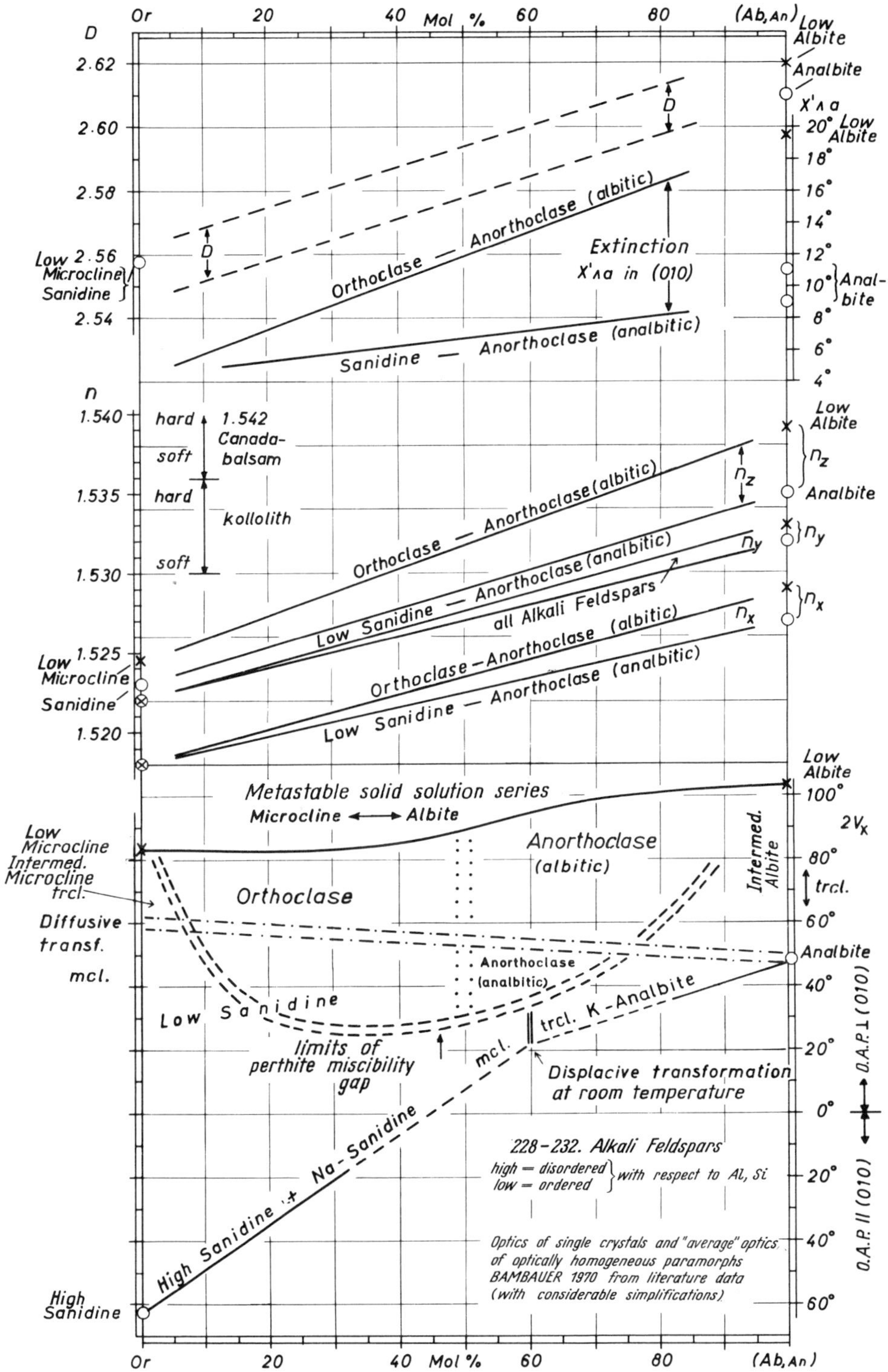

Or
20
40 Mol % 60
80
(Ab,An)
D
2.62
2.60
2.58
Low 2.56
Microcline
Sanidine
2.54
Low Albite
Analbite
X'∧a
20° Low Albite
18°
16°
14°
12°
10° Anal- bite
8°
6°
4°
D
D
Orthoclase — Anorthoclase (albitic)
Extinction
X'∧a in (010)
Sanidine — Anorthoclase (analbitic)
n
1.540
hard
1.542
Canada-balsam
soft
1.535
hard
kollolith
soft
1.530
Low
Microcline
1.525
Sanidine
1.520
Low Albite
$n_z$
Analbite
$n_y$
$n_x$
Orthoclase — Anorthoclase (albitic)
Low Sanidine — Anorthoclase (analbitic)
all Alkali Feldspars
Orthoclase-Anorthoclase (albitic)
Low Sanidine — Anorthoclase (analbitic)
Low Albite
Metastable solid solution series
Microcline ⟷ Albite
100°
$2V_x$
Low Microcline
Intermed. Microcline
trcl.
Anorthoclase
(albitic)
Intermed. Albite
80°
trcl.
Orthoclase
60°
Diffusive transf.
Analbite
mcl.
Anorthoclase
(analbitic)
40°
Low Sanidine
trcl. K-Analbite
20°
limits of perthite miscibility gap
mcl.
Displacive transformation at room temperature
0°
O.A.P.⊥(010)
228–232. Alkali Feldspars
high = disordered
low = ordered
with respect to Al, Si
20°
High Sanidine + Na-Sanidine
Optics of single crystals and "average" optics of optically homogeneous paramorphs BAMBAUER 1970 from literature data (with considerable simplifications)
40°
O.A.P. II (010)
High Sanidine
60°
Or
20
40 Mol % 60
80
(Ab,An)


(96)

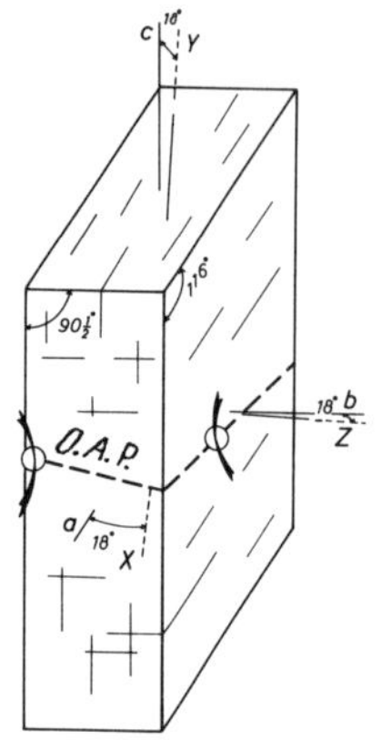


230-1. Low microcline
(LAVES 1965)

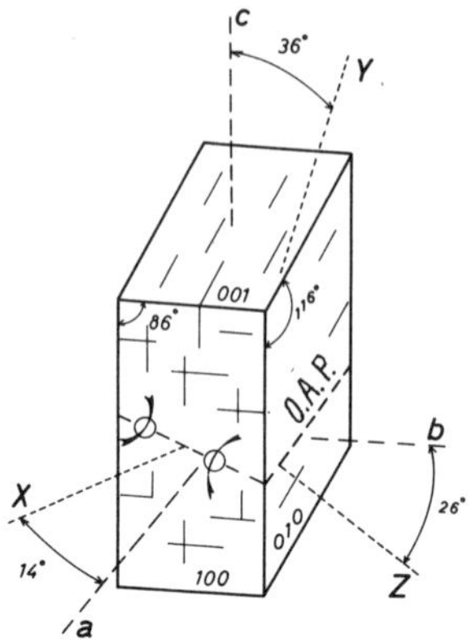


231-1. Analbite (synth.)
(SCHNEIDER 1957)

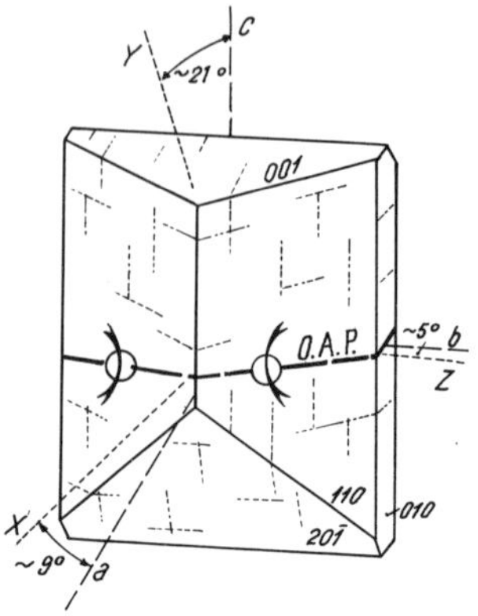


232-1. Anorthoclase
(mcl. "average" optics)

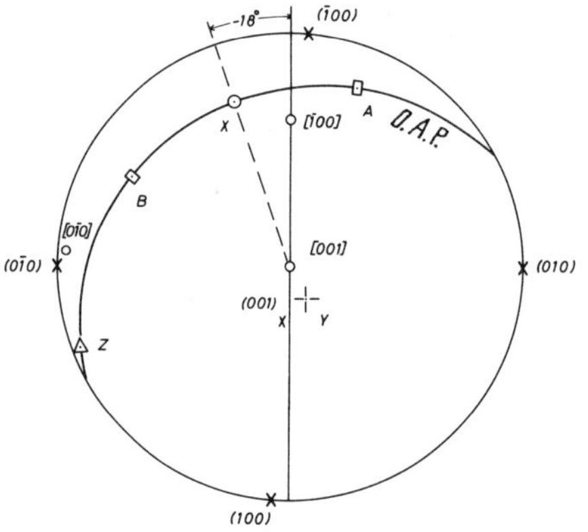


230-2. Low microcline
LAVES (1965)

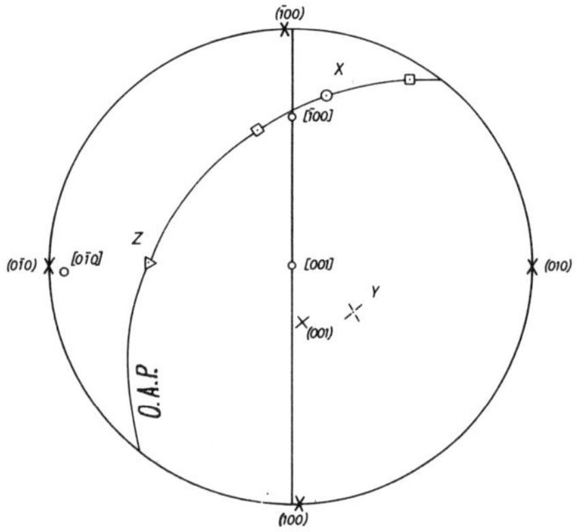


231-2. Analbite (synth.)
SCHNEIDER (1957)

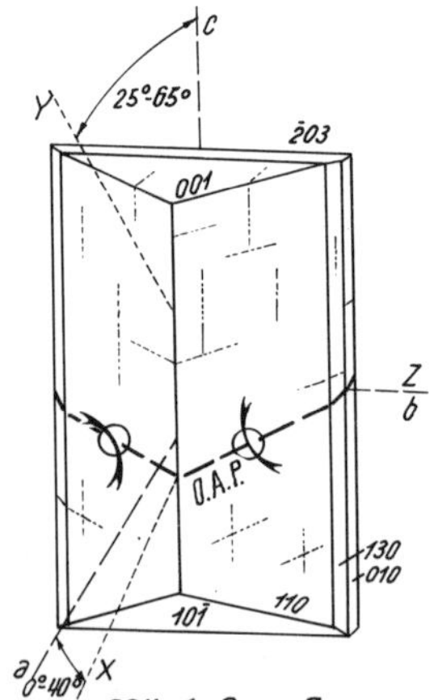


234-1. Or > Cn

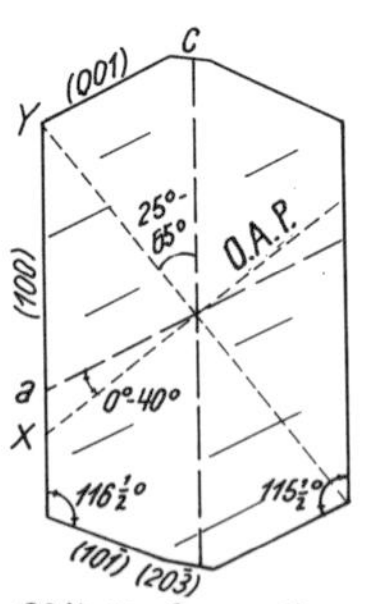


234-2. Or > Cn
section // (010)

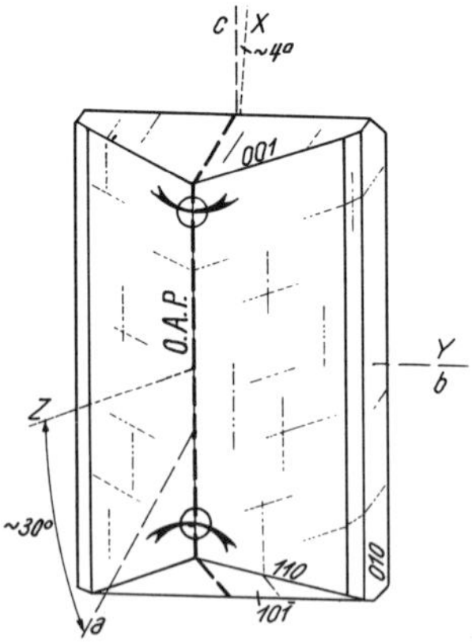


234-3. Cn-rich

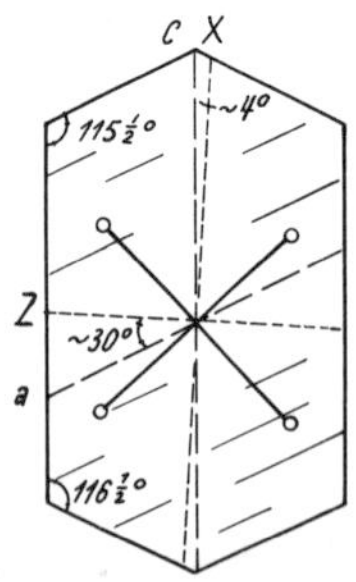


234-4. Cn-rich
section // (010)

## 234. Hyalophane group

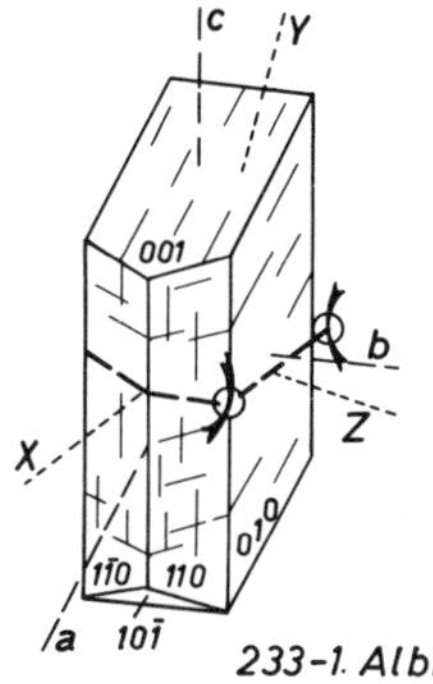


233-1. Albite

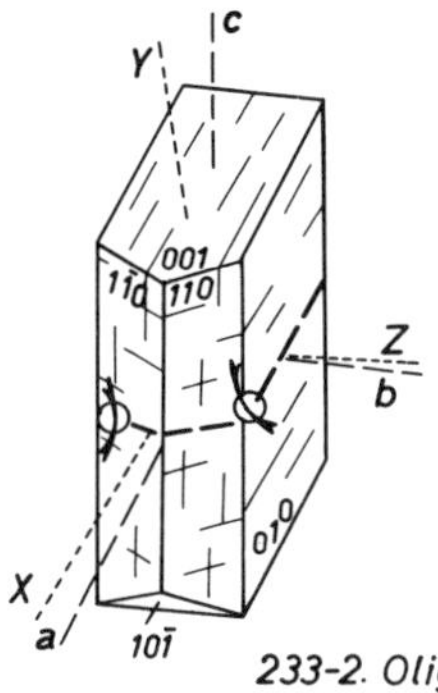


233-2. Oligoclase

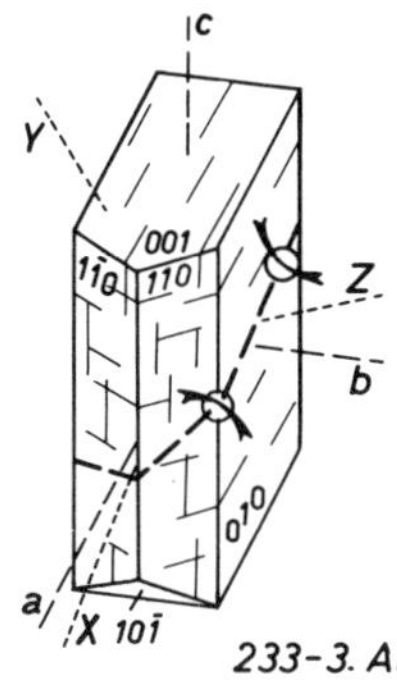


233-3. Andesine

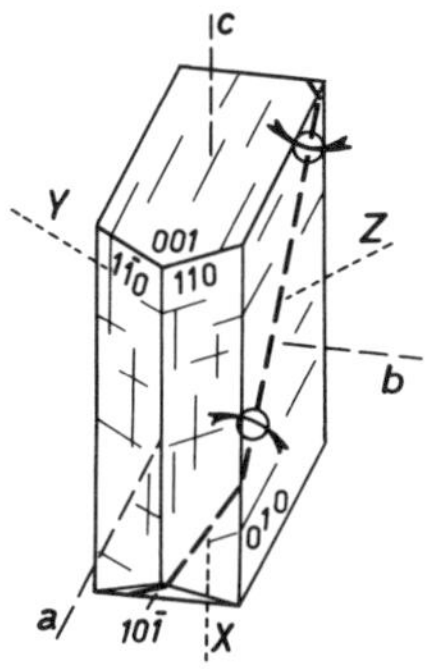


233-4. Labradorite

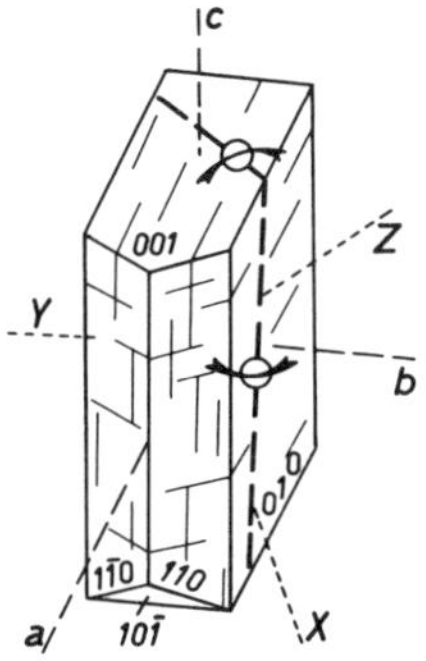


233-5. Bytownite

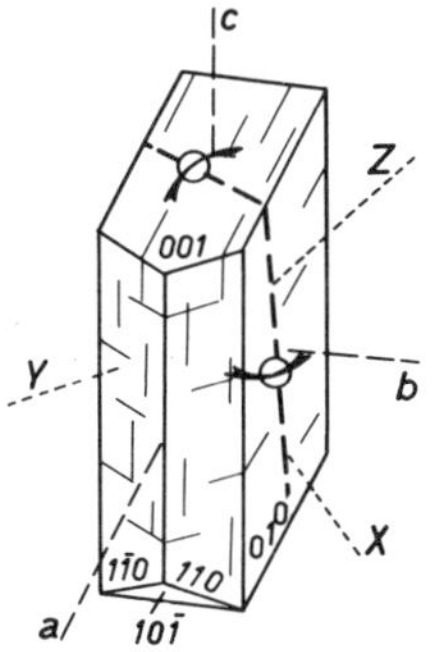


233-6. Anorthite

233. *Optics of low plagioclases*

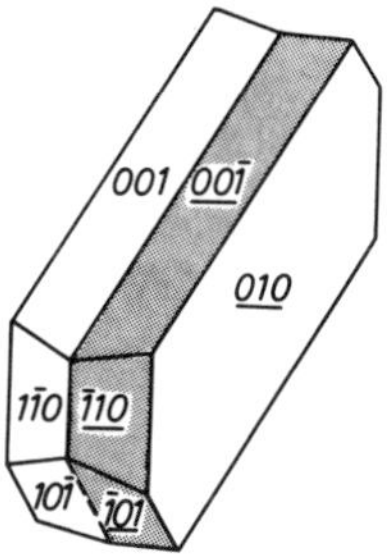


233-7. ◊ ⊥ (010) *Albite twin*

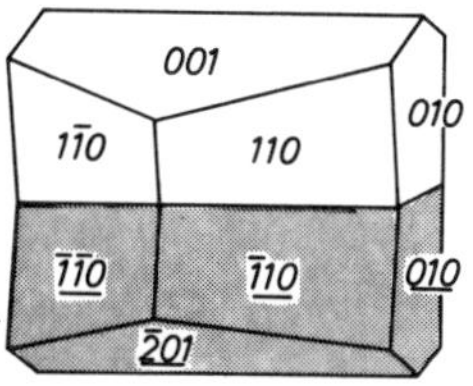


233-8. ◊ [010] *Pericline twin (simplified)*

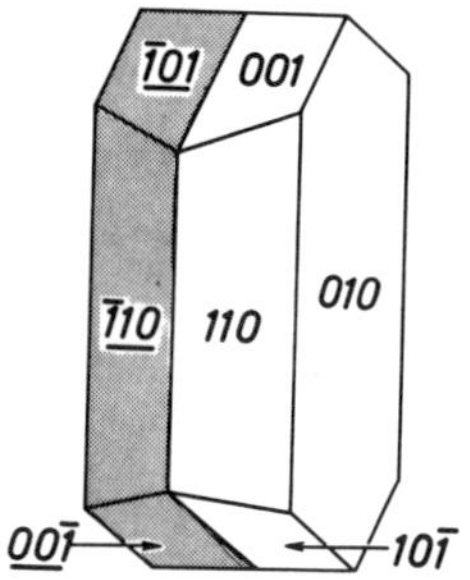


233-9. ◊ [001] *Carlsbad twin (contact-twin)*

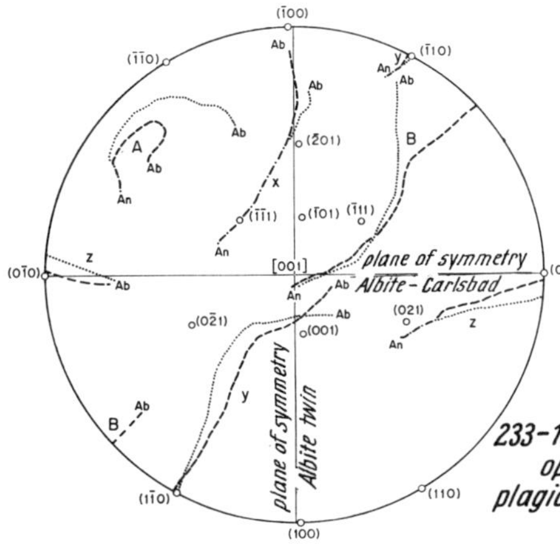


*233-10. Stereogram of the optic elements in plagioclase on a plane ⊥c*
see vol. 2, p. 734

*233-11. Sign of plagioclase extinction angles*
*in (100): rule of MICHEL-LÉVY*
*in (001) and (010): rule of SCHUSTER*
*—— = crystallogr. directions*
*– – – = directions of extinction X'*
see vol. 2, p. 743

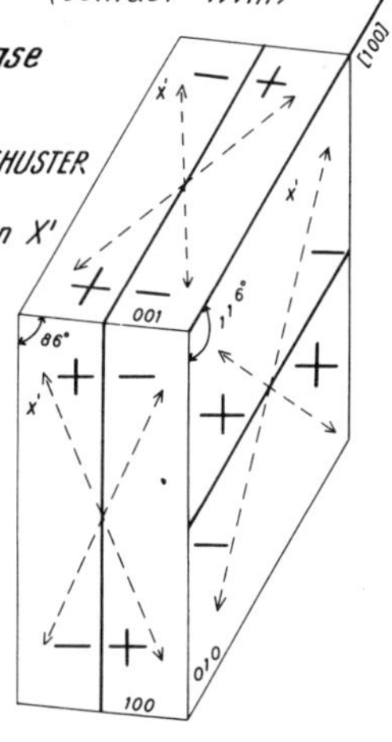

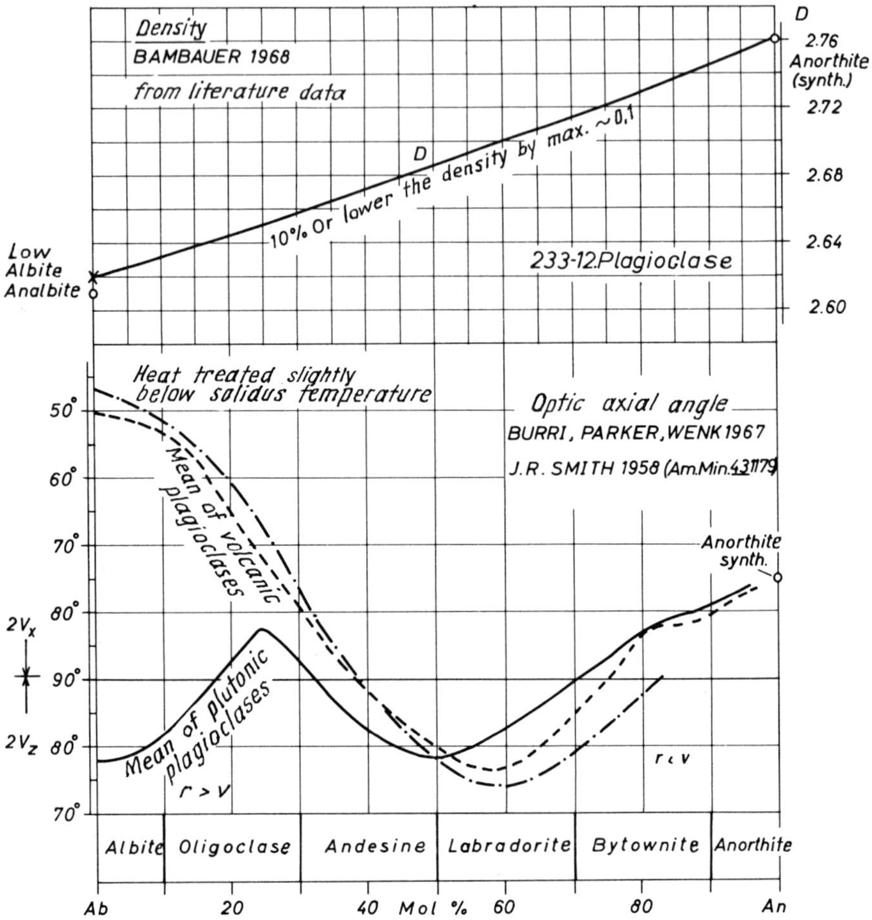

Density
BAMBAUER 1968
from literature data
D
10% Or lower the density by max. ~0,1
233-12 Plagioclase
D
2.76
Anorthite
(synth.)
2.72
2.68
2.64
2.60
Low
Albite
Analbite
Heat treated slightly
below solidus temperature
Optic axial angle
BURRI, PARKER, WENK 1967
J.R. SMITH 1958 (Am.Min.43 1179)
Mean of volcanic plagioclases
Mean of plutonic plagioclases
r > v
r < v
Anorthite synth.
50°
60°
70°
80°
90°
80°
70°
2Vx
2Vz
Albite
Oligoclase
Andesine
Labradorite
Bytownite
Anorthite
Ab
20
40 Mol % 60
80
An

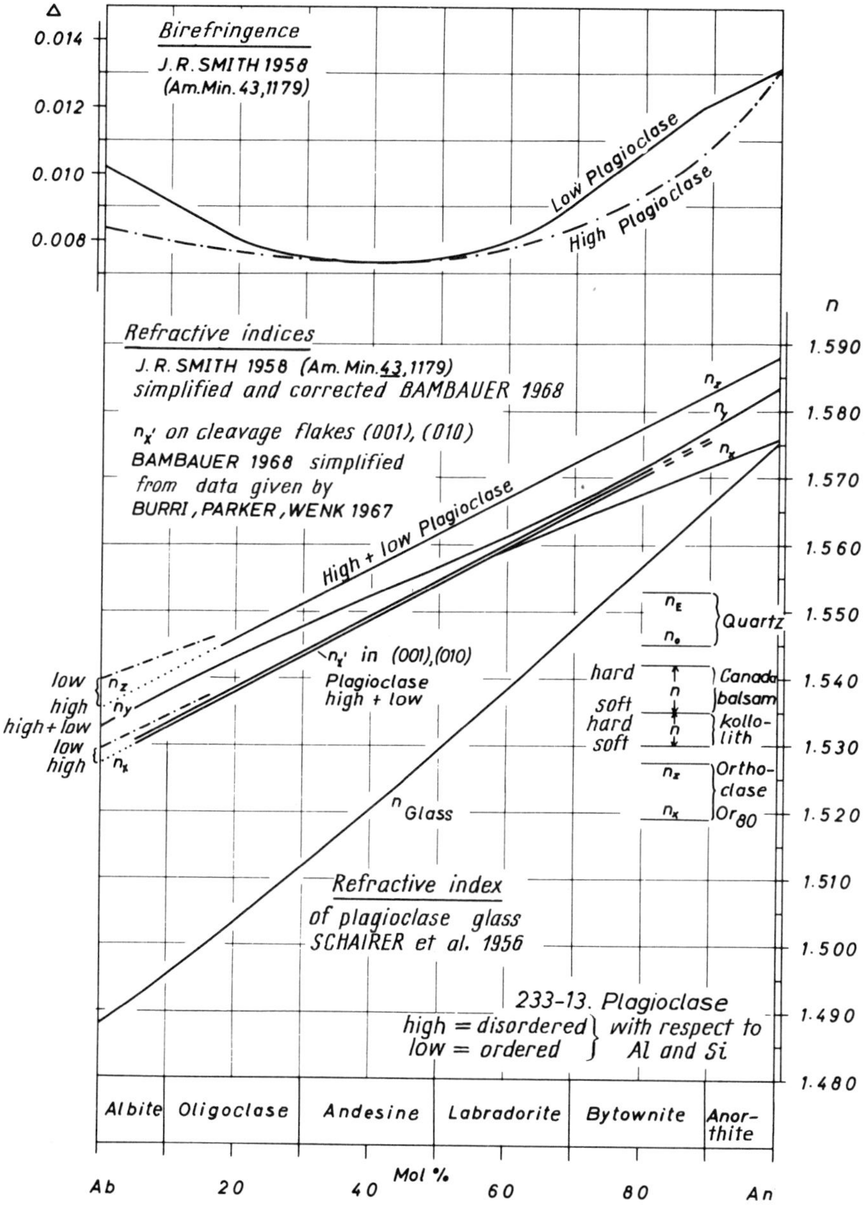


233-13. Plagioclase
high = disordered } with respect to
low = ordered } Al and Si

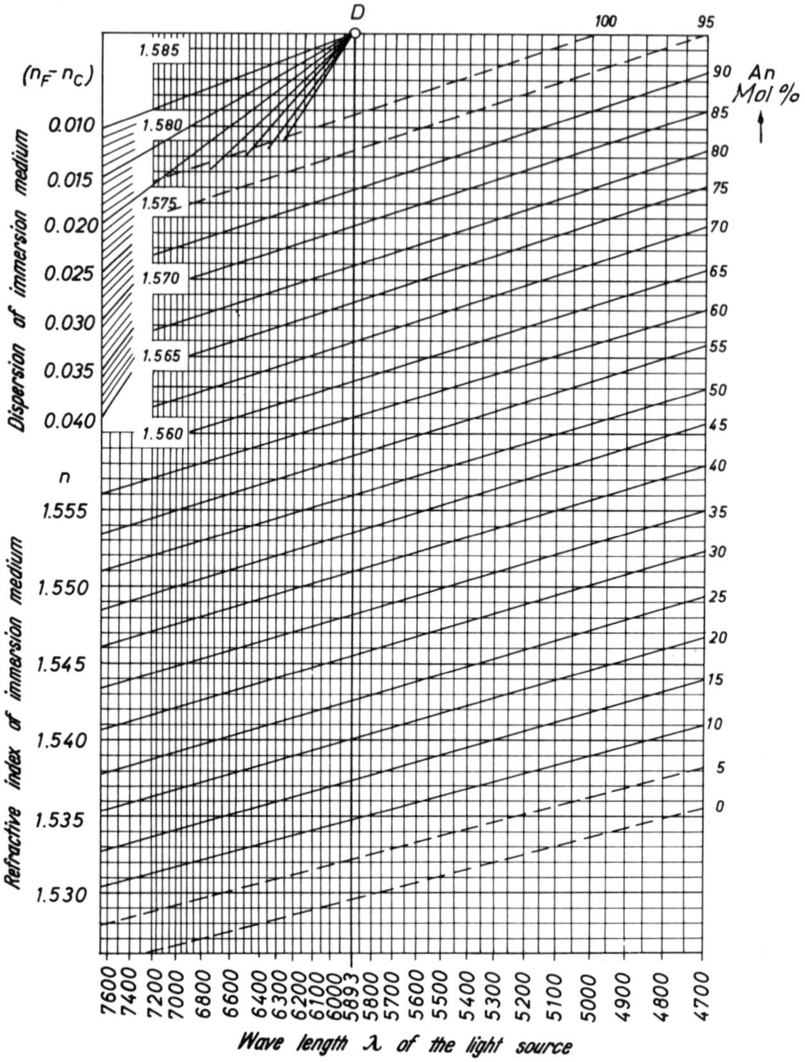


# 233 – 14. Low Plagioclase

determination of refractive indices with cleavage flakes after TSUBOI 1934 (Jap. J. Geol. Geogr. 11, 325): measurement of $n_{x'}$ in (001) by comparison with an immersion oil and variable monochromatic light.

Data from MORSE (1968) Am. Min. 53, 105 best results in the range $An_{10} - An_{80}$

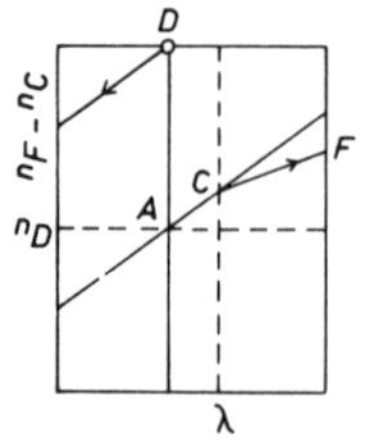


## explanation:

1.) point A has abscissa = 5893 Å, ordinate = $n_D$ of imm. oil

2.) look up the ray D ⟶ $(n_F - n_C)$ of immersion oil

3.) draw a line parallel to this ray through A (= dispersion line of immersion oil)

4.) find intersection C of this line with the line of the wave length at which the Becke line disappears

5.) line C ⟶ F gives anorthite content at F

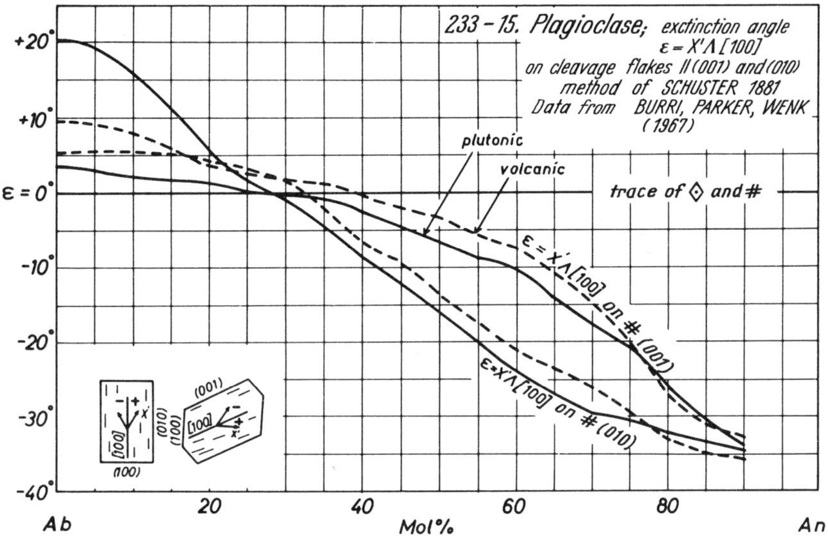
233 - 15. Plagioclase; extinction angle
ε = X'Λ [100]
on cleavage flakes ll (001) and (010)
method of SCHUSTER 1881
Data from BURRI, PARKER, WENK
(1967)
plutonic
volcanic
trace of ◊ and #
ε = X'Λ [100] on # (001)
ε = X'Λ [100] on # (010)
+20°
+10°
ε = 0°
-10°
-20°
-30°
-40°
(001)
(010)
(100)
[100]
(100)
X'
Ab
20
40
60
80
An
Mol%

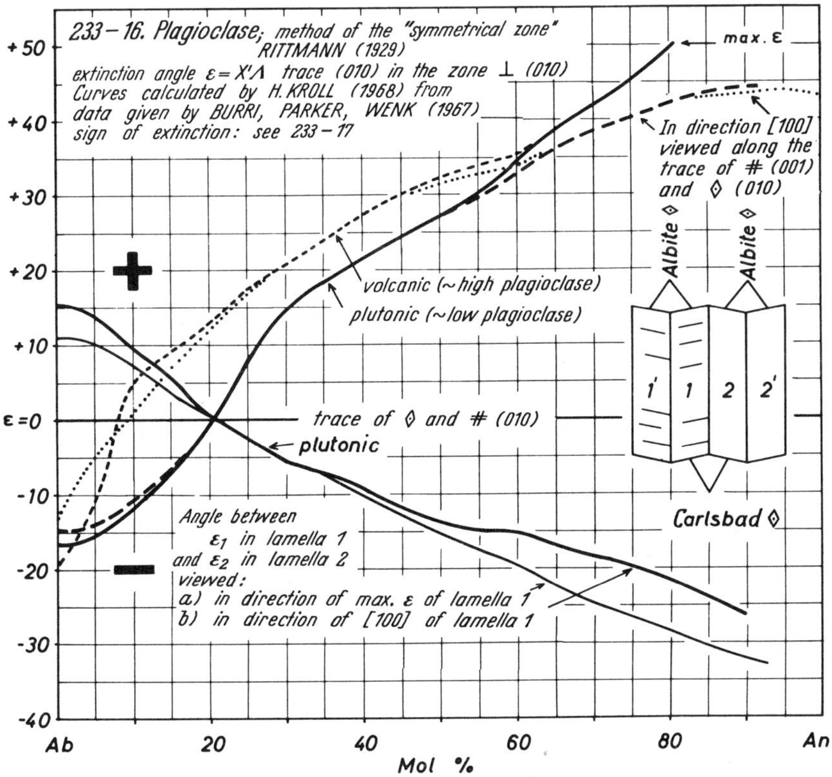
233 - 16. Plagioclase; method of the "symmetrical zone"
RITTMANN (1929)
extinction angle ε = X'Λ trace (010) in the zone ⊥ (010)
Curves calculated by H. KROLL (1968) from
data given by BURRI, PARKER, WENK (1967)
sign of extinction: see 233 - 17
max. ε
In direction [100]
viewed along the
trace of # (001)
and ◊ (010)
Albite ◊
Albite ◊
volcanic (~high plagioclase)
plutonic (~low plagioclase)
1'
1
2
2'
trace of ◊ and # (010)
plutonic
Carlsbad ◊
Angle between
ε1 in lamella 1
and ε2 in lamella 2
viewed:
a) in direction of max. ε of lamella 1
b) in direction of [100] of lamella 1
+50
+40
+30
+20
+10
ε = 0
-10
-20
-30
-40
Ab
20
40
60
80
An
Mol %

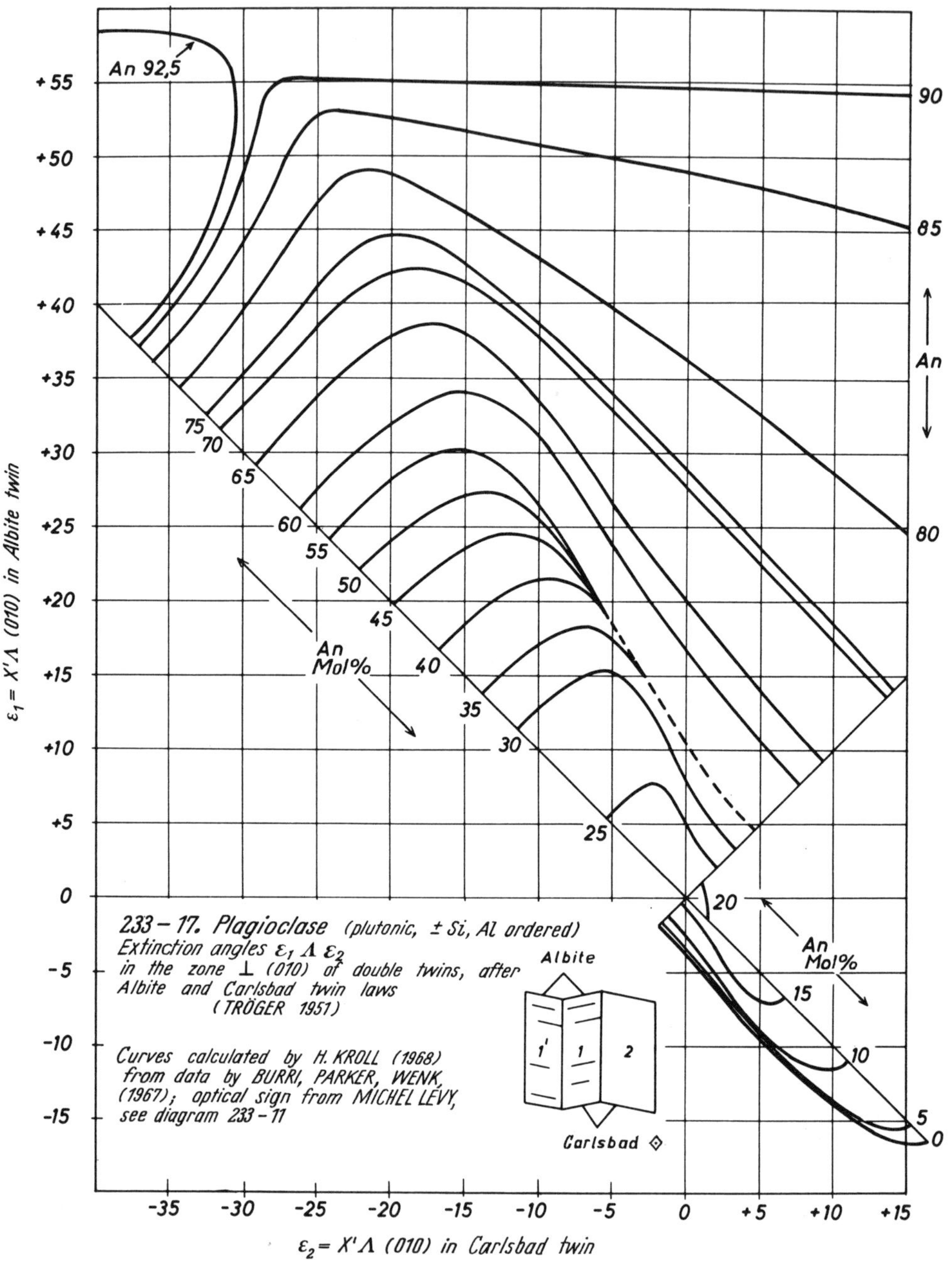

233 – 17. Plagioclase (plutonic, ± Si, Al ordered)
Extinction angles $\varepsilon_1 \Lambda \varepsilon_2$
in the zone ⊥ (010) of double twins, after
Albite and Carlsbad twin laws
(TRÖGER 1951)
Curves calculated by H. KROLL (1968)
from data by BURRI, PARKER, WENK,
(1967); optical sign from MICHEL LÉVY,
see diagram 233 – 11
$\varepsilon_1$ = X' Λ (010) in Albite twin
$\varepsilon_2$ = X' Λ (010) in Carlsbad twin
An 92,5
An Mol%
An
Albite
Carlsbad ◇

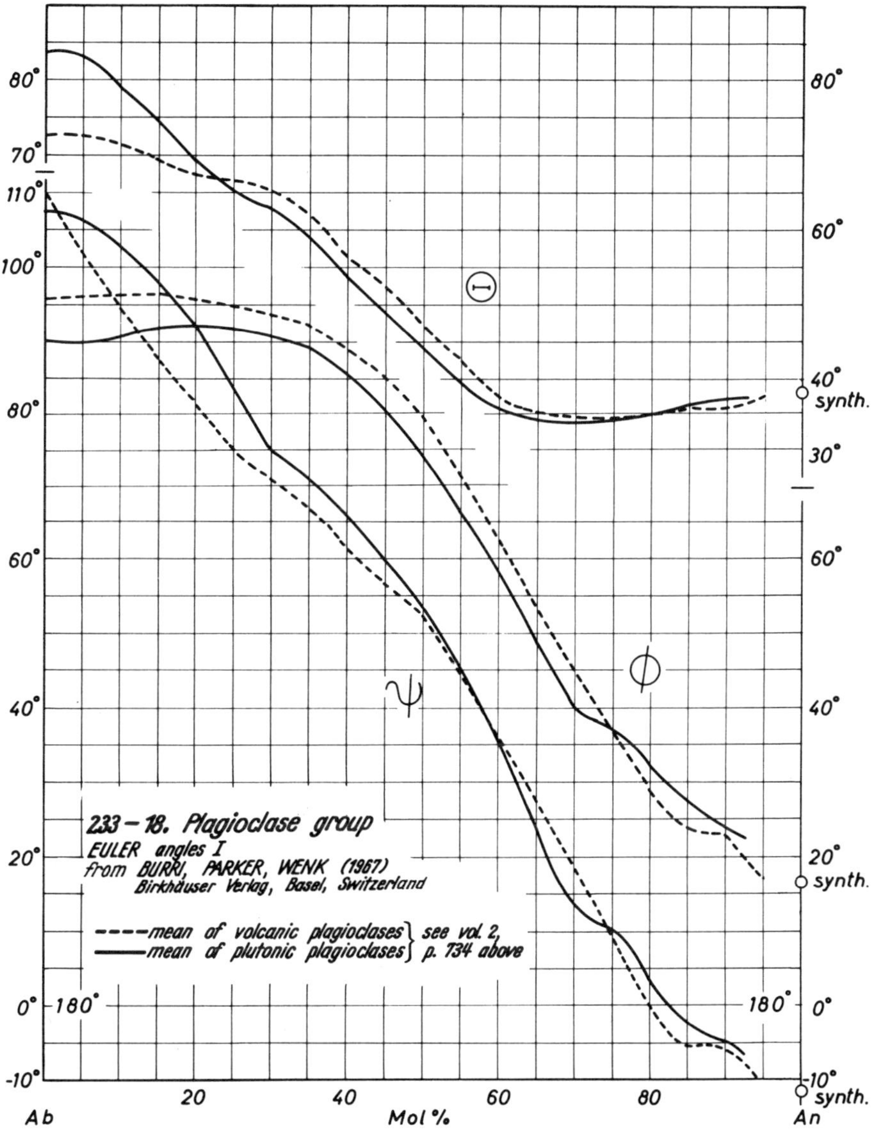
233 – 18. Plagioclase group
EULER angles I
from BURRI, PARKER, WENK (1967)
Birkhäuser Verlag, Basel, Switzerland
mean of volcanic plagioclases} see vol. 2,
mean of plutonic plagioclases} p. 734 above
Θ
Φ
ψ
80°
70°
110°
100°
80°
60°
40°
20°
0°
-10°
180°
60°
40°
synth.
30°
20°
-10°
Ab
20
40
60
80
Mol %
An

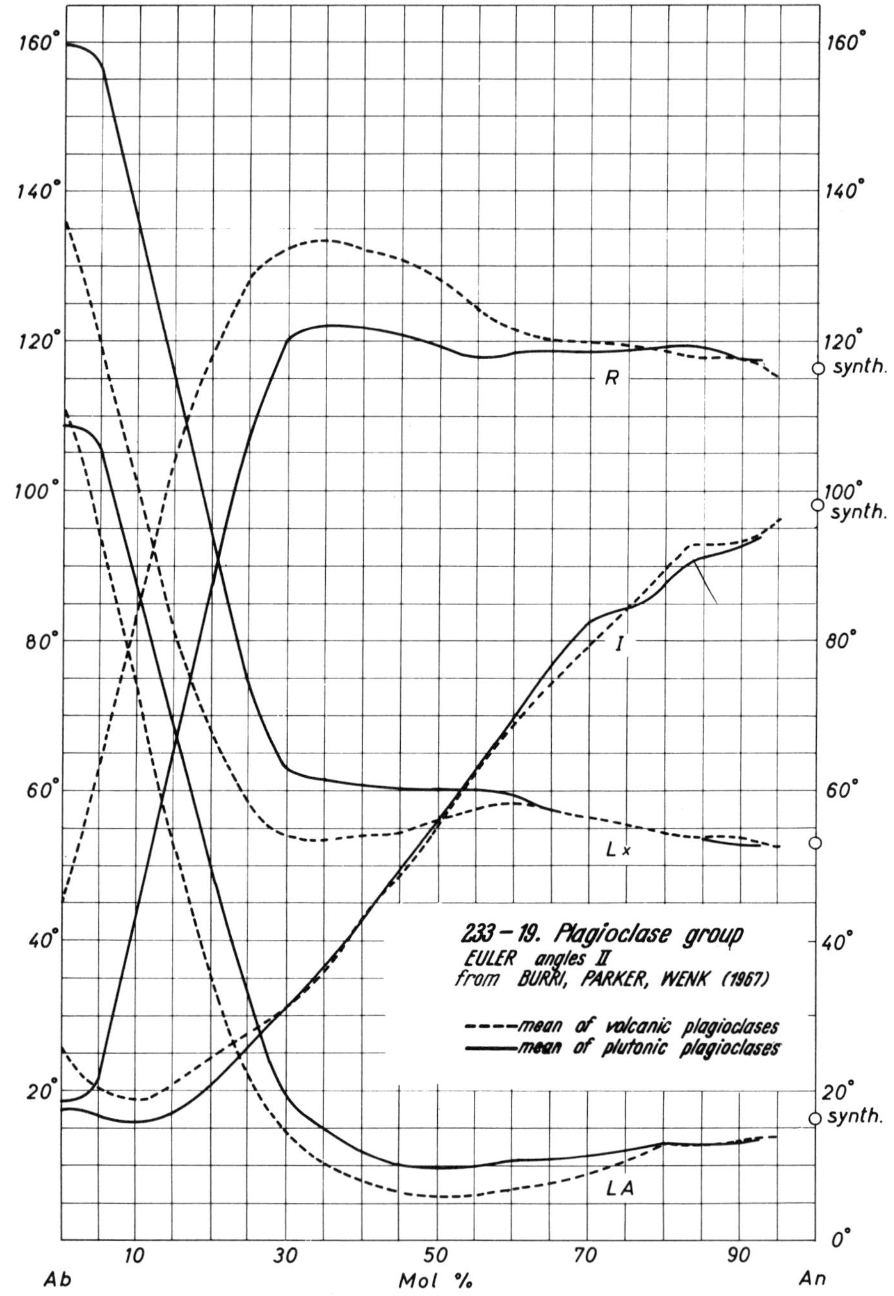
160°
140°
120°
100°
80°
60°
40°
20°
0°
synth.
synth.
synth.
R
I
Lx
LA
233 – 19. Plagioclase group
EULER angles II
from BURRI, PARKER, WENK (1967)
mean of volcanic plagioclases
mean of plutonic plagioclases
10
30
50
70
90
Ab
Mol %
An

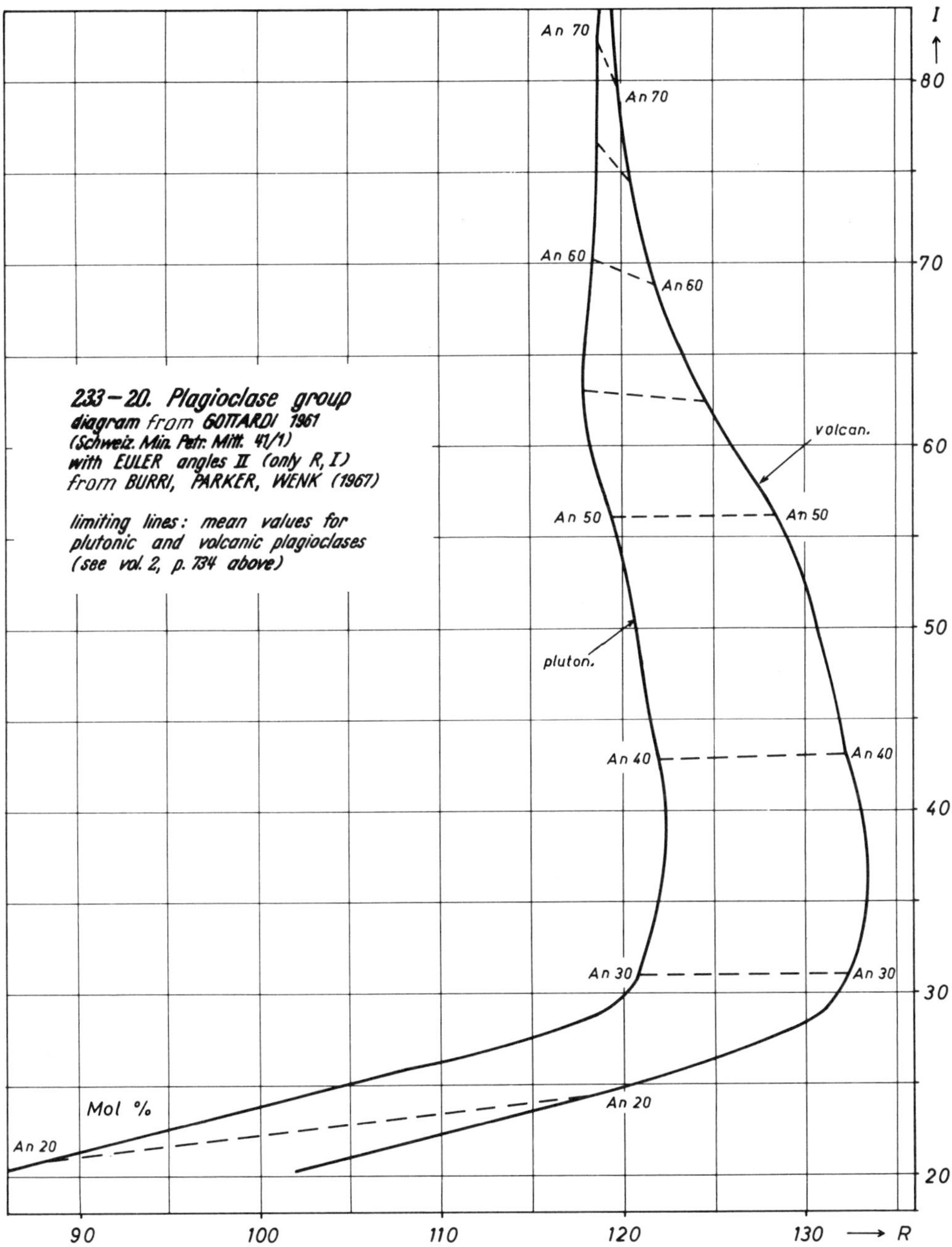
233-20. Plagioclase group
diagram from GOTTARDI 1961
(Schweiz. Min. Petr. Mitt. 41/1)
with EULER angles II (only R, I)
from BURRI, PARKER, WENK (1967)
limiting lines: mean values for
plutonic and volcanic plagioclases
(see vol. 2, p. 734 above)
An 70
An 70
An 60
An 60
volcan.
An 50
An 50
pluton.
An 40
An 40
An 30
An 30
An 20
An 20
Mol %
I
80
70
60
50
40
30
20
90
100
110
120
130
R

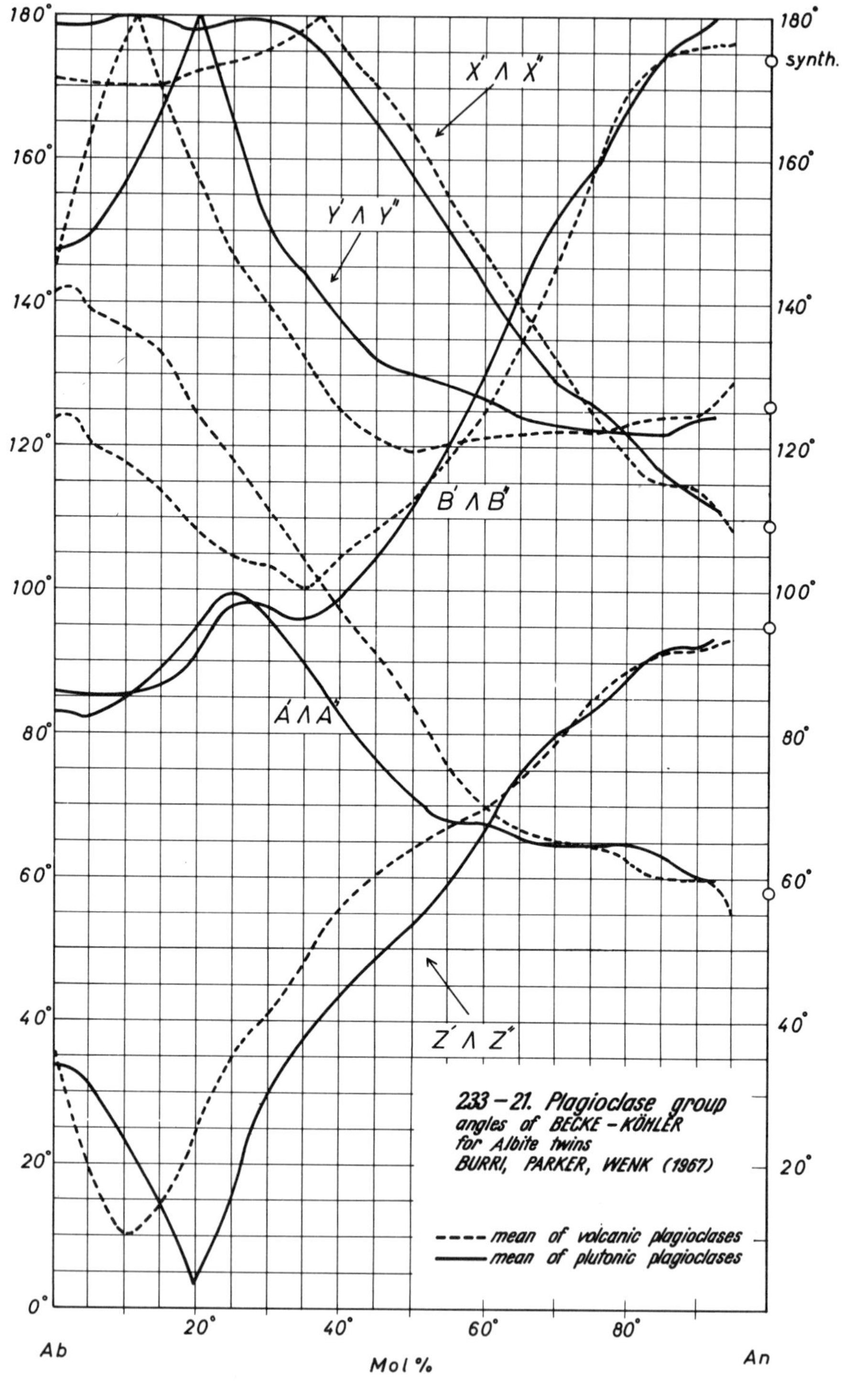
180°
160°
140°
120°
100°
80°
60°
40°
20°
0°
180°
synth.
160°
140°
120°
100°
80°
60°
40°
20°
X' Λ X"
Y' Λ Y"
B' Λ B"
A' Λ A"
Z' Λ Z"
233 – 21. Plagioclase group
angles of BECKE – KÖHLER
for Albite twins
BURRI, PARKER, WENK (1967)
mean of volcanic plagioclases
mean of plutonic plagioclases
20°
40°
60°
80°
Ab
Mol %
An

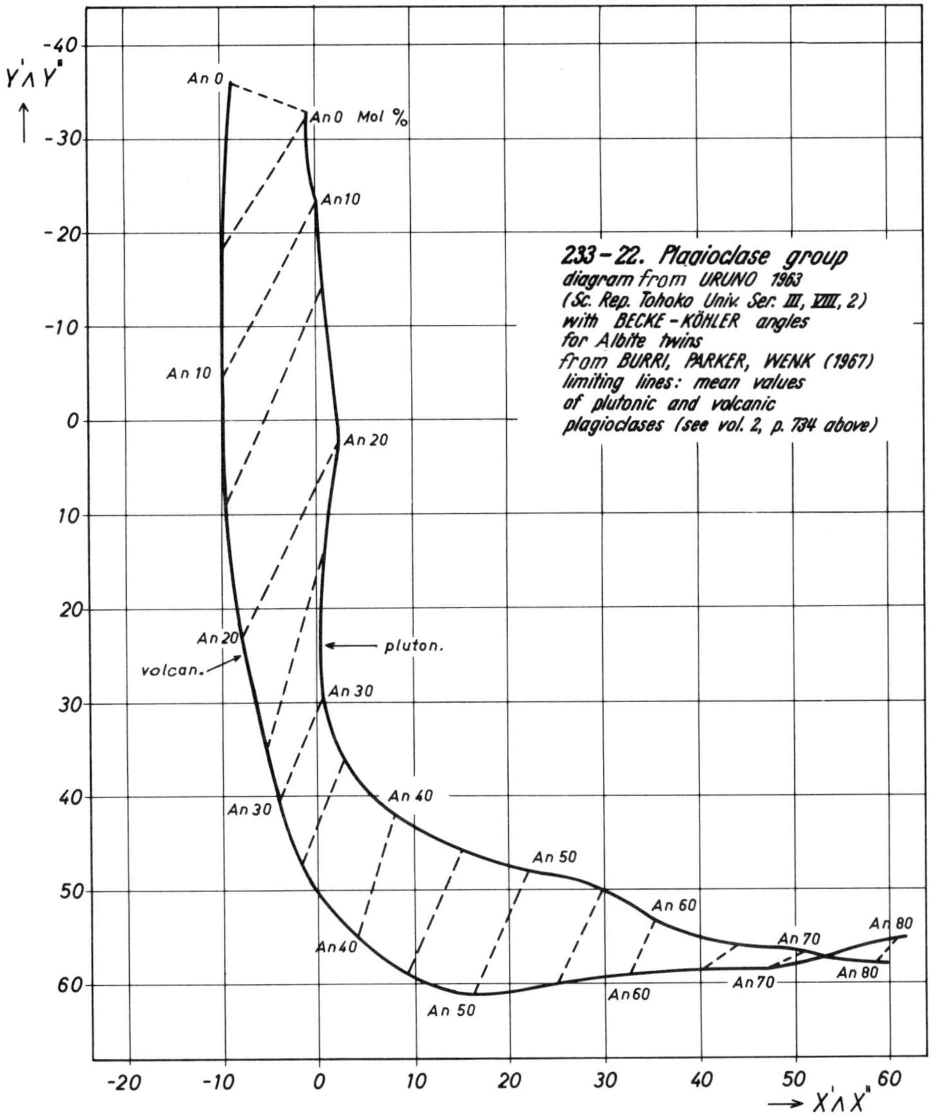
233-22. Plagioclase group
diagram from URUNO 1963
(Sc. Rep. Tohoko Univ. Ser. III, VIII, 2)
with BECKE-KÖHLER angles
for Albite twins
from BURRI, PARKER, WENK (1967)
limiting lines: mean values
of plutonic and volcanic
plagioclases (see vol. 2, p. 734 above)
Y'∧Y"
X'∧X"
-40
-30
-20
-10
0
10
20
30
40
50
60
-20
-10
0
10
20
30
40
50
60
An 0
An 0 Mol %
An 10
An 10
An 20
An 20
pluton.
volcan.
An 30
An 30
An 40
An 40
An 50
An 50
An 60
An 60
An 70
An 70
An 80
An 80

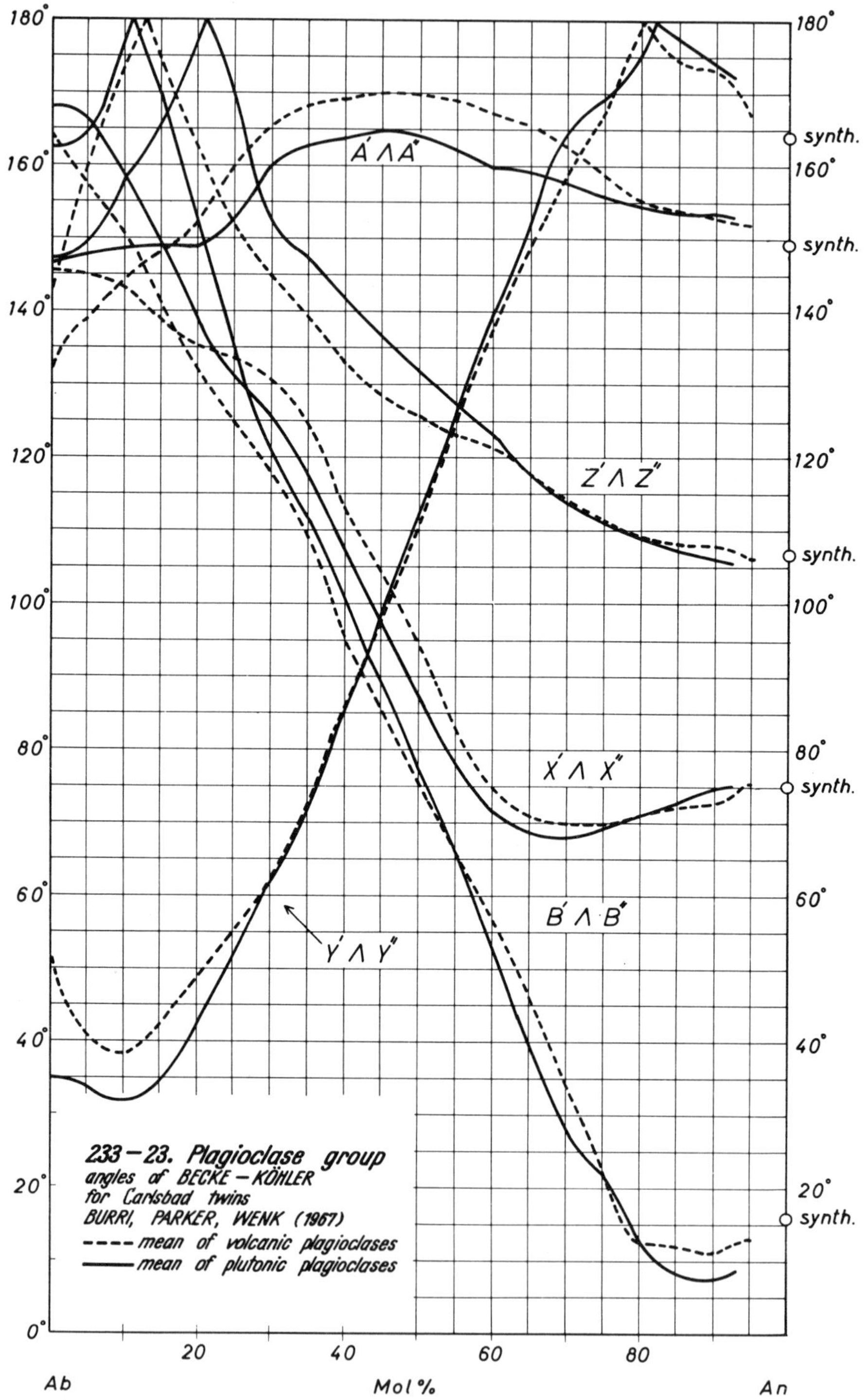
180°
160°
140°
120°
100°
80°
60°
40°
20°
0°
synth.
A'∧A''
Z'∧Z''
X'∧X''
B'∧B''
Y'∧Y''
233-23. Plagioclase group
angles of BECKE-KÖHLER
for Carlsbad twins
BURRI, PARKER, WENK (1967)
---- mean of volcanic plagioclases
—— mean of plutonic plagioclases
20
40
60
80
Ab
Mol %
An

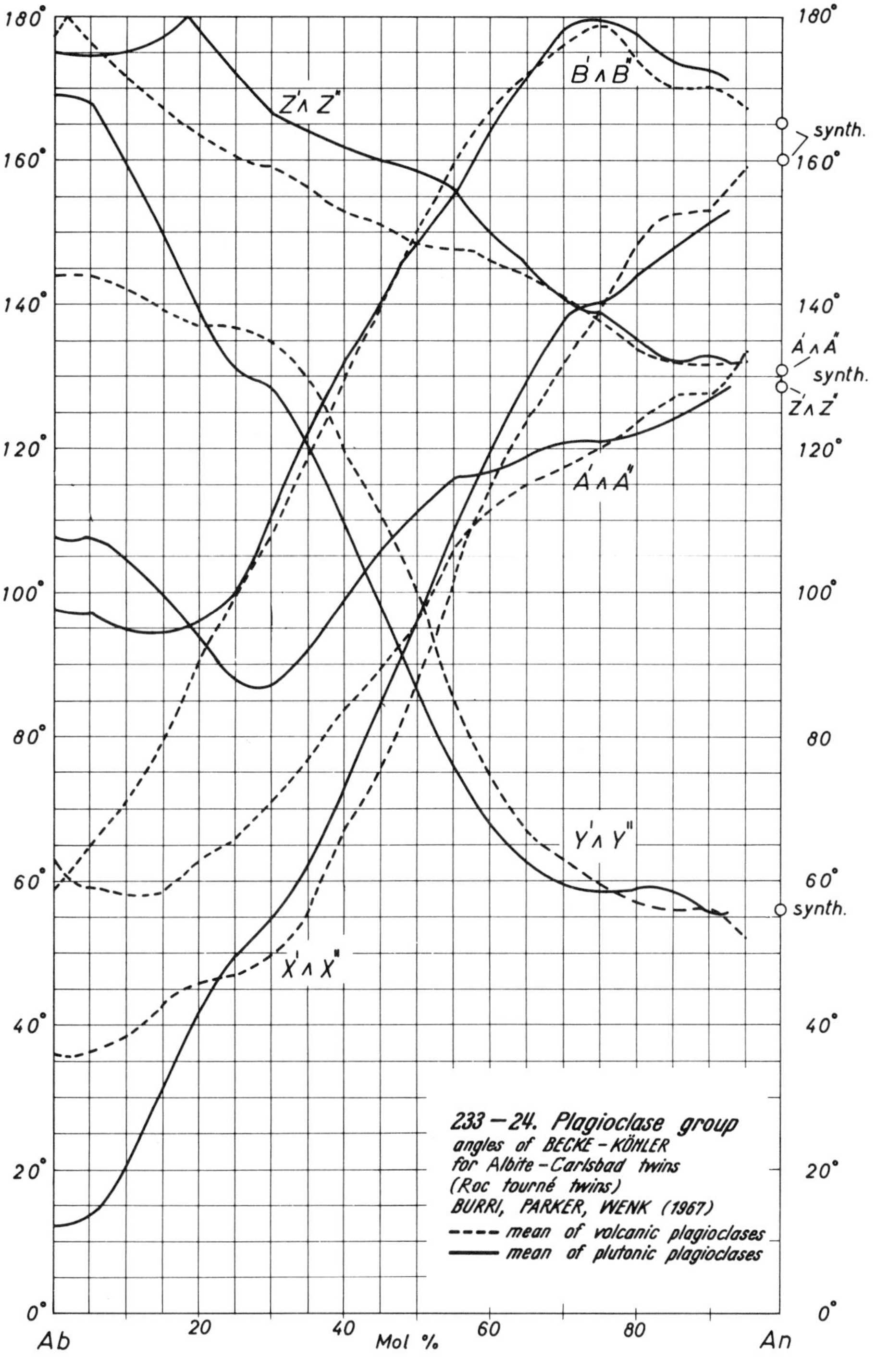
180°
160°
140°
120°
100°
80°
60°
40°
20°
0°
Z'∧Z"
B'∧B"
synth.
A'∧A"
synth.
Z'∧Z"
A'∧A"
Y'∧Y"
synth.
X'∧X"
233 – 24. Plagioclase group
angles of BECKE - KÖHLER
for Albite - Carlsbad twins
(Roc tourné twins)
BURRI, PARKER, WENK (1967)
mean of volcanic plagioclases
mean of plutonic plagioclases
Ab
20
40
Mol %
60
80
An

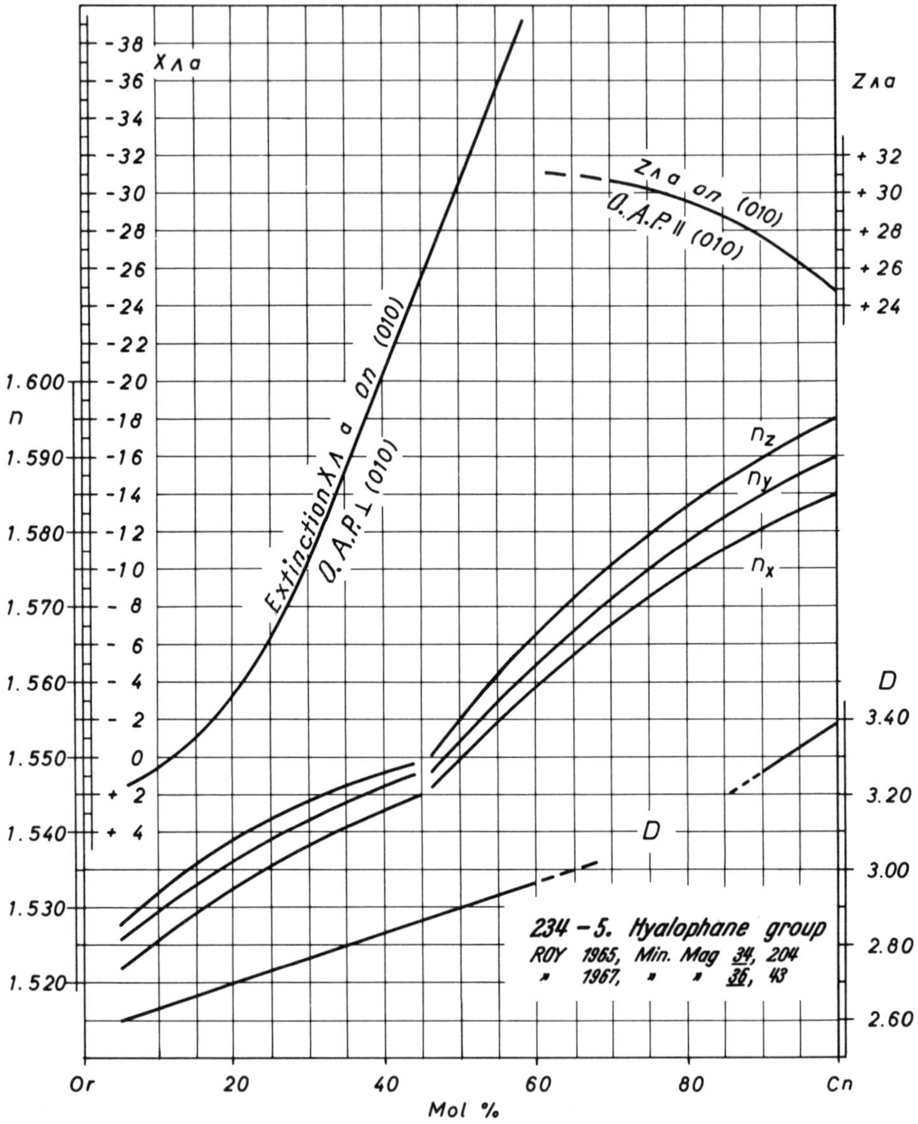
X∧a
Z∧a
Z∧a on (010)
O.A.P. ∥ (010)
Extinction X∧a on (010)
O.A.P. ⊥ (010)
n
nz
ny
nx
D
D
234 – 5. Hyalophane group
ROY 1965, Min. Mag 34, 204
" 1967, " " 36, 43
Or
20
40
60
80
Cn
Mol %

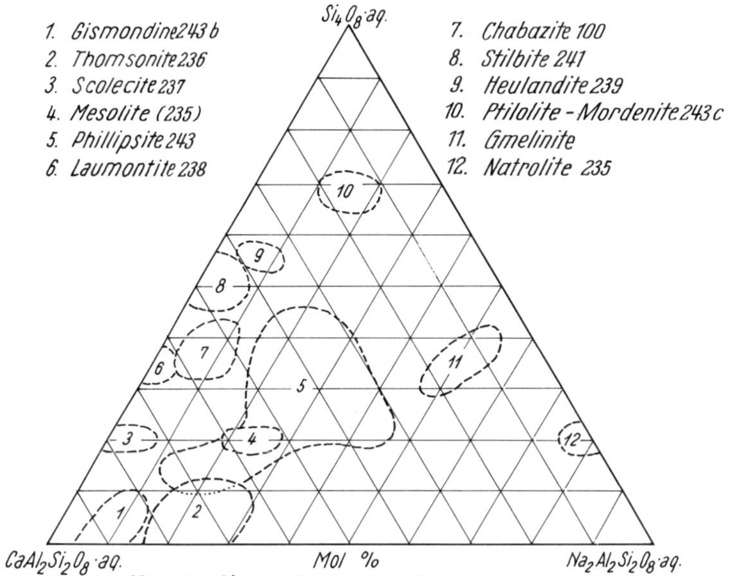


**235-243. Chemical variation of the zeolites**
WINCHELL 1937 (Am. Min. 22, 90-92)

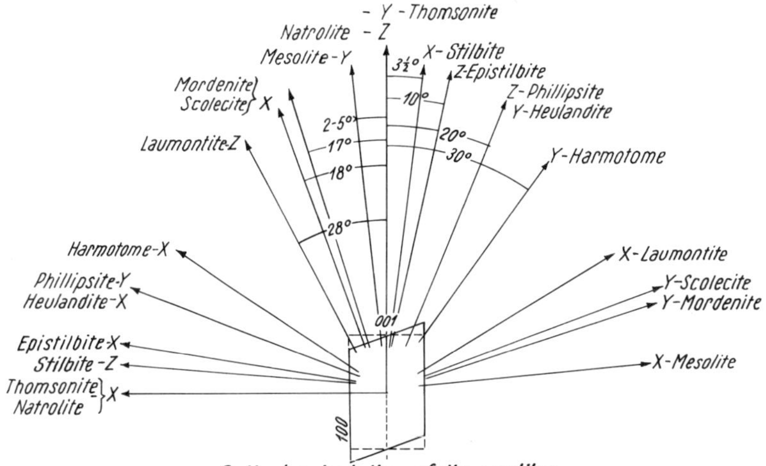


**Optical orientation of the zeolites**
in section II (010)
TRÖGER 1951 from WINCHELL 1927 (Opt. Min. II 389)

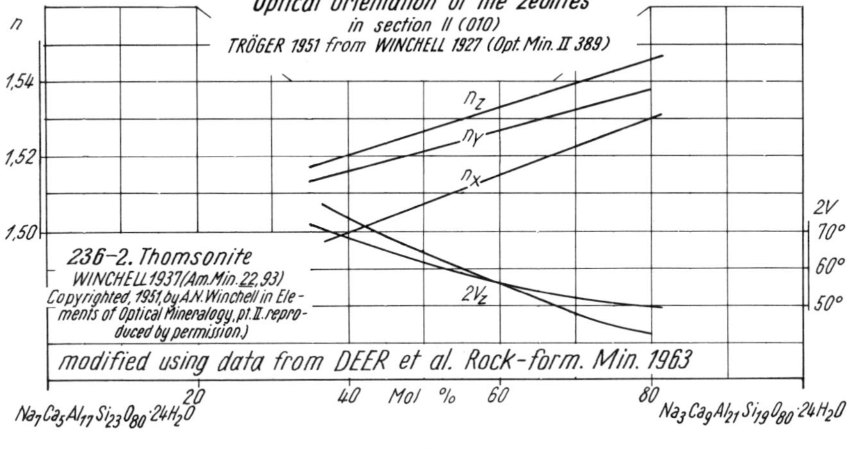


**236-2. Thomsonite**
WINCHELL 1937 (Am. Min. 22, 93)
(Copyrighted, 1951, by A.N. Winchell in Elements of Optical Mineralogy, pt. II. reproduced by permission.)
modified using data from DEER et al. Rock-form. Min. 1963

| No. | Name<br>Symmetry<br>Chem. Composition | Common Crystal Forms, Twinning<br>Crystal Habit<br>Aggregates | Cleavage<br>Translation Gliding<br>Deform. Twinning<br>Hardness, Density | Optic Orientation | Refractive Indices<br>Birefringence |
|---|---|---|---|---|---|
| 235 | Natrolite orh<br>$Na_2[Al_2Si_3O_{10}]$<br>$\cdot 2H_2O$ | {110},{111},{010},<br>{100}, pseudotetr.;<br>◇ {110} rare;<br>acicular ‖ c,<br>fibrous, stellate | # {110} good, 89°,<br>(010) parting;<br>H = 5 – 5½ D = 2.2± | X ‖ a<br>Y ‖ b, O.A.P. (010)<br>Z ‖ c, bi;<br>l = (+)<br>fig. p. 142 | $n_x$ = 1.473 – 1.489<br>$n_y$ = 1.476 – 1.491<br>$n_z$ = 1.485 – 1.501<br>⊕Δ = 0.012 – 0.012 |
| 236 | Thomsonite orh<br>$NaCa_2[Al_5Si_5O_{20}]$<br>$\cdot 6H_2O$ | {110},{001},{100},<br>{010};<br>◇ {110};<br>columnar ‖ c, fibrous,<br>flaky ‖ (010), radiated | # (010) perf.,<br># (100) good;<br>H = 5 – 5½ D = 2.3± | X ‖ a<br>Y ‖ c, O.A.P. (001)<br>Z ‖ b, bi;<br>l = (±)<br>fig. p. 142 | $n_x$ = 1.497 – 1.530<br>$n_y$ = 1.513 – 1.532<br>$n_z$ = 1.518 – 1.545<br>⊕Δ = 0.020 – 0.015<br>diagr. p. 139 |
| 237 | Scolecite mcl<br>$Ca[Al_2Si_3O_{10}]$<br>$\cdot 3H_2O$ | {110},{010},{hkl};<br>β = 90¾°, pseudo-orh.;<br>◇ (100) always present;<br>acicular ‖ c, fibrous | # {110} good, 88½°;<br>H = 5 – 5½ D = 2.3± | XΛc = +15° to +18°, bi;<br>YΛa = –14° to –17°,<br>O.A.P. ⊥ (010)<br>Z ‖ b<br>l' = (–)<br>fig. p. 142/3 | $n_x$ = 1.509 – 1.514<br>$n_y$ = 1.516 – 1.520<br>$n_z$ = 1.521 – 1.525<br>⊕Δ = 0.012 – 0.011 |
| 238 | Laumontite mcl<br>$Ca[Al_2Si_4O_{12}]$<br>$\cdot 4H_2O$ | {110},{100},{010},<br>{101},{$\bar{2}$03};<br>β = 111½°;<br>◇ (100);<br>columnar ‖ c, acicular | # (010) perf.,<br># {110} perf., 94°<br># (100) poor;<br>H = 3 – 3½ D = 2.3± | XΛa = 10° to 26°, bi;<br>Y ‖ b, O.A.P. (010)<br>ZΛc = 8° to 11°<br>l' = (+)<br>fig. p. 142/3 | $n_x$ = 1.502 – 1.513<br>$n_y$ = 1.512 – 1.524<br>$n_z$ = 1.514 – 1.525<br>⊖Δ = 0.012 – 0.012 |
| 239 | Heulandite mcl<br>$Ca[Al_2Si_7O_{18}]\cdot 6H_2O$ | {010},{101},{$\bar{1}$01},<br>{001},{110};<br>β = 91½°; pseudo-orh.;<br>◇ (001);<br>flaky ‖ (010), subparallel aggreg., rarely radiated | # (010) perf.;<br>H = 3½ – 4 D = 2.20± | XΛa = 0° – 34°<br>YΛc = 0° to +32°,<br>O.A.P. ⊥ (010)<br>Z ‖ b, bi;<br>l' = (+)<br>fig. p. 142/3 | $n_x$ = 1.496 – 1.499<br>$n_y$ = 1.497 – 1.500<br>$n_z$ = 1.501 – 1.505<br>⊕Δ = 0.005 – 0.006 |
| 240 | Epistilbite mcl<br>$Ca[Al_2Si_6O_{16}]\cdot 5H_2O$ | {100},{010},{110},<br>{11$\bar{1}$},{001};<br>β = 90¾°;<br>◇ (100),{110},{102},<br>penetration ◇;<br>columnar ‖ c,<br>spherulitic | # (010) perf.;<br>H = 4 D = 2.25± | XΛa = +11°, bi;<br>Y ‖ b, O.A.P. ‖ (010)<br>ZΛc = –10°<br>l' = (+)<br>fig. p. 142/3 | $n_x$ = 1.502 – 1.505<br>$n_y$ = 1.510 – 1.515<br>$n_z$ = 1.512 – 1.519<br>⊖Δ = 0.010 – 0.014 |
| 241 | Stilbite mcl<br>(Desmine)<br>$Ca[Al_2Si_7O_{18}]\cdot 7H_2O$ | {100},{010},{10$\bar{1}$},<br>{001};<br>β = 92°; almost always penetration-◇ ‖ (100),<br>{102}, {110};<br>tabular ‖ (010), elongated ‖ c, sheaf-like | # (010) perf.,<br># {10$\bar{1}$} poor;<br>H = 3½ – 4 D = 2.15± | XΛc = –3° to –12°, bi;<br>Y ‖ b, O.A.P. (010)<br>ZΛa = +3 to –13°<br>l' = (±)<br>fig. p. 142/3 | $n_x$ = 1.486 – 1.498<br>$n_y$ = 1.494 – 1.507<br>$n_z$ = 1.496 – 1.509<br>⊖Δ = 0.010 – 0.011 |
| 242 | Harmotome mcl<br>$Ba[Al_2Si_6O_{16}]\cdot 6H_2O$ | {100},{010},{011};<br>β = 124¾°; penetration-◇ ‖ (100), {102},<br>{110}; columnar ‖ c,<br>tabular ‖ (010) | # (010) distinct,<br># (100) poor;<br>H = 4½ D = 2.4± | XΛa = +63° to 67°<br>YΛc = –28° to –82°,<br>O.A.P. ⊥ (010)<br>Z ‖ b, bi;<br>l' = (±)<br>fig. p. 142/3 | $n_x$ = 1.503 – 1.506<br>$n_y$ = 1.505 – 1.509<br>$n_z$ = 1.508 – 1.514<br>⊕Δ = 0.005 – 0.008 |
| 243 | Phillipsite orh<br>(½ <u>Ca</u>,K,Na)$_5$<br>$[Al_5Si_{11}O_{32}]\cdot 10H_2O$ | {100},{010},{011};<br>β = 125¾°; often pseudo-cubic due to penetration-◇ ‖ (100),<br>‖ {110},{102};<br>columnar ‖ c,<br>bladed ‖ (010) | # (010) distinct,<br># (100) distinct;<br>H = 4 – 4½ D = 2.15± | X ‖ b<br>YΛa = +65° to 46°,<br>O.A.P. ⊥ (010)<br>ZΛc = –11° to –30°, bi;<br>l' = (+)<br>fig. p. 142/3 | Si-rich Si-poor<br>$n_x$ = 1.481 – 1.504<br>$n_y$ = 1.484 – 1.509<br>$n_z$ = 1.486 – 1.514<br>⊕Δ = 0.003 – 0.010 |
| 243b | Gismondite mcl<br>$Ca[Al_2Si_2O_8]\cdot 4H_2O$ | ◇ (110), (001);<br>radiated, stellate | # (101) good;<br>H = 4½ D = 2.25± | Z ~ ⊥ (100)<br>O.A.P. ⊥ (010) | $n_x$ = 1.531 – 1.540<br>$n_y$ = 1.540 – 1.543<br>$n_z$ = 1.548 – 1.548<br>⊖Δ = 0.017 – 0.008 |
| 243c | Mordenite orh<br>(<u>Na$_2$</u>,<u>Ca</u>,K$_2$Mg)<br>$[AlSi_5O_{12}]_2\cdot 6H_2O$ | fibrous, radiated,<br>massive | # (100) good,<br># (010) distinct;<br>H = 5 D = 2.1 | XΛc = (4°) – 0°<br>Z ‖ b;<br>Y ‖ a, O.A.P. (100) | $n_x$ = 1.472 – 1.483<br>$n_y$ = 1.475 – 1.485<br>$n_z$ = 1.477 – 1.487<br>⊕/⊖Δ = 0.005 – 0.004 |
| 244 | Whewellite mcl<br>$Ca[C_2O_4]\cdot H_2O$ | {001},{011},{010},<br>{110},{112},{$\bar{1}$01};<br>β = 107¼°;<br>◇ {10$\bar{1}$} heart-shaped;<br>columnar ‖ a,<br>tabular ‖ (010) | # {$\bar{1}$01} perf.,<br># (010) distinct,<br># (001), {110} poor;<br>brittle;<br>H = 2½ – 3 D = 2.2± | X ‖ b;<br>YΛa = –14° to –7°,<br>O.A.P. ⊥ (010)<br>ZΛc = +28° to +31°, bi;<br>fig. p. 143 | λ: Li Na Tl<br>$n_x$ = 1.488 – 1.491 – 1.494<br>$n_y$ = 1.551 – 1.555 – 1.560<br>$n_z$ = 1.645 – 1.655 – 1.657<br>⊕Δ = 0.157 – 0.164 – 0.163 |

| Angle and Dispersion of Optic Axes | Color in Thin Section Pleochroism | Chemical Reactions Special Remarks | Distinguishing Features of Similar Minerals | Occurrences | |
|---|---|---|---|---|---|
| $2V_Z = 58^o - 64^o$<br>$r < v$ weak | colorless | gelatinizes easily with HCl, then stainable with aniline colors; pseudomorph after nepheline, sodalite, plagioclase | thomsonite: gr. n; other zeolites: oblique extinction; scolecite: l = (–) | hydrothermal in vesicles and fissures of alkaline rocks (phonolite, basalts, also Na-syenites); rarely autometasomatic in the groundmass of volcanics | 235 |
| $2V_Z = 47^o - 75^o$<br>$r > v$ distinct | colorless | gelatinizes easily with HCl, then stainable with aniline colors | natrolite: sm. n; other zeolites: oblique extinction; chabasite: uniaxial | hydrothermal in vacuoles of effusive rocks (basalts, phonolites); pseudomorph in contact-metam. rocks | 236 |
| $2V_X = 36^o - 58^o$<br>$r \ll v$ | colorless | gelatinizes with HCl, then stainable with aniline colors | natrolite, thomsonite: l' = (+) and straight ext.; stilbite: sm. X Λ c and sm. n; other fibrous zeolites: gr. n | hydrothermal in fissures and vesicles of volcanic rocks (basalts, phonolites); in contact rocks, also in alpine fissures | 237 |
| $2V_X = 32^o - 47^o$<br>$r < v$ strong | colorless | gelatinizes with HCl, then stainable with aniline colors; weathers to powder in open air, abnormal interf. colors | scolecite: sm. Δ; heulandite: sm. Δ; O.A.P. ⊥ (010); epistilbite: one #; natrolite: ⊕ | hydrothermal in fissures and vesicles of basic magmatites and schists; also in ore lodes; typical of zeolite facies | 238 |
| $2V_Z = (0^o) - 34^o$ $-(55^o)$<br>$r > v$ | colorless | decomposed by HCl: $SiO_2$-powder, not stainable; cleavage flakes parallel # (010) show crossed dispersion | scolecite: gr. n; laumontite: gr. Δ, O.A.P. ∥ (010); analcime: sm. n, ± isotropic; stilbite: gr. Δ, O.A.P. ∥ (010); | hydrotherm. filling of vesicles in basaltic rocks; in fissures of metamorphic rocks, in hydrotherm. ore lodes; typical of zeolite facies | 239 |
| $2V_X = 44^o$<br>$r < v$ | colorless | decomposed by HCl: $SiO_2$-powder, not stainable | stilbite: l' = (+); phillipsite: gr. Z Λ c, opt. ⊕; heulandite: ⊕; laumontite: diff. # | with stilbite, hydrothermal in basalt vesicles and in tuffs | 240 |
| $2V_X = 30^o - 49^o$<br>$r < v$ | colorless | decomposed by HCl: $SiO_2$-powder, not stainable | harmotome, phillipsite: O.A.P. ⊥ (010), opt. ⊕; epistilbite: l' = (+); heulandite: opt. ⊕ | hydrotherm. in basalt vesicles and granite cavities; in fissures of schists, in hydrotherm. ore lodes | 241 |
| $2V_Z = 79^o$<br>(may decrease to $32^o$ after immersion in hot medium) | colorless | decomposed by HCl: $SiO_2$-powder, not stainable; not as frequent as phillipsite; Ba-color in flame | phillipsite: X ∥ b; stilbite: O.A.P. ∥ (010) | hydrotherm. in vesicles of basaltic volcanics; in hydrotherm. ore lodes | 242 |
| $2V_Z = 60^o - 80^o$<br>$r < v$ | colorless | gelatinizes with HCl, then stainable with aniline colors | harmotome: Z ∥ b; stilbite: O.A.P. ∥ (010) | hydroth. in vesicles of basalts; in precipitates of hot springs, also in deep-sea clays, recent in tuffs | 243 |
| $2V_X = 82^o - 86^o$ | colorless | soluble in acids; characteristic fern-like forms | thomsonite, phillipsite: sm. n, sm. Δ | usually in vesicles of basalts and ankaratrites, tephrites; also in altered granites | 243b |
| $2V_{Z,X} = 76^o - 90^o$ | colorless, also reddish due to $Fe_2O_3$ | insoluble in HCl | natrolite: sm. 2V; scolecite and thomsonite: gr. n, sm. 2V | in vesicles and fissures of basic, also of acid volcanics and associated tuffs; authigen. in sediments, in zeolite facies rocks | 243c |
| $2V_Z = 83^o - 85^o$<br>$r < v$ weak | colorless | easily soluble in acids, slightly soluble in $H_2O$ | calcite: uniaxial, ⊖; aragonite: ⊖; anhydrite: sm. Δ; hydrargillite: sm. Δ; other similar minerals: uniaxial | diagenetic with calcite in fissures, in coals and shales; constituent of bituminous marls; in hydrotherm. ore lodes; in the organs of plants and animals | 244 |

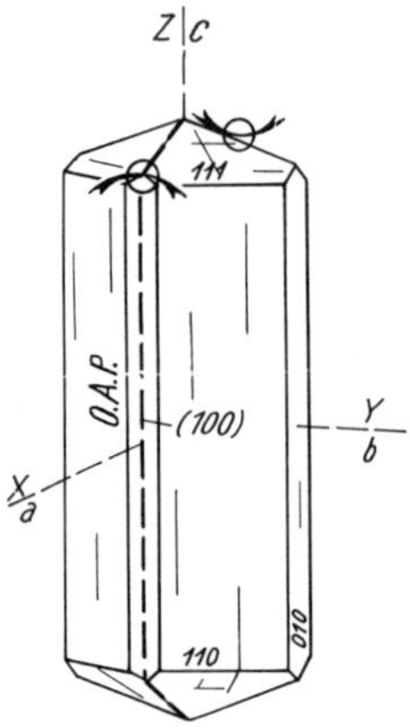


235. Natrolite

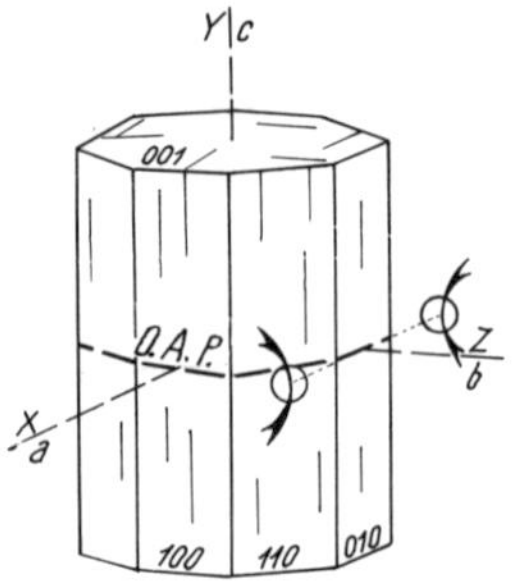


236-1. Thomsonite

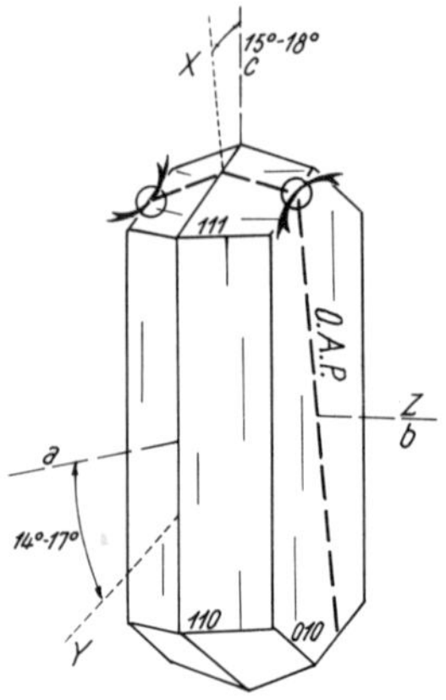


237-1. Scolecite

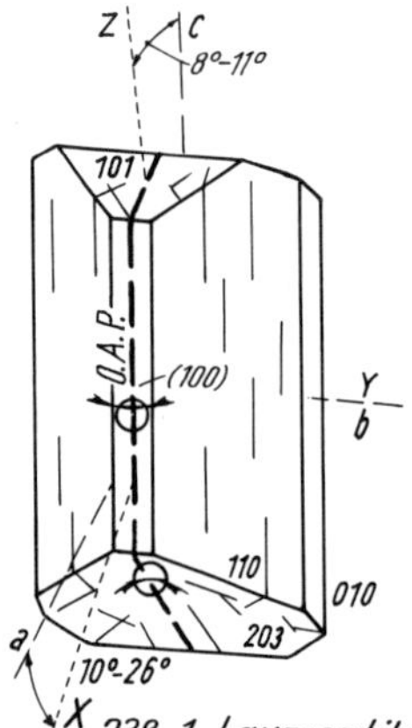


238-1. Laumontite

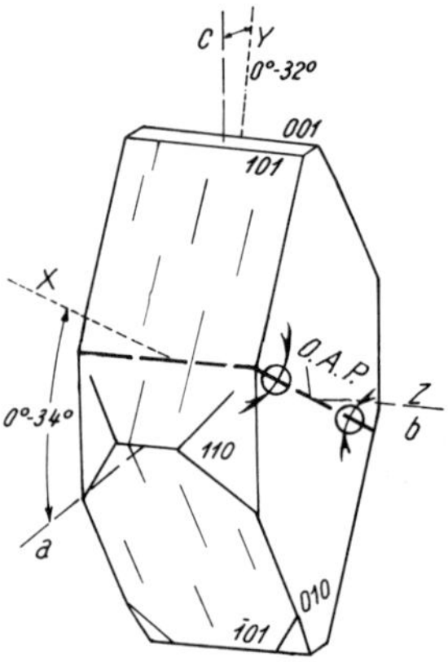


239-1. Heulandite

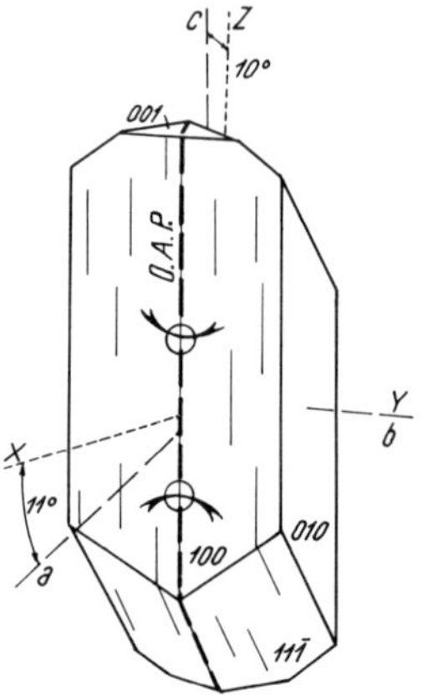


240-1. Epistilbite

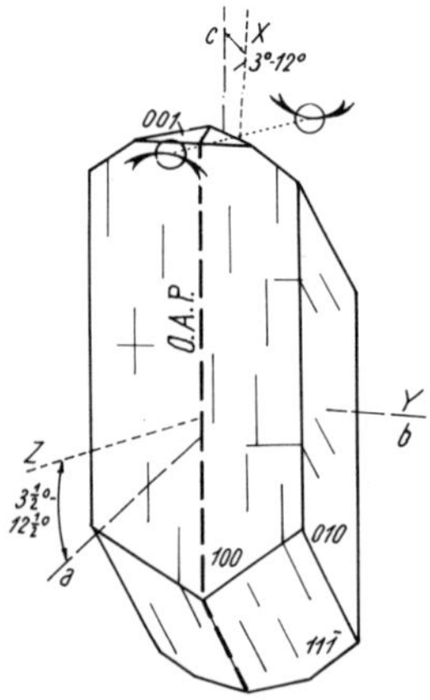


241-1. Stilbite

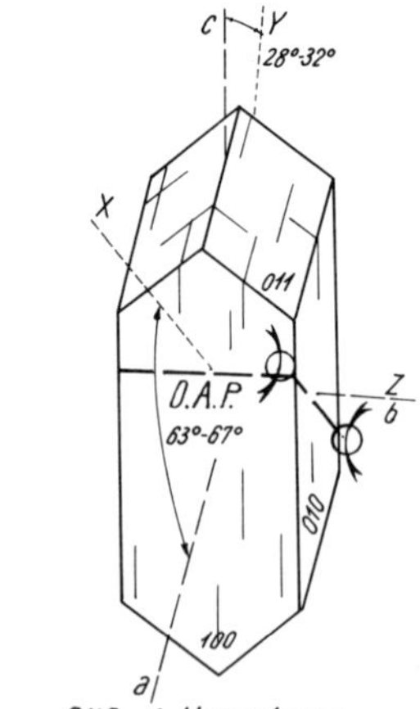


242-1. Harmotome

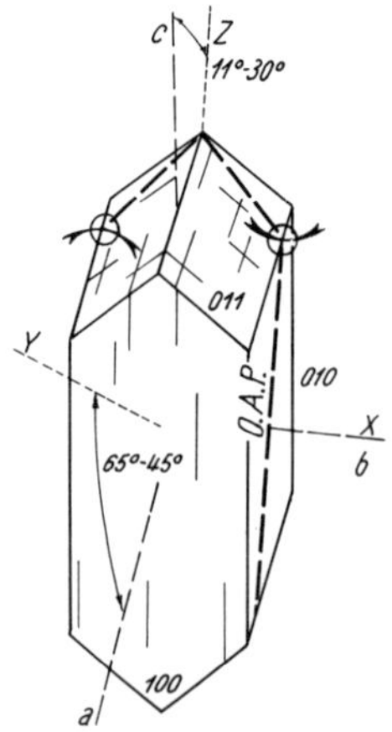


243-1. Phillipsite

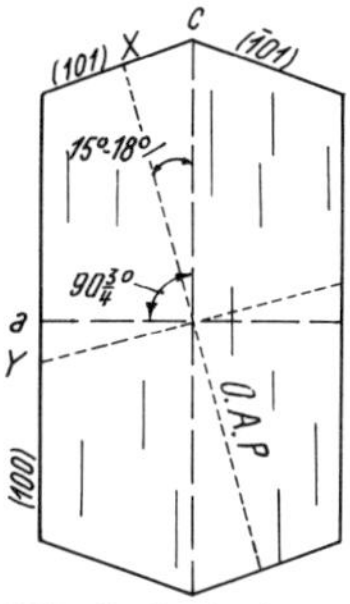


237-2. Scolecite
section II (010)

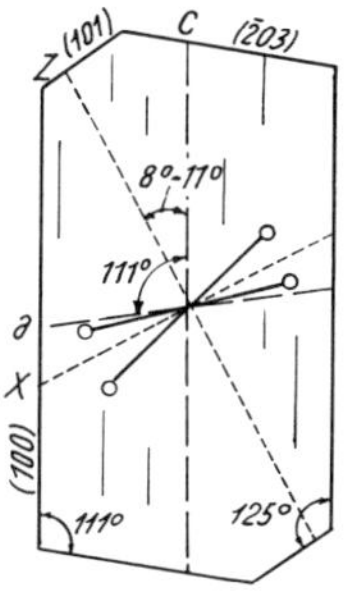


238-2. Laumontite
section II (010)

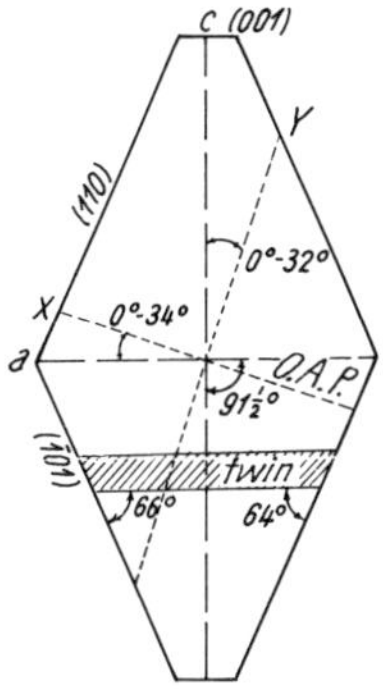


239-2. Heulandite
section II (010)

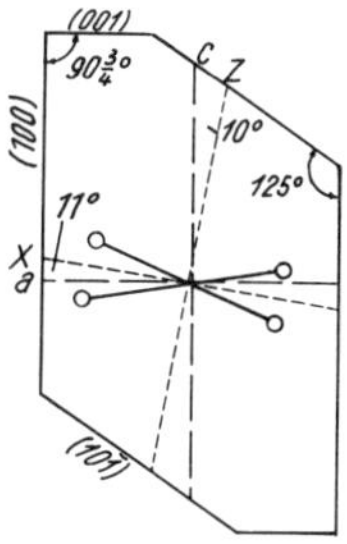


240-2. Epistilbite
section II (010)

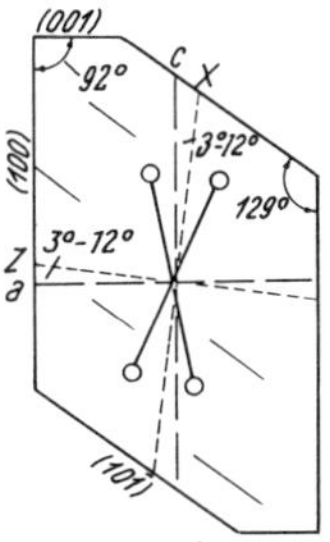


241-2. Stilbite
section II (010)

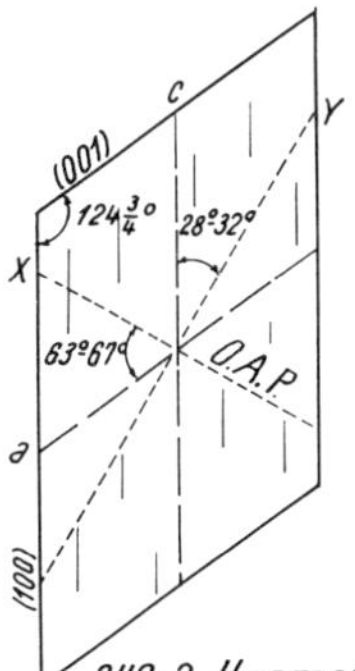


242-2. Harmotome
section II (010)

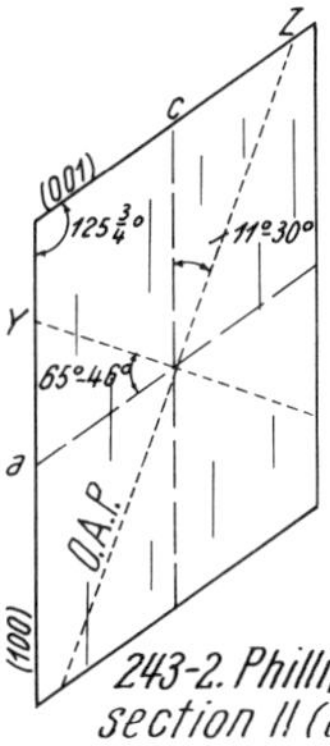


243-2. Phillipsite
section II (010)

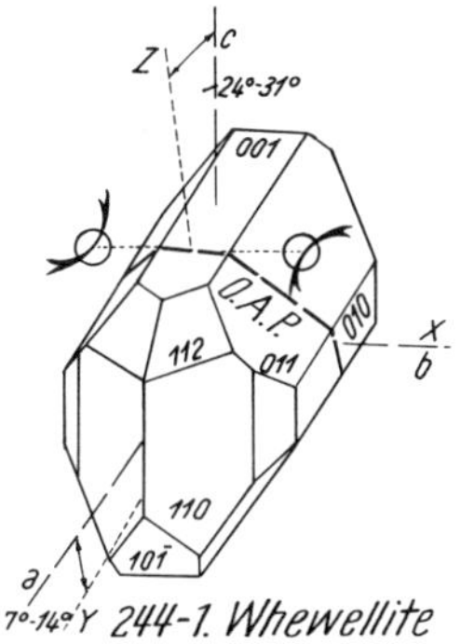


244-1. Whewellite

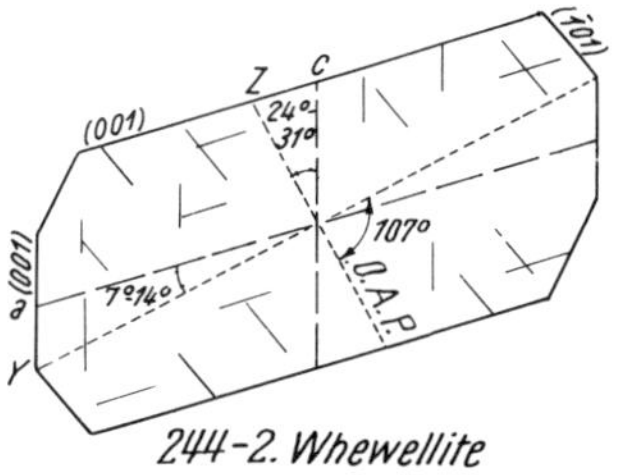


244-2. Whewellite
section II (010)

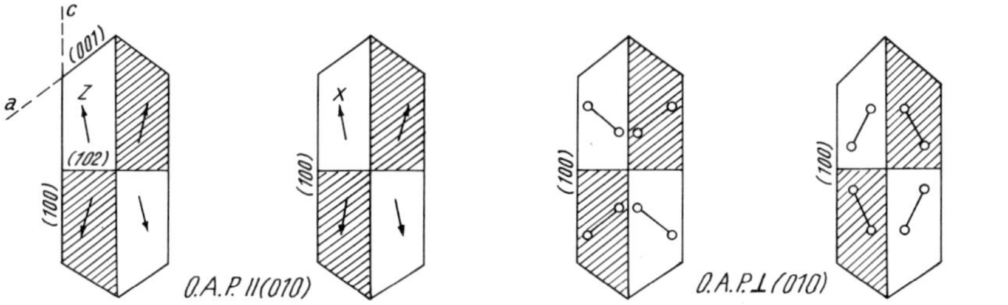


240-3. Epistilbite 241-3. Stilbite 242-3. Harmotome 243-3. Phillipsite

penetration twins
(cleavage flakes II (010)

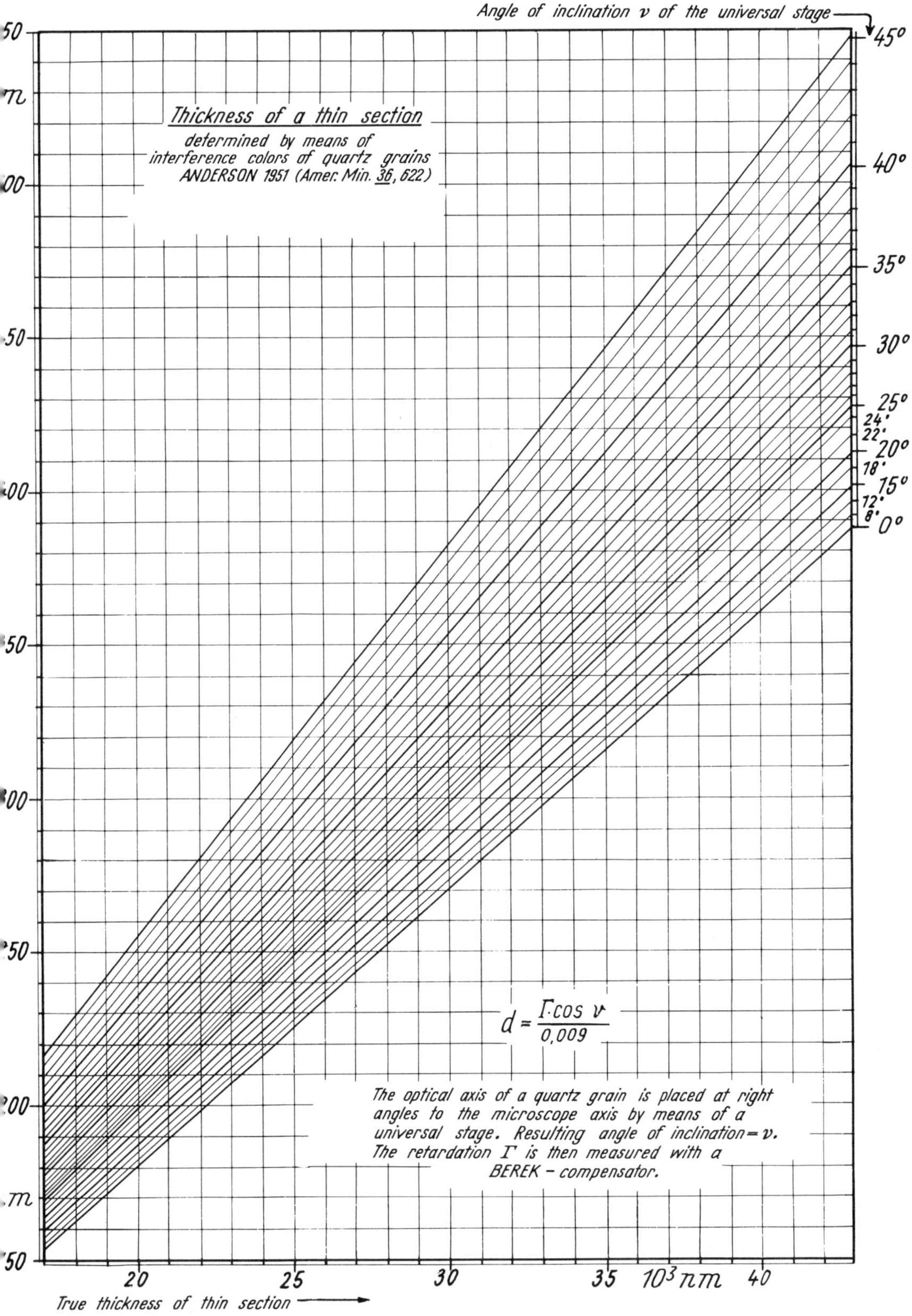
Angle of inclination ν of the universal stage
45°
40°
35°
30°
25°
24°
22°
20°
18°
15°
12°
8°
0°
Thickness of a thin section
determined by means of
interference colors of quartz grains
ANDERSON 1951 (Amer. Min. 36, 622)
d = Γ·cos ν / 0,009
The optical axis of a quartz grain is placed at right angles to the microscope axis by means of a universal stage. Resulting angle of inclination = ν. The retardation Γ is then measured with a BEREK – compensator.
20
25
30
35
10³ nm
40
True thickness of thin section

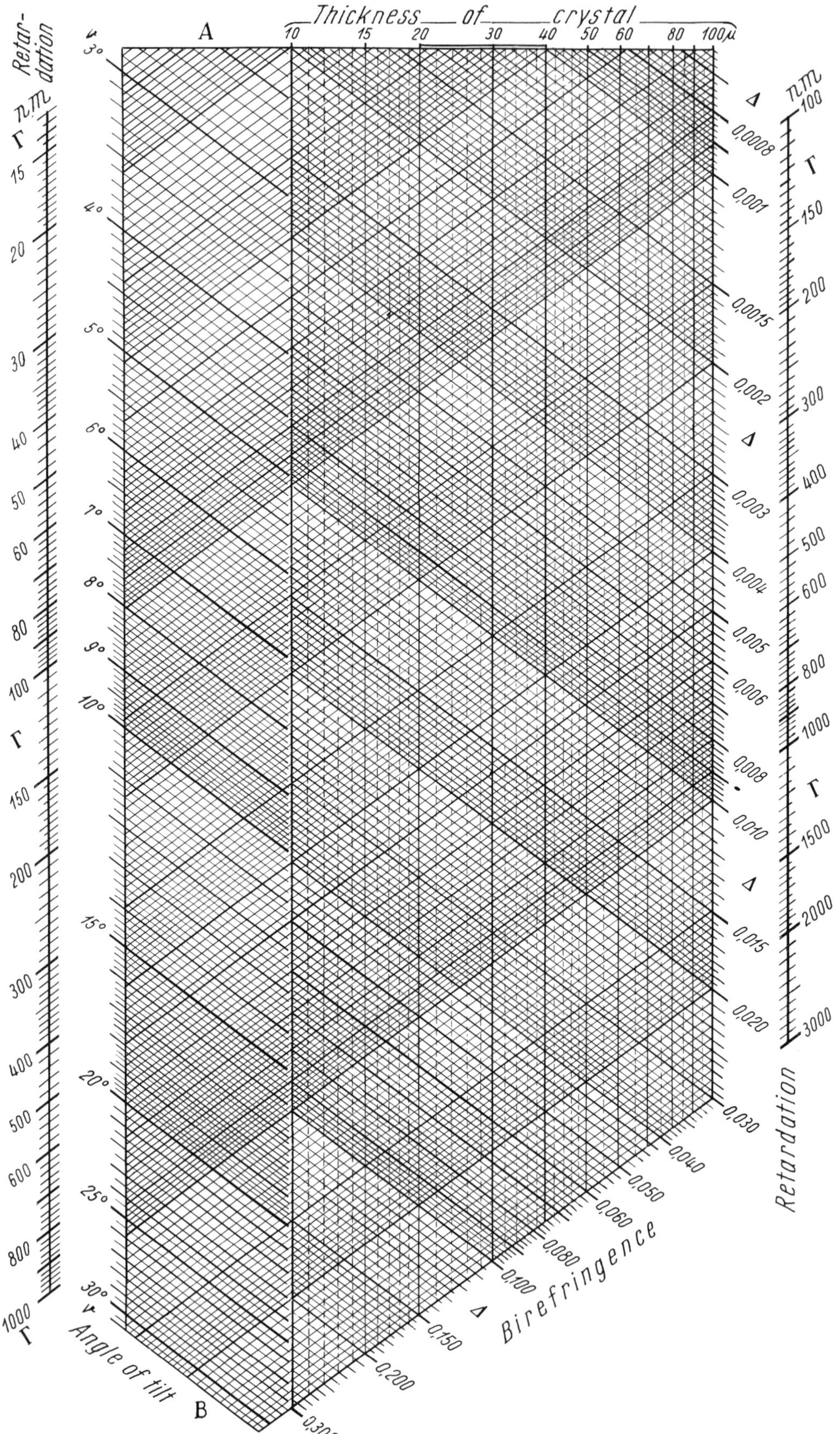
Thickness of crystal
10
15
20
30
40
50
60
80
100μ
A
B
Retar- dation
nm
Γ
15
20
30
40
50
60
80
100
150
200
300
400
500
600
800
1000
3°
4°
5°
6°
7°
8°
9°
10°
15°
20°
25°
30°
Angle of tilt
Δ
0,0008
0,001
0,0015
0,002
0,003
0,004
0,005
0,006
0,008
0,010
0,015
0,020
0,030
0,040
0,050
0,060
0,080
0,100
0,150
0,200
0,300
Birefringence
100
150
200
300
400
500
600
800
1000
1500
2000
3000
Retardation

# Birefringence

## Nomogram by E. TRÖGER 1951
## after a design by PETROV 1940
## (Iswest. Akad. Nauk. S.S.S.R., Ser. geol., 1940, Nr. 2, 151)

Graphical solutions to the following 3 problems are possible

1) Determine retardation $\Gamma$ from the angle of tilt $\nu$ of the BEREK - compensator

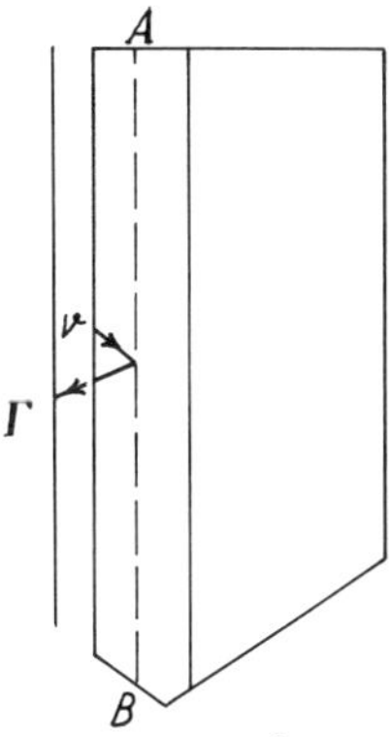


2) Determine relations between retardation $\Gamma$, thickness of section d, and birefringence $\Delta$ of the crystal

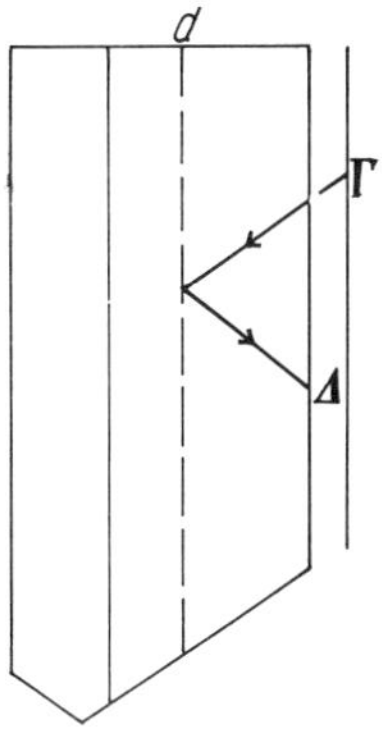


3) Determine birefringence $\Delta$ on a section of known thickness by measuring the angle of tilt $\nu$ of the BEREK - compensator

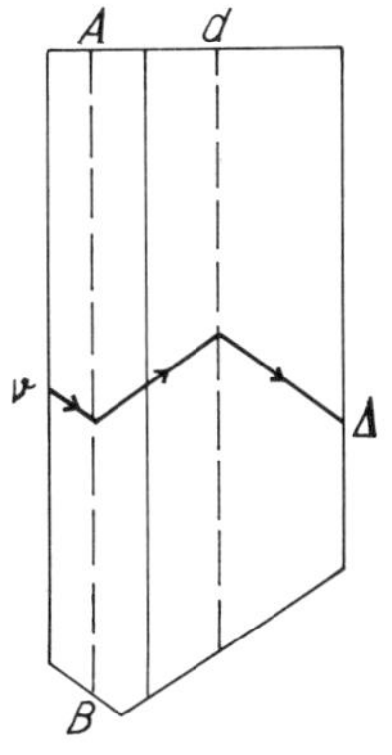


**Calibration:** the conversion function of the compensator used has to be drawn as a vertical line A–B. The position of this line has to be determined from data given by the manufacturer of the instrument.

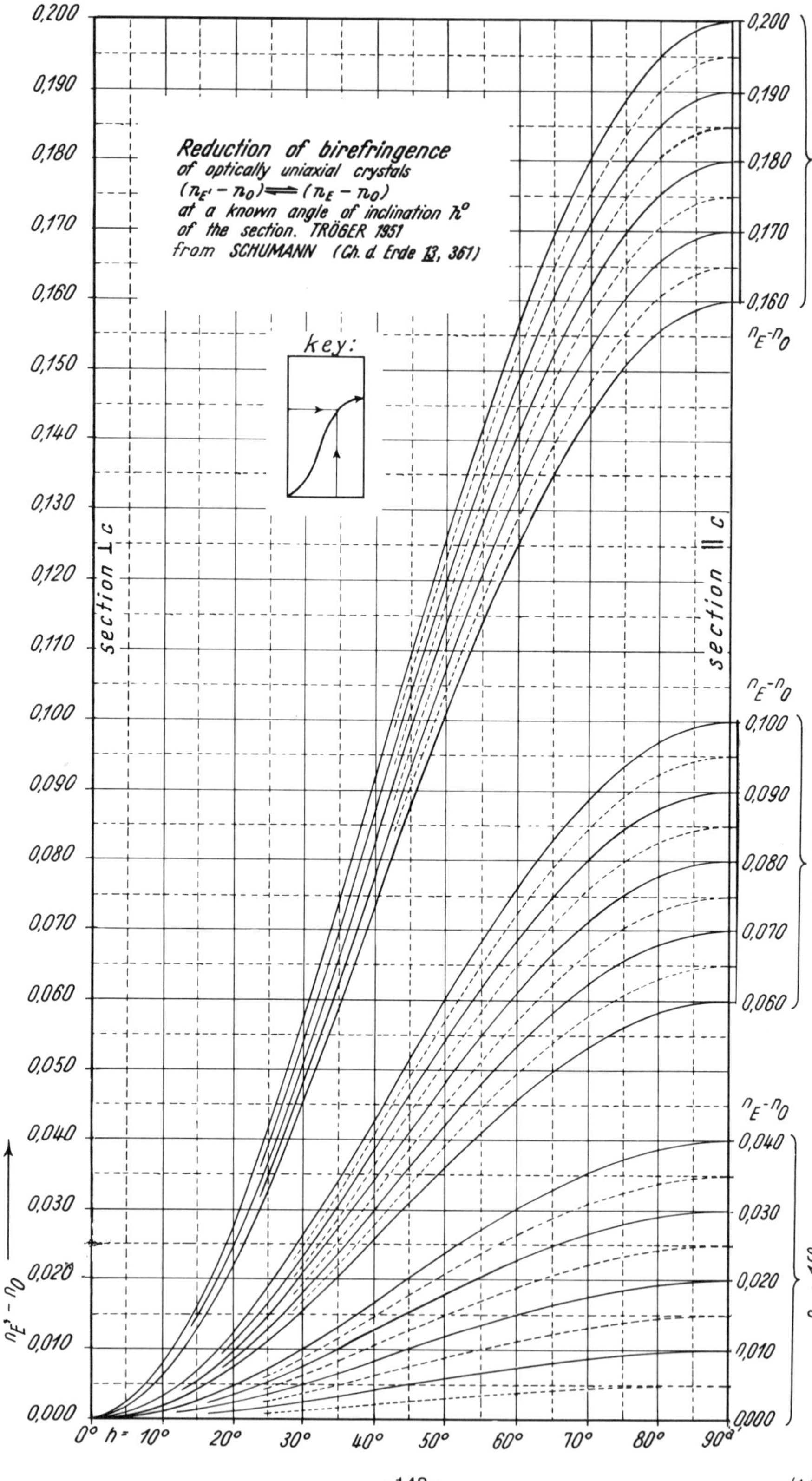
Reduction of birefringence
of optically uniaxial crystals
$(n_{E'} - n_O) \rightleftharpoons (n_E - n_O)$
at a known angle of inclination h°
of the section. TRÖGER 1951
from SCHUMANN (Ch. d. Erde 13, 361)
key:
section ⊥ c
section ∥ c
$n_{E'} - n_O$
$n_E - n_O$
$n_O \sim 1,70$
$n_O \sim 2,00$
$n_O \sim 1,50$
0,000
0,010
0,020
0,030
0,040
0,050
0,060
0,070
0,080
0,090
0,100
0,110
0,120
0,130
0,140
0,150
0,160
0,170
0,180
0,190
0,200
0° h = 10° 20° 30° 40° 50° 60° 70° 80° 90°

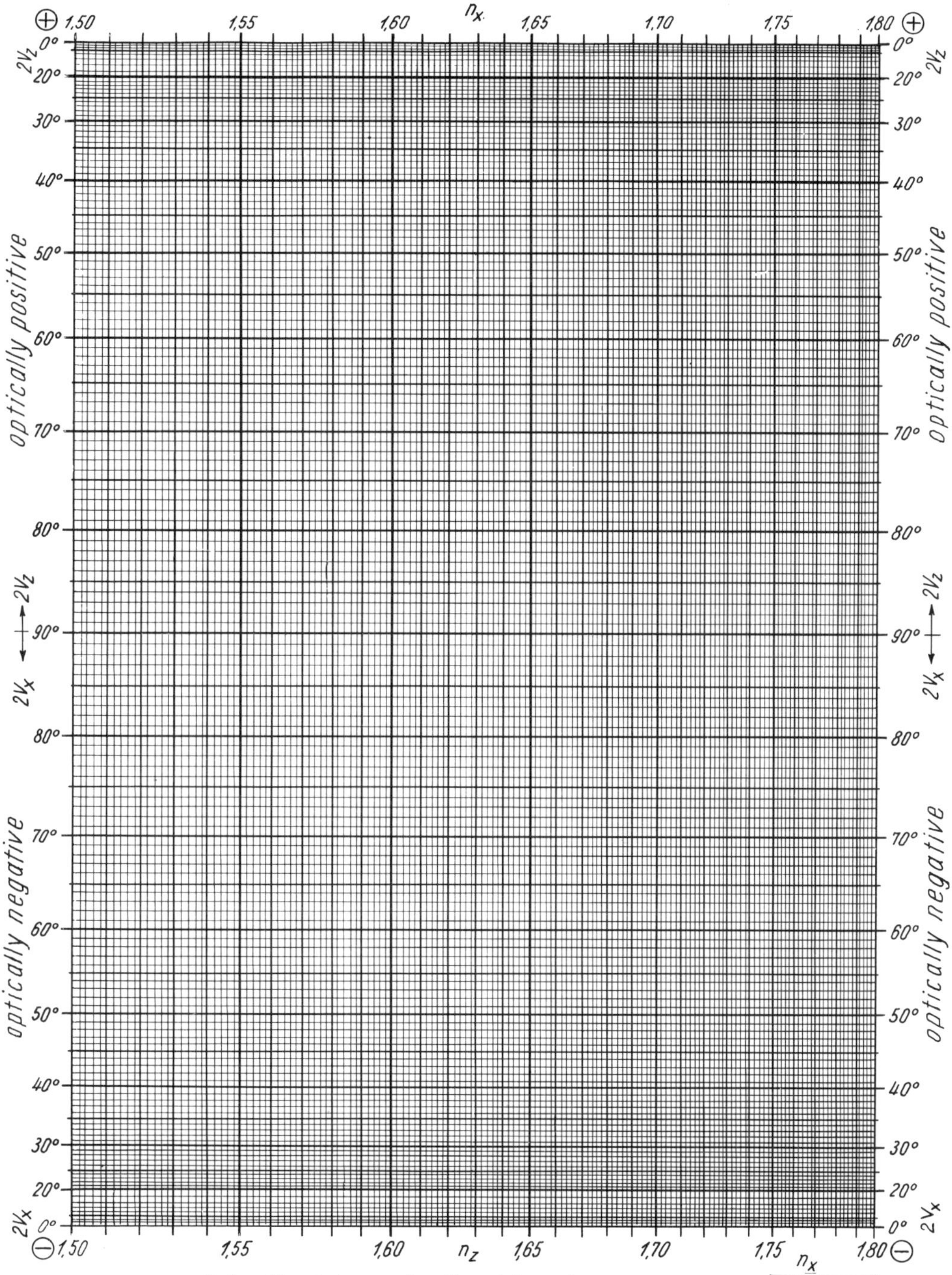


Refractive index and optic axial angle

Nomogram after MERTIE 1942 (Amer. Min. 27, 542)
for solution of the following equations:

$$\sin^2 V_X = \frac{\frac{1}{n_Y^2} - \frac{1}{n_Z^2}}{\frac{1}{n_X^2} - \frac{1}{n_Z^2}} \qquad \sin^2 V_Z = \frac{\frac{1}{n_X^2} - \frac{1}{n_Y^2}}{\frac{1}{n_X^2} - \frac{1}{n_Z^2}}$$

(Use a plate of acrylate glass with engraved line for easy reading!)

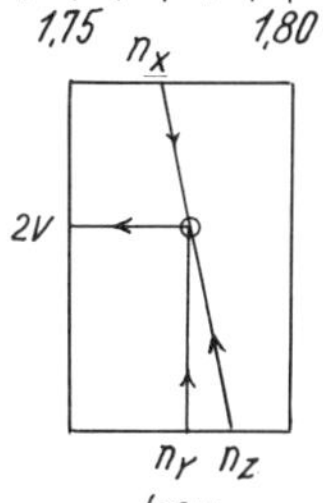


key

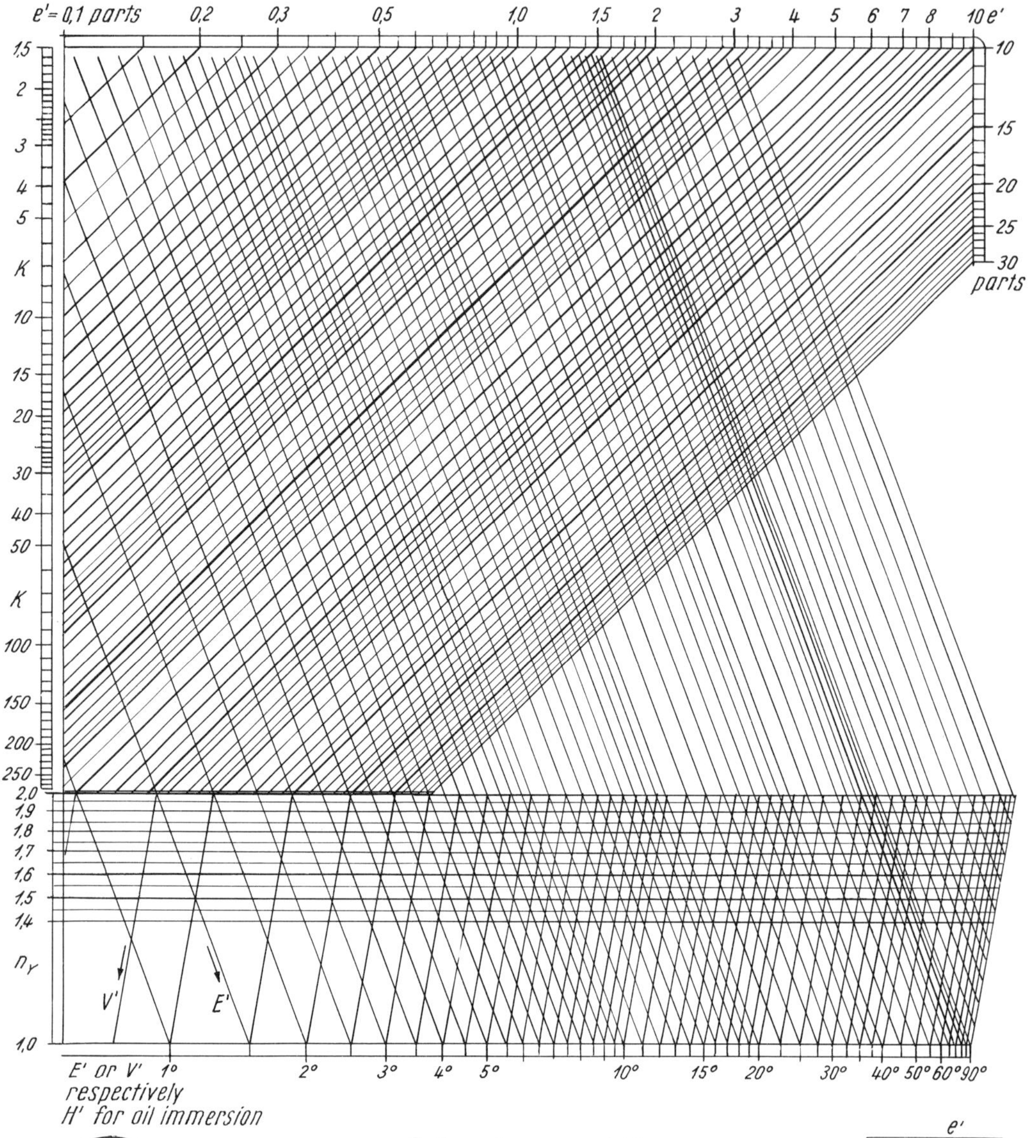


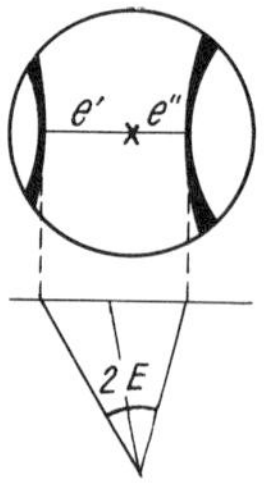


## Optic axial angle 2 V from conoscopic measurement

Nomogram after H. WINCHELL 1946, (Am. Min. 31, 44)

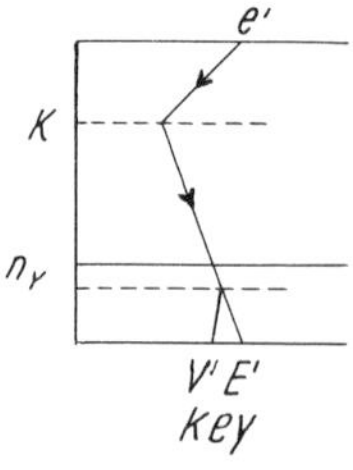


The distance of the melatopes is to be measured in two parts from the center to avoid errors due to asymmetry

$$e' = K \cdot \sin E' = n_Y \cdot K \cdot \sin V'$$

$$E' + E'' = 2E; \qquad V' + V'' = 2V$$

Calibration: MALLARD's constants K have to be determined empirically with crystals of known 2E or with a universal stage for the common ocular and objective combinations. The constants K are then to be marked as horizontal lines in the nomogram.

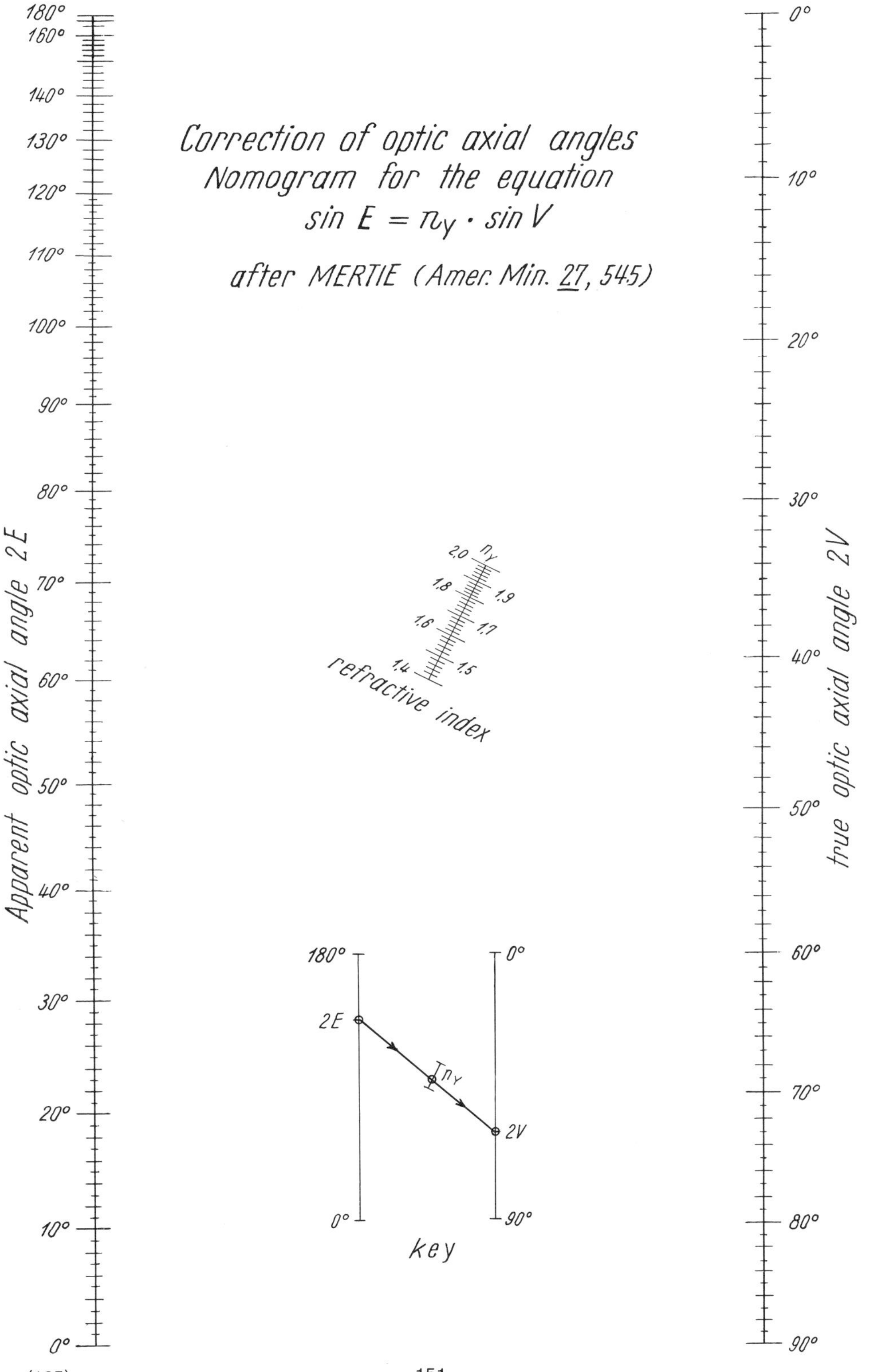
Correction of optic axial angles
Nomogram for the equation
sin E = n_Y · sin V
after MERTIE (Amer. Min. 27, 545)
Apparent optic axial angle 2E
180° 160° 140° 130° 120° 110° 100° 90° 80° 70° 60° 50° 40° 30° 20° 10° 0°
true optic axial angle 2V
0° 10° 20° 30° 40° 50° 60° 70° 80° 90°
refractive index
n_Y
1,4 1,5 1,6 1,7 1,8 1,9 2,0
key
180° 0° 2E n_Y 2V 0° 90°

## *Angle of inclination*

*Correction of universal stage readings*

*Nomogram by TRÖGER 1939 (Zbl. Min. 1939 A, 179)*

*for the segment 1,55*

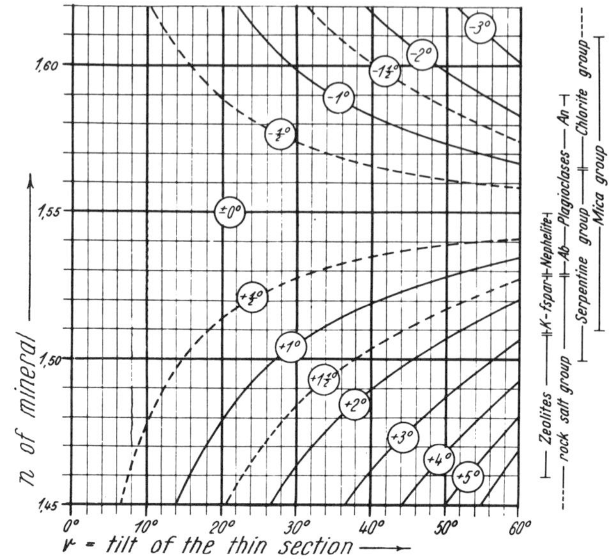


$h = \nu \pm k$

*for the segment 1,65*

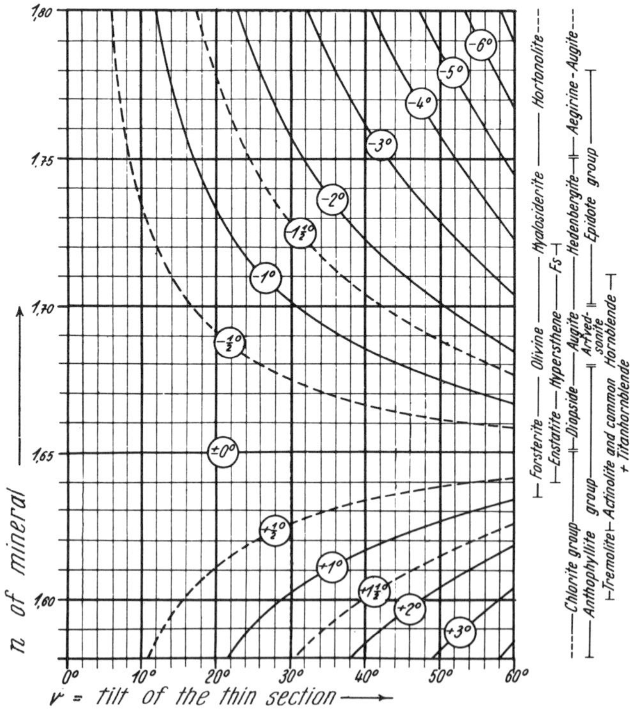

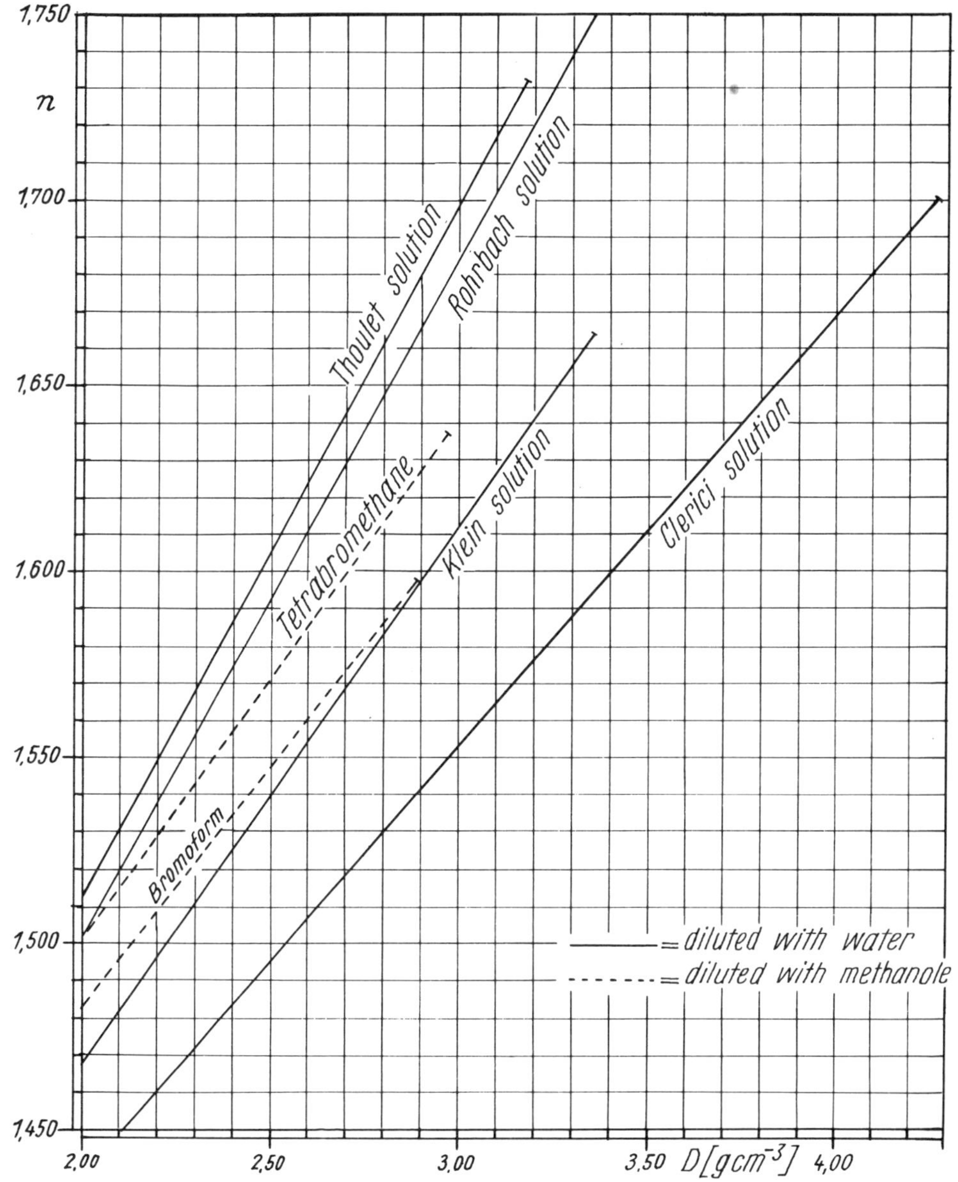


**Heavy liquids**

*Control of density by measuring the refractive indices at 20°C.*
*TRÖGER 1956 and FOSTER 1947 (Amer. Min. 32, 464)*

# Densities of Rock-Forming Minerals
(except water-soluble minerals)

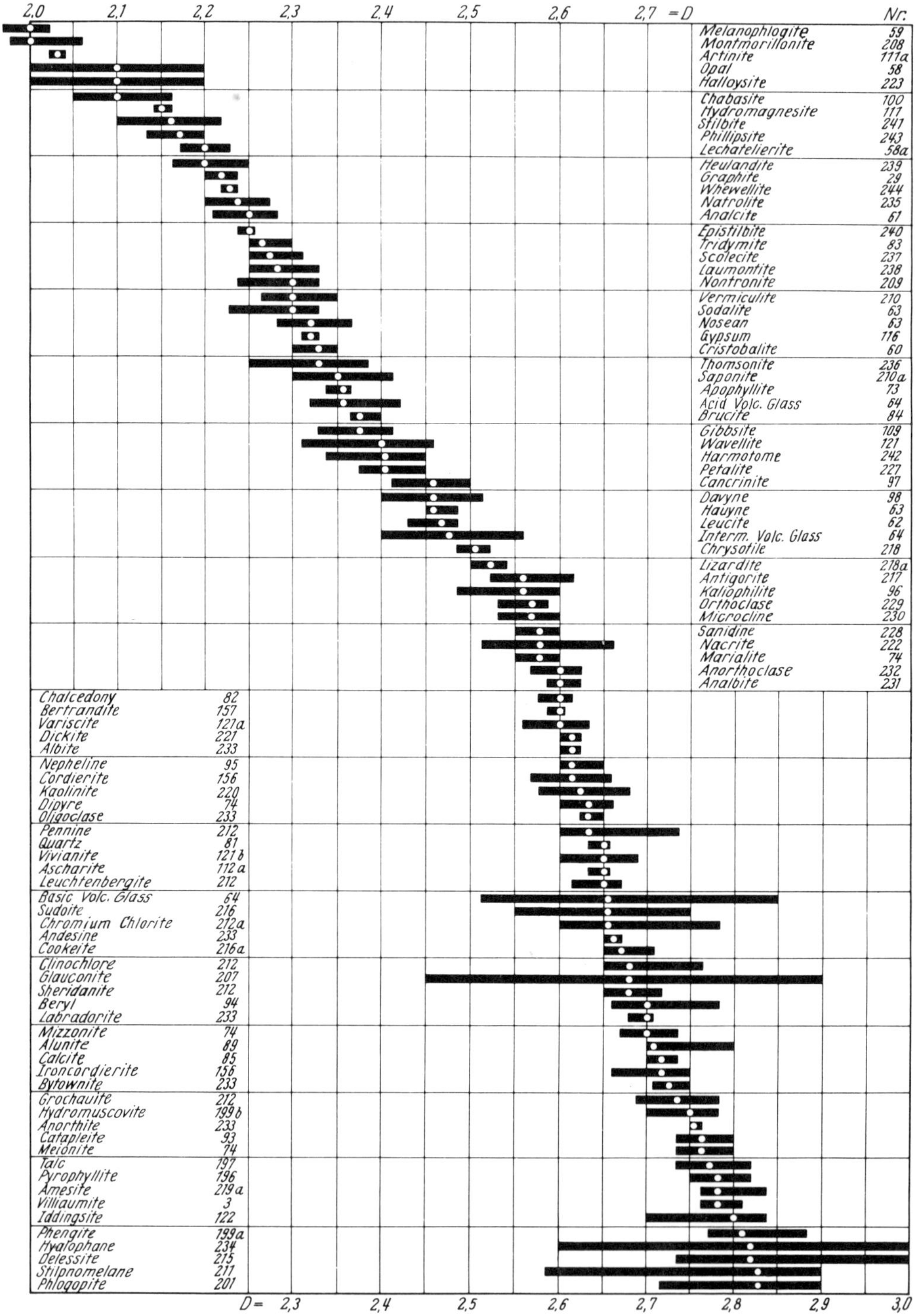

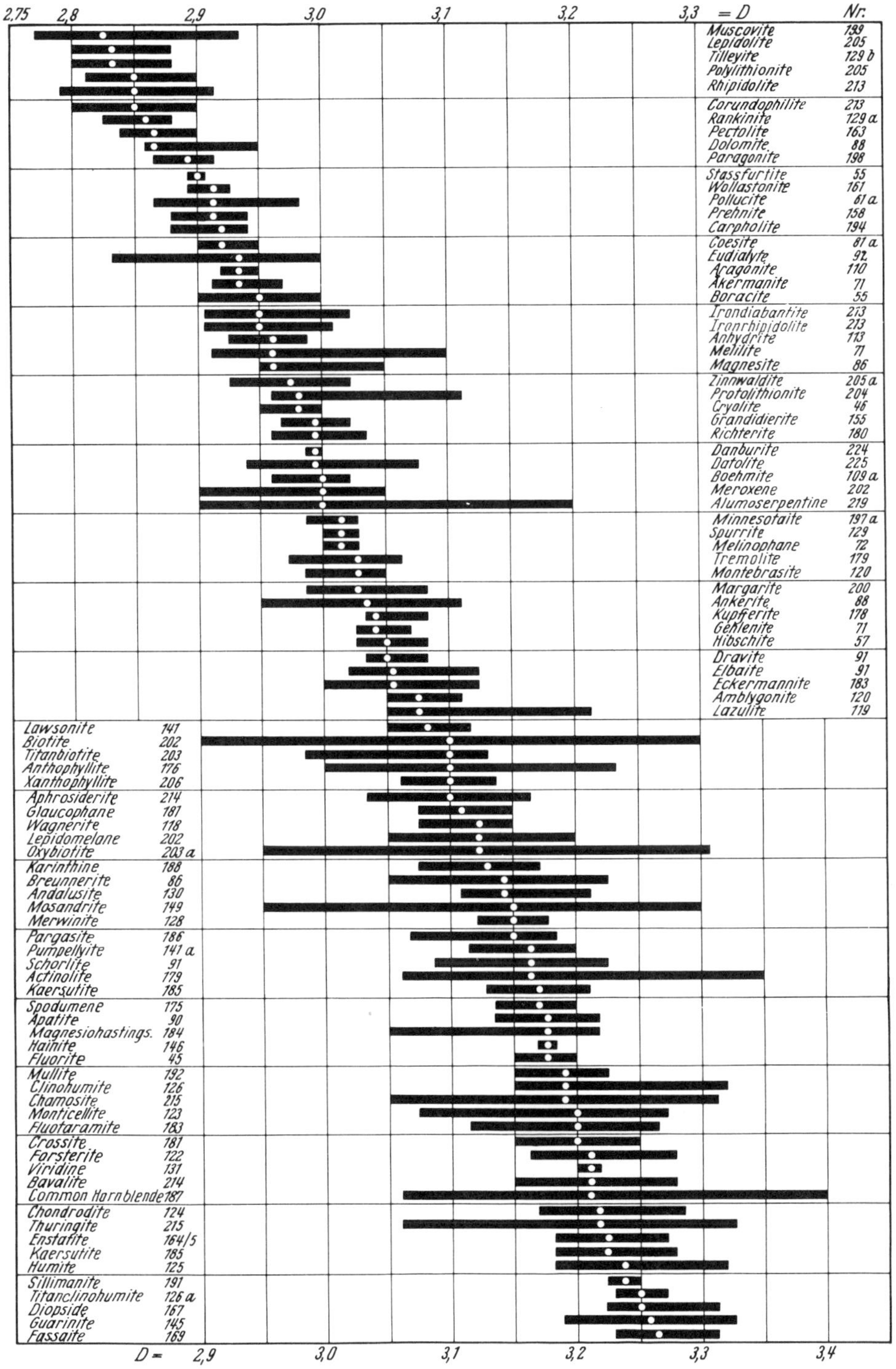
2,75 2,8 2,9 3,0 3,1 3,2 3,3 = D
Nr.
Muscovite 199
Lepidolite 205
Tilleyite 129 b
Polylithionite 205
Rhipidolite 213
Corundophilite 213
Rankinite 129 a
Pectolite 163
Dolomite 88
Paragonite 198
Stassfurtite 55
Wollastonite 161
Pollucite 61 a
Prehnite 158
Carpholite 194
Coesite 81 a
Eudialyte 92
Aragonite 110
Åkermanite 71
Boracite 55
Irondiabantite 213
Ironrhipidolite 213
Anhydrite 113
Melilite 71
Magnesite 86
Zinnwaldite 205 a
Protolithionite 204
Cryolite 46
Grandidierite 155
Richterite 180
Danburite 224
Datolite 225
Boehmite 109 a
Meroxene 202
Alumoserpentine 219
Minnesotaite 197 a
Spurrite 129
Melinophane 72
Tremolite 179
Montebrasite 120
Margarite 200
Ankerite 88
Kupfferite 178
Gehlenite 71
Hibschite 57
Dravite 91
Elbaite 91
Eckermannite 183
Amblygonite 120
Lazulite 119
Lawsonite 141
Biotite 202
Titanbiotite 203
Anthophyllite 176
Xanthophyllite 206
Aphrosiderite 214
Glaucophane 187
Wagnerite 118
Lepidomelane 202
Oxybiotite 203 a
Karinthine 188
Breunnerite 86
Andalusite 130
Mosandrite 149
Merwinite 128
Pargasite 186
Pumpellyite 141 a
Schorlite 91
Actinolite 179
Kaersutite 185
Spodumene 175
Apatite 90
Magnesiohastings. 184
Hainite 146
Fluorite 45
Mullite 192
Clinohumite 126
Chamosite 215
Monticellite 123
Fluotaramite 183
Crossite 181
Forsterite 122
Viridine 131
Bavalite 214
Common Hornblende 187
Chondrodite 124
Thuringite 215
Enstatite 164/5
Kaersutite 185
Humite 125
Sillimanite 191
Titanclinohumite 126 a
Diopside 167
Guarinite 145
Fassaite 169
D = 2,9 3,0 3,1 3,2 3,3 3,4

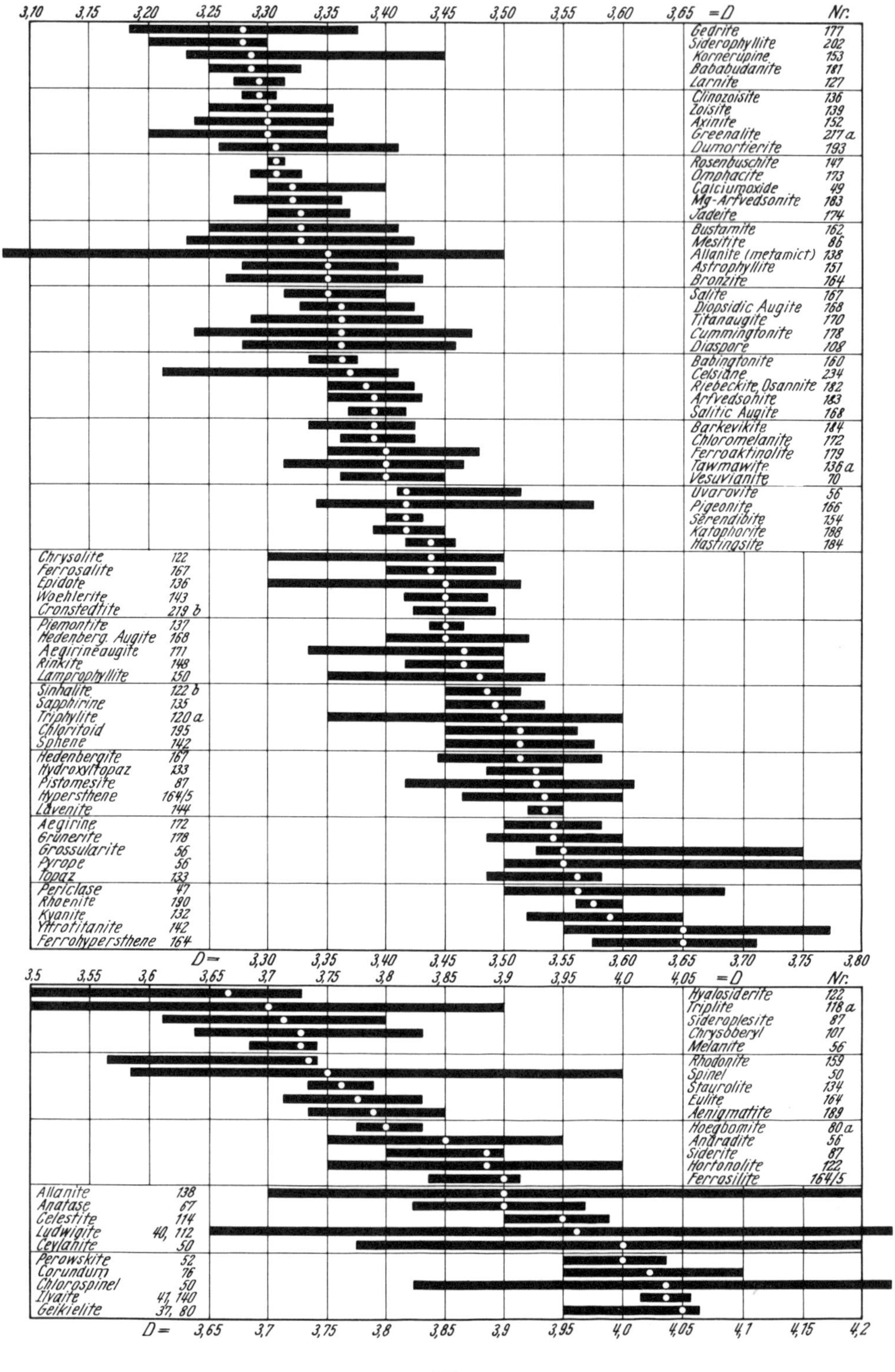
3,10 3,15 3,20 3,25 3,30 3,35 3,40 3,45 3,50 3,55 3,60 3,65 =D
Nr.
Gedrite 177
Siderophyllite 202
Kornerupine 153
Bababudanite 181
Larnite 127
Clinozoisite 136
Zoisite 139
Axinite 152
Greenalite 217 a
Dumortierite 193
Rosenbuschite 147
Omphacite 173
Calciumoxide 49
Mg-Arfvedsonite 183
Jadeite 174
Bustamite 162
Mesitite 86
Allanite (metamict) 138
Astrophyllite 151
Bronzite 164
Salite 167
Diopsidic Augite 168
Titanaugite 170
Cummingtonite 178
Diaspore 108
Babingtonite 160
Celsiane 234
Riebeckite, Osannite 182
Arfvedsonite 183
Salitic Augite 168
Barkevikite 184
Chloromelanite 172
Ferroaktinolite 179
Tawmawite 136 a
Vesuvianite 70
Uvarovite 56
Pigeonite 166
Serendibite 154
Kataphorite 188
Hastingsite 184
Chrysolite 122
Ferrosalite 167
Epidote 136
Woehlerite 143
Cronstedtite 219 b
Piemontite 137
Hedenberg. Augite 168
Aegirineaugite 171
Rinkite 148
Lamprophyllite 150
Sinhalite 122 b
Sapphirine 135
Triphylite 120 a
Chloritoid 195
Sphene 142
Hedenbergite 167
Hydroxyltopaz 133
Pistomesite 87
Hypersthene 164/5
Låvenite 144
Aegirine 172
Grünerite 178
Grossularite 56
Pyrope 56
Topaz 133
Periclase 47
Rhoenite 190
Kyanite 132
Yttrotitanite 142
Ferrohypersthene 164
D= 3,30 3,35 3,40 3,45 3,50 3,55 3,60 3,65 3,70 3,75 3,80
3,5 3,55 3,6 3,65 3,7 3,75 3,8 3,85 3,9 3,95 4,0 4,05 =D
Nr.
Hyalosiderite 122
Triplite 118 a
Sideroplesite 87
Chrysoberyl 101
Melanite 56
Rhodonite 159
Spinel 50
Staurolite 134
Eulite 164
Aenigmatite 189
Hoegbomite 80 a
Andradite 56
Siderite 87
Hortonolite 122
Ferrosilite 164/5
Allanite 138
Anatase 67
Celestite 114
Ludwigite 40, 112
Ceylanite 50
Perowskite 52
Corundum 76
Chlorospinel 50
Ilvaite 41, 140
Geikielite 37, 80
D= 3,65 3,7 3,75 3,8 3,85 3,9 3,95 4,0 4,05 4,1 4,15 4,2

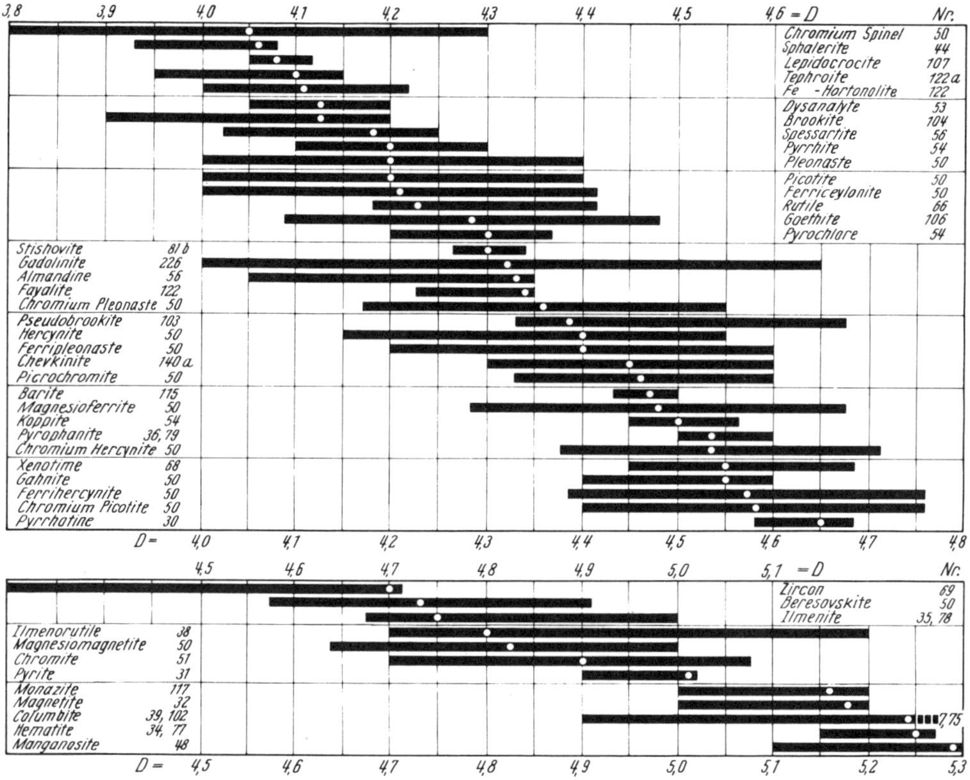
3,8 3,9 4,0 4,1 4,2 4,3 4,4 4,5 4,6 = D
Nr.
Chromium Spinel 50
Sphalerite 44
Lepidocrocite 107
Tephroite 122 a
Fe - Hortonolite 122
Dysanalyte 53
Brookite 104
Spessartite 56
Pyrrhite 54
Pleonaste 50
Picotite 50
Ferriceylonite 50
Rutile 66
Goethite 106
Pyrochlore 54
Stishovite 81 b
Gadolinite 226
Almandine 56
Fayalite 122
Chromium Pleonaste 50
Pseudobrookite 103
Hercynite 50
Ferripleonaste 50
Chevkinite 140 a
Picrochromite 50
Barite 115
Magnesioferrite 50
Koppite 54
Pyrophanite 36, 79
Chromium Hercynite 50
Xenotime 68
Gahnite 50
Ferrihercynite 50
Chromium Picotite 50
Pyrrhotine 30
D = 4,0 4,1 4,2 4,3 4,4 4,5 4,6 4,7 4,8
4,5 4,6 4,7 4,8 4,9 5,0 5,1 = D
Nr.
Zircon 69
Beresovskite 50
Ilmenite 35, 78
Ilmenorutile 38
Magnesiomagnetite 50
Chromite 51
Pyrite 31
Monazite 117
Magnetite 32
Columbite 39, 102
Hematite 34, 77
Manganosite 48
7,75
D = 4,5 4,6 4,7 4,8 4,9 5,0 5,1 5,2 5,3

# Dispersion of the Refractive Indices

**of Rock-Forming Minerals**

(Abstracted from a table by A. N. WINCHELL, Am. Mineral. 14, 128, 1929)

$n_D = n_O$ of optical uniaxial minerals, $n_D = n_Y$ of optical biaxial minerals

| Mineral | $n_D$ |
|---|---|
| $n_F - n_C \sim 0.004$ | |
| Fluorite | 1.434 |
| $n_F - n_C \sim 0.006$–7 | |
| Ice | 1.309 |
| Scapolite | 1.581 |
| Tremolite | 1.618 |
| $n_F - n_C \sim 0.008 \pm 1/2$ | |
| Leucite | 1.508 |
| Cancrinite | 1.516 |
| Adularia | 1.523 |
| Gypsum | 1.523 |
| Sanidine | 1.521 |
| Albite | 1.533 |
| Nepheline | 1.538 |
| Cordierite | 1.537 |
| Oligoclase | 1.543 |
| Quartz | 1.544 |
| Grothite | 1.554 |
| Anhydrite | 1.575 |
| Muscovite | 1.599 |
| Topaz | 1.618 |
| Melilite | 1.634 |
| Anthophyllite | 1.637 |
| $n_F - n_C \sim 0.009 \pm 1/2$ | |
| Natrolite | 1.479 |
| Sodalite | 1.483 |
| Cancrinite | 1.525 |
| Anorthoclase | 1.628 |
| Apophyllite | 1.534 |
| Nepheline | 1.538 |
| Cordierite | 1.546 |
| Andesine | 1.552 |
| Labradorite | 1.563 |
| Bytownite | 1.573 |
| Beryl | 1.574 |
| Anorthite | 1.583 |
| Scapolite | 1.587 |
| Barite | 1.637 |
| Tourmaline | 1.646 |

| Mineral | $n_D$ |
|---|---|
| Sillimanite | 1.660 |
| $n_F - n_C \sim 0.010 \pm 1/2$ | |
| Sodalite | 1.484 |
| Anorthite | 1.583 |
| Muscovite | 1.587 |
| Tourmaline | 1.635 |
| Apatite | 1.638 |
| Sillimanite | 1.658 |
| Spodumene | 1.665 |
| Spinel | 1.721 |
| Diaspore | 1.772 |
| $n_F - n_C \sim 0.011 \pm 1/2$ | |
| Microcline | 1.522 |
| Brucite | 1.566 |
| Scapolite | 1.590 |
| Tourmaline | 1.646 |
| Diopside | 1.678 |
| Clinozoisite | 1.715 |
| Kyanite | 1.722 |
| Chrysoberyl | 1.747 |
| Corundum | 1.768 |
| Spessartine | 1.814 |
| $n_F - n_C \sim 0.012 \pm 1/2$ | |
| Wavellite | 1.526 |
| Brucite | 1.566 |
| Beryl | 1.570 |
| Clinochlore | 1.596 |
| Tourmaline | 1.652 |
| Sillimanite | 1.660 |
| Monticellite | 1.662 |
| Boracite | 1.667 |
| Olivine | 1.670 |
| Aragonite | 1.681 |
| Triphylite | 1.688 |
| Pyrope | 1.705 |
| Olivine | 1.713 |
| Spinel | 1.719 |
| Staurolite | 1.745 |
| Epidote | 1.763 |

| Mineral | $n_D$ |
|---|---|
| $n_F - n_C \sim 0.013 \pm 1/2$ | |
| Halite | 1.544 |
| Scapolite | 1.570 |
| Beryl | 1.582 |
| Clinochlore | 1.590 |
| Calcite | 1.658 |
| Forsterite | 1.659 |
| Dolomite | 1.680 |
| Aragonite | 1.681 |
| Axinite | 1.685 |
| Olivine | 1.701 |
| Augite | 1.704 |
| Clinozoisite | 1.717 |
| Periclase | 1.738 |
| Grossularite | 1.744 |
| Pyrope | 1.744 |
| Staurolite | 1.750 |
| Spessartite | 1.804 |
| $n_F - n_C \sim 0.014 \pm 1/2$ | |
| Tourmaline | 1.654 |
| Calcite | 1.659 |
| Rinkite | 1.668 |
| Axinite | 1.685 |
| Dolomite | 1.692 |
| Clinozoisite | 1.717 |
| Pyroxene | 1.726 |
| Periclase | 1.738 |
| $n_F - n_C \sim 0.015 \pm 1/2$ | |
| Xenotime | 1.721 |
| Pyrope | 1.744 |
| Grossularite | 1.744 |
| Almandine | 1.796 |
| Spessartite | 1.815 |
| $n_F - n_C \sim 0.016$ | |
| Datolite | 1.649 |
| Hedenbergite | 1.737 |
| Epidote | 1.742 |
| Pyrope | 1.746 |
| Grossularite | 1.763 |
| Monazite | 1.791 |

| Mineral | $n_D$ |
|---|---|
| $n_F - n_C \sim 0.019–21$ | |
| Garnet | 1.742 |
| Tephroite | 1.786 |
| Gahnite | 1.807 |
| Spessartite | 1.811 |
| Uvarovite | 1.838 |
| $n_F - n_C \sim 0.022–24$ | |
| Siderite | 1.872 |
| Sphene | 1.909 |
| Zircon | 1.925 |
| $n_F - n_C \sim 0.025–27$ | |
| Mullite | 1.652 |
| Epidote | 1.757 |
| Grossularite | 1.771 |
| Diamond | 2.418 |
| $n_F - n_C \sim 0.035 \pm 3$ | |
| Acmite | 1.799 |
| Andradite | 1.895 |
| Sphene | 1.905 |
| Cassiterite | 1.999 |
| $n_F - n_C \sim 0.080 \pm 10$ | |
| Sulfur | 2.038 |
| Wurtzite | 2.356 |
| Sphalerite | 2.368 |
| $n_F - n_C \sim 0.150 \pm 30$ | |
| Goethite | 2.390 |
| Brookite | 2.586 |
| Rutile | 2.616 |
| $n_F - n_C \sim 0.500$ | |
| Hematite | 3.220 |

# Color and Pleochroism

## in thin sections

### 1. Red and Brown Colors

**Rose-colored, pink:** Spinel, pyrope, almandine, nosean, xenotime O, eudialyte, chrysoberyl X, andalusite X, sapphirine X, piemontite Y, rhodonite Y, hypersthene X, pigeonite XY, mullite X, lepidolite YZ, kaemmererite X, cookeite XY, dumortierite YZ, tephroite Y

**Carmine-red:** Piemontite Z, thulite, bustamite Y

**Dark red:** Villiaumite O, titanclinohumite X, astrophyllite X

**Purple:** Fluorite, corundum O, geikielite, piemontite Z, babingtonite Y

**Lilac:** Piemontite Y, glaucophane Y, dumortierite Y, kaemmererite X

**Lilac to brown:** Schorlite, enstatite X, titanaugite YZ, karinthine Y

**Pale lilac to brown:** Apatite, schorlite, clinohypersthene X

**Pale brown:** Periclase, volcanic glasses, anatase O, xenotime, zircon, vesuvianite, magnesite, siderite, apatite, sphene Z, acmite Y, cummingtonite Y, phlogopite YZ, sinhalite Y

**Pale brownish yellow:** Gedrite Z, richterite X, osannite Y, common hornblende X, chrysolite X, zinnwaldite X, xanthophyllite YZ

**Yellowish brown:** Sphalerite, cassiterite, rutile O or E, anatase E, catapleite, goethite Y, allanite X, ilvaite X, chevkinite X, gedrite Z, biotite X, wodanite X, oxybiotite X, kaersutite X

**Pale olive-brown:** Schorlite, muscovite YZ

**Olive-brown:** Schorlite, goethite Z, common hornblende Z, biotite YZ, stilpnomelane YZ, kaersutite Z

**Yellowish red:** Loeweite, pyrochlore, hematite E, pyrophanite, eudialyte, lepidocrocite Y, titanclinohumite Y, astrophyllite X, iddingsite YZ, katophorite X, triplite (variable)

**Brownish red:** Wodanite Z, aenigmatite Y, kaersutite YZ, katophorite YZ, tephroite X, oxybiotite YZ

**Reddish brown:** Koenenite O, perovskite, pyrochlore, cassiterite, rutile O, hematite O, pyrophanite, geikielite, niobite X, tantalite Y, pseudobrookite Y, brookite, baddeleyite Z, lepidocrocite Z, allanite Y, chevkinite YZ, sphene Z, kaersutite YZ, aenigmatite XY, rhoenite Y, biotite YZ, wodanite Z, xanthophyllite X, cronstedtite X

**Brown:** Melanite, volcanic glasses, cassiterite, tawmawite Y, allanite Z, ilvaite X, acmite X, aenigmatite Y, rhoenite X, biotite YZ, protolithionite YZ, sinhalite X, bavalite

**Dark brown:** Rinneite, chromite, melanite, ilmenite, niobite Y, tantalite Z, ludwigite Z, allanite Z, ilvaite Y, aenigmatite Z, rhoenite Z, biotite YZ, katophorite YZ

### 2. Yellow Colors

**Pale yellow:** Tachydrite, loeweite, sphalerite, spessartite, hibschite, beckelite, volcanic glasses, xenotime O, melilite, melinophane O, scapolite, dolomite, magnesite, dravite, catapleite, beryl, chrysoberyl Y, brookite, lepidocrocite X, monazite, chondrodite YZ, humite YZ, clinohumite YZ, titanclinohumite YZ, viridine X, staurolite XY, sapphirine X, woehlerite, låvenite XY, guarinite, hainite, rosenbuschite, rinkite, mosandrite, lamprophyllite XY, cornerupine, cordierite X, bertrandite, rhodonite Z, pigeonite Z, jadeite Z, cummingtonite Y, richterite Y, hornblende X, chrysotile X, minnesotaite X, muscovite YZ, phlogopite YZ, hydrobiotite YZ, wodanite X, protolithionite X, pennine, delessite X, stilpnomelane X, nontronite X, cookeite Z, dumortierite YZ

**Yellow:** Villiaumite E, picotite, pyrochlore, dravite, tantalite X, goethite X, lepidocrocite X, chondrodite X, humite X, clinohumite X, tawmawite Y, piemontite X, thulite, astrophyllite Y, chamosite X, stilpnomelane X, katophorite X

**Dark yellow:** Dravite, pseudobrookite Z, viridine Z, staurolite Z

**Pale greenish yellow:** Vesuvianite, melinophane E, monazite, olivine XZ, epidote X, thulite, sphene Y, cornerupine, serendibite X, hypersthene Y, osannite Y, jenkinsite X, pennine, vermiculite YZ, iddingsite X

**Greenish yellow:** Rutile E, anatase O, vesuvianite, epidote Y, aegirine YZ, acmite Z, riebeckite Z, rhipidolite XY, glauconite X, gedrite XY, Na-tremolite Y

**Orange:** Goethite Z, olivine Y, titanclinohumite X, piemontite X, låvenite Z, lamprophyllite Z, astrophyllite XZ, rhodonite X, bustamite XZ, richterite YZ, wodanite Y, xanthophyllite X, iddingsite

**Greyish yellow:** Melilite, hematite, dravite, staurolite Y, allanite XY, lamprophyllite XY, hypersthene Y, titanaugite X, aegirine-augite Z, gedrite XY, cummingtonite Z, biotite X, xanthophyllite YZ, glauconite YZ

**Brownish yellow:** Picotite, dysanalyte, melanite, volcanic glasses, xenotime E, dravite, brookite, baddeleyite X, goethite X, chondrodite X, staurolite Z, babingtonite Z, titanaugite Z, aegirine Z, arfvedsonite Y, wodanite Y, glauconite YZ

**Dark brownish yellow:** Picotite, melanite

## 3. Green Colors

**Pale green:** Boracite, nosean, xenotime E, vesuvianite, corundum E, chrysoberyl Z, andalusite YZ, sapphirine Y, hypersthene Z, clinohypersthene Z, pigeonite Z, hedenbergite YZ, augite XY, fassaite YZ, aegirine Y, jadeite X, omphacite, hiddenite, cummingtonite Z, actinolite Y, pargasite Y, minnesotaite YZ, muscovite YZ, xanthophyllite YZ, jenkinsite YZ, chamosite YZ, dumortierite YZ, sinhalite Y, sudoite, cookeite XY, greenalite, alumoserpentine YZ

**Greyish green:** Pleonaste, dysanalyte, hypersthene Z, augite Z, fluotaramite Y, chrysotile X, delessite Z, rhipidolite Z

**Pale bluish green:** Pumpellyite Y, grandidierite Z, hedenbergite XZ, augite X, aegirine-augite Y, actinolite Z, eckermannite Y, pargasite Z, pennine, dumortierite X

**Pale yellowish green:** Chlorospinel, grossularite, andradite, vesuvianite, corundum E, beryl, epidote Z, clinohypersthene Y, hedenbergite XY, augite Y, fassaite Y, aegirine-augite Z, actinolite X, glaucophane X, eckermannite Z, common hornblende X, carpholite XY, chrysotile Z, chloritoid X, pennine, delessite X, vermiculite YZ, glauconite X, alumoserpentine X, vivianite YZ

**Yellowish green:** Chlorospinel, andradite, viridine X, epidote Y, allanite Z, aegirine Y, glaucophane X, common hornblende Y, carpholite XY, biotite X, chloritoid Z, rhipidolite XY, iddingsite, nontronite Y, chloromelanite Z

**Olive-green:** Cassiterite, brookite, baddeleyite Y, tawmawite X, augite Z, common hornblende Z or Y, biotite X, xanthophyllite YZ, delessite Z, rhipidolite Z, glauconite YZ, nontronite Z, gadolinite X, cronstedtite YZ

**Grass-green:** Viridine Y, aegirine-augite X, aegirine X, common hornblende Y, gadolinite YZ, chlorite

**Emerald-green:** Manganosite, hercynite, gahnite, uvarovite, viridine Y, tawmawite XZ, aegirine X, chloromelanite XY, cronstedtite X

**Dark green:** Hercynite, ludwigite XY, ilvaite Z, babingtonite X, fluotaramite Z, biotite YZ, chloritoid X, stilpnomelane YZ, thuringite YZ, bavalite YZ

**Blue-green:** Anatase O or E, grandidierite X, eckermannite X, fluotaramite X, common hornblende Z, chloritoid X, glauconite YZ, gedrite Z

**Dark bluish green:** Serendibite YZ, arfvedsonite Z, glauconite X

### 4. Blue and Violet Colors

**Pale blue:** Spinel, hauyne, anatase O, corundum E, catapleite, kyanite YZ, chloritoid Z, sapphirine Y, dumortierite YZ
**Pale bluish violet:** Hauyne, corundum, fluorite, cordierite, glaucophane Y, dumortierite
**Azure-blue:** Anatase E, corundum O, lazulite YZ, sapphirine Z, riebeckite Y
**Greenish blue:** Barroisite Z, tephroite Z
**Greyish blue:** Anatase E, schorlite, sapphirine Y, osannite Z, arfvedsonite Z, chloritoid Y, Na-tremolite Z
**Violet-blue:** Glaucophane Z, babingtonite Z, dumortierite X
**Deep blue:** Corundum O, sapphirine Z, serendibite Z, glaucophane Z, babingtonite X, riebeckite X, osannite X, vivianite X
**Pale violet:** Fluorite, rinneite, schorlite, kyanite Y, axinite, titanaugite Y, kunzite, lepidolite YZ
**Greyish violet:** Perovskite, schorlite, babingtonite Y, titanaugite Y, glaucophane Y
**Brownish violet:** Perovskite, ilmenite, brookite, titanaugite YZ
**Deep violet:** Fluorite

## Other Features

### 1. Pleochroitic Haloes

**within** the following **host minerals:**
Amphibole, andalusite, chlorite, micas, cordierite, pyroxene (?), brittle micas, staurolite, tourmaline
**around** the following **inclusions:**
Dumortierite, allanite, rutile, sphene, xenotime, cassiterite, zircon; (topaz, pleonaste, pyrrhotine, apatite, ilmenite, monazite, epidote)

### 2. Isomorphous zonal Structure

**Colored:** Melanite, tourmaline, pyroxene, amphibole, micas
**Colorless:** Feldspars, nepheline, leucite, zircon
**Hour-glass structure:** Titanaugite, sanidine, aegirine, brookite, chloritoid

### 3. Abnormal Inference Colors

occur frequently with the following minerals:
Chrysoberyl, clinozoisite, zoisite, pennine, melilite, vesuvianite, brucite, dumortierite, prehnite, calcic garnets, chloritoid, titanaugite, fassaite, sodic amphiboles, sanidine, laumontite, apophyllite, $Fe^{2+}$-chlorites

### 4. Abnormally birefringent "Panelling"

occurs frequently with the following minerals:
Leucite, grossularite, andradite, eudialyte, apatite, perovskite, chabasite and other zeolites

### 5. Twin Lamellae

occur frequently with the following minerals:
Leonite, (polyhalite), cryolite, (cassiterite), rutile, ilmenite, hematite, calcite, siderite, dolomite, baddeleyite, (aragonite), gibbsite, hydromagnesite, anhydrite, (monazite), amblygonite, humite group, merwinite, spurrite, kyanite, lawsonite, woehlerite, låvenite, guarinite, rinkite, mosandrite,

serendibite, wollastonite, clinopyroxene (partly), cummingtonite, actinolite, common hornblende, aenigmatite, rhoenite, dumortierite, xanthophyllite, chloritoid, microcline, analbite, plagioclase group

are rarely detectable because of low extinction angles:

Epidote group, mica group, chlorite group

## 6. Magnetism

The following guidelines for a sequence in the FRANTZ-isodynamic separator may be given (for a transversal angle of inclination of 15° and a right-left angle of inclination of 25°)

**hand magnet:** Magnetite, metallic iron, pyrrhotine (partly)

up to 0.5 A: Ilmenite, hematite, chromite, ilvaite, garnet (almandine), olivine (partly), biotite, chloritoid (?), siderite, augite (partly), hypersthene

up to 0.75 A: Hornblendes, augite (partly), actinolite, garnets (pyrope), staurolite (partly), epidote (partly), biotite (partly), chlorites, schorlite, olivine (partly), allanite, dolomite (partly)

up to 1.0 A: Diopside, tremolite, enstatite, spinel, staurolite (partly), phlogopite, muscovite (partly), zoisite, light colored tourmalines, vesuvianite, dolomite (partly)

up to 1.5 A: Sphene, monazite, xenotime, rutile, corundum (partly), beryl (partly)

above 1.5 A: Apatite, zircon, pyrite (fresh), topaz, fluorite, kyanite, sillimanite, corundum (partly), quartz, feldspars

## 7. Solubility

Insoluble in HF:

Magnetite, chromite, hematite, ilmenite, graphite, coal, perovskite, spinel, kornerupine, astrophyllite, fluorite, rutile, anatase, cassiterite, zircon, xenotime, chrysoberyl, corundum, baddeleyite, tourmaline, serendibite, beryl, brookite, staurolite, diaspore, kyanite, chloritoid, xanthophyllite, sapphirine, axinite, sillimanite, dumortierite, andalusite, topaz, bertrandite

Slightly soluble in HCl:

Hermatite, ilmenite, sphalerite, boracite, garnet, vesuvianite, pseudobrookite, goethite, sphene, allanite, ottrelite, låvenite, carpholite, lazulite, gibbsite, labradorite

Silica powder forms with HCl:

Analcite, leucite, meionite, apophyllite, rinkite, mosandrite, anorthite, stilbite, epistilbite

Gelatinizes with HCl (staining reactions):

Sodalite group, gehlenite, melilite, eudialyte, nepheline, hydronephelite, ilvaite, forsterite, olivine, fayalite, monticellite, datolite, rosenbuschite, wollastonite, pectolite, humite group, chlorite group, serpentine, heulandite, natrolite, chabasite, phillipsite, harmotome, laumontite, thomsonite, mesolite, scolecite

Soluble in HCl:

Pyrrhotite, wurtzite, magnetite, periclase, apatite, calcite, dolomite, magnesite, siderite, brucite, cancrinite, monazite (with insoluble residue), wagnerite, wavellite, aragonite, hydromagnesite

Soluble in KOH:

Opal, chalcedony, tridymite, sulfur, kaolinite (with insoluble residue)

Dissociates in $H_2O$ (alkalic reaction of finest powder):

Albite, analcite, andesine, anorthite, apophyllite, axinite, biotite, bytownite, clinochlore, clinozoisite, datolite, epidote, heulandite, hydronephelite, grossularite, labradorite, leucite, lithionite, margarite, microcline, muscovite, natrolite, nepheline, oligoclase, pennine, prehnite, serpentine, stilbite, talc, tremolite, vesuvianite, wollastonite, zoisite

Table of d-spacings

of the four strongest peaks in the X-ray powder diagram of rock-forming minerals arranged according to decreasing d-spacings of the strongest peak; (for estimated intensities: vst = very strong, st = strong, m = medium, w = weak, vw = very weak). Full references to the literature cited here are to be found in the corresponding chapters of Vol. 2.

a) Minerals translucent in thin section

| Name | No. | d-spacings I | II | III | IV | intensity I | II | III | IV | Literature |
|---|---|---|---|---|---|---|---|---|---|---|
| Saponite group (Fe-rich) | 210a | 15,6 | 1,53 | 4,56 | 2,63 | 100 | 50 | 40 | 30 | MIYAMOTO, 1957 |
| Montmorillonite | 208 | 15,0 | 4,50 | 5,01 | 3,02 | 100 | 80 | 60 | 60 | ROSENQUIST, 1959 |
| Faujasite | 61c | 14,3 | 5,71 | 8,75 | 7,51 | 100 | 100 | 80 | 80 | BARRER, 1959 |
| Guarinite | 145 | 14,0 | 3,89 | 3,01 | 2,90 | 100 | 80 | 80 | 80 | PIERROT, 1964 |
| Vermiculite | 210 | 14,0 | 3,45 | 1,52 | 4,57 | 100 | 60 | 60 | 40 | WALKER, 1951 |
| Montmorillonite | 208 | 13,6 | 4,47 | 3,34 | 3,23 | 100 | 18 | 10 | 10 | MOLLOY & KERR, 1961 |
| Nontronite | 209 | 13,4 | 4,51 | 3,49 | 1,79 | 100 | 60 | 50 | 40 | " " 1961 |
| Stilpnomelane group | 211 | 12,1 | 2,56 | 4,02 | 2,71 | 100 | 50 | 35 | 35 | BLAKE, 1965 |
| Erionite | 100a | 11,57 | 2,84 | 3,80 | 6,64 | 100 | 100 | 90 | 80 | STAPLES & CARD, 1959 |
| Erionite | 100a | 11,4 | 4,32 | 2,86 | 6,60 | 100 | 90 | 75 | 60 | DEFFEYES, 1959 |
| Hydrobiotite | 202a | 11,4 | 3,41 | 2,62 | 1,53 | 100 | 80 | 80 | 60 | WALKER, 1951 |
| Astrophyllite | 151 | 10,6 | 3,51 | 2,77 | 2,64 | 100 | 80 | 60 | 60 | NEUMANN, 1957 |
| Siderophyllite | 202 | 10,26 | 3,38 | 2,65 | 2,46 | 100 | 80 | 70 | 40 | EUGSTER & WONES, 1962 |
| Hydromuscovite | 199b | 10,1 | 4,48 | 3,34 | 2,56 | st | st | st | st | LEVINSON, 1955 |
| Halloysite (hydrated) | 223 | 10,1 | 4,42 | 3,34 | 2,56 | 100 | 100 | 90 | 80 | MACEWAN & AMOROS, 1950 |
| Phlogopite (2M) | 201 | 10,1 | 3,36 | 2,62 | 2,02 | 100 | 100 | 100 | 65 | YODER & EUGSTER, 1954 |
| Lepidomelane | 202 | 10,01 | 3,35 | 2,63 | 2,44 | 100 | 45 | 50 | 25 | OVERKOTT, 1961 |
| Muscovite | 199 | 10,01 | 3,35 | 2,56 | 2,01 | 100 | 100 | 90 | 75 | YODER & EUGSTER, 1955 |
| Phlogopite (3T) | 201 | 10,1 | 3,35 | 2,01 | 2,51 | 100 | 100 | 100 | 50 | SMITH & YODER, 1956 |
| Mica group | 198-207 | 10,00 | 3,35 | 2,60 | 2,01<br>-2,50 | 100 | 95 | 85 | 10<br>-75 | Arith. mean of micas |
| Phlogopite (1M) | 201 | 9,94 | 3,35 | 2,61 | 2,01 | 100 | 100 | 30 | 30 | SMITH & YODER, 1956 |
| Cryophyllite | 205 | 9,9 | 3,30 | 2,58 | 3,62 | 100 | 70 | 35 | 25 | FOSTER & EVANS, 1962 |
| Meroxene | 202 | 9,82 | 3,32 | 2,00 | 2,49 | 100 | 70 | 15 | 10 | PAPENFUSS, 1963 |
| Ferrierite | 243d | 9,61 | 3,99 | 3,54 | 3,49 | 100 | 90 | 80 | 80 | STAPLES, 1955 |
| Paragonite (3H) | 198 | 9,60 | 4,82 | 3,22 | 5,34 | vst | vst | vst | m | DIETRICH, 1956 |
| Minnesotaite | 197a | 9,6 | 2,52 | 3,17 | 3,33 | 100 | 70 | 50 | 25 | BLAKE, 1965 |
| Chabasite | 100 | 9,46 | 5,56 | 5,03 | 4,32 | 100 | 100 | 100 | 100 | WEIBEL, 1963 |
| Talc | 197 | 9,34 | 3,12 | 4,66 | 2,48 | 100 | 100 | 90 | 65 | STEMPLE & BRINDLEY, 1960 |
| Osannite | 182 | 8,78 | 3,13 | 2,77 | 1,44 | vst | vst | vst | vst | JAKOB & al. 1931 |
| Meso Cordierite | 156 | 8,58 | 3,04 | 3,38 | 4,11 | 100 | 90 | 90 | 80 | CLARINGBULL, Br. Mus.<br>MIYASHIRO & al., 1955 |
| Richterite | 180 | 8,55 | 2,71 | 3,38 | 3,27 | 100 | 80 | 70 | 60 | ASTM-[10-456] |
| High Cordierite | 156a | 8,48 | 3,03 | 3,14 | 3,38 | 100 | 85 | 65 | 55 | Nat. Bur. Standards, 1961 |
| Low Cordierite | 156 | 8,45 | 8,52 | 3,04 | 3,03 | 100 | 95 | 65 | 65 | SCHREYER & SCHAIRER, 1961 |
| Magnesioriebeckite | 181 | 8,45 | 3,14 | 2,72 | 3,42 | 100 | 90 | 80 | 65 | ERNST, 1960 |
| Riebeckite | 182 | 8,42 | 2,72 | 3,1 | 3,43 | 100 | 100 | 80 | 60 | CILLIERS & al., 1961 |

| | d-spacings | | | | | intensity | | | | Literature |
|---|---|---|---|---|---|---|---|---|---|---|
| | No. | I | II | III | IV | I | II | III | IV | |
| Wavellite | 121 | 8,39 | 3,44 | 3,20 | 2,56 | 100 | 80 | 80 | 20 | ASTM-[2-0075] |
| "Basaltic" Hornblende | 185 | 8,39 | 3,13 | 2,55 | 3,28 | 100 | 68 | 35 | 30 | PAPENFUSS, 1963 |
| Tremolite | 179 | 8,38 | 3,12 | 2,71 | 3,27 | 100 | 100 | 90 | 75 | STEMPLE & BRINDLEY, 1960 |
| Grunerite | 178 | 8,33 | 2,77 | 3,07 | 2,64 | 100 | 90 | 80 | 70 | KLEIN, 1964 |
| Harmotome | 242 | 8,24 | 7,17 | 6,26 | 4,03 | 100 | 100 | 100 | 100 | WATERSON, 1953 |
| Ferrogedrite | 177 | 8,24 | 3,04 | 3,22 | 8,95 | 100 | 76 | 21 | 12 | SEKI & YAMASAKI, 1957 |
| Holmquistite | 177a | 8,12 | 3,00 | 4,43 | 3,34 | 100 | 90 | 70 | 60 | NICKEL & al. 1960 |
| Aenigmatite | 189 | 8,11 | 3,14 | 2,70 | 2,54 | 100 | 100 | 80 | 80 | KELSEY & McKIE, 1964 |
| Gypsum | 116 | 7,56 | 3,06 | 4,27 | 2,68 | 100 | 57 | 51 | 28 | F.H.GILLERY, [6-0046] |
| Lizardite | 218a | 7,36 | 3,66 | 2,50 | 1,54 | vst | vst | st | st | ZUSSMAN, 1956 |
| Chrysotile | 218 | 7,36 | 3,66 | 2,45 | 1,54 | vst | vst | st | st | ZUSSMAN, 1956 |
| Lizardite | 218a | 7,31 | 3,66 | 2,50 | 1,54 | vst | st | vst | st | RUCKLIDGE & ZUSSMANN, 1965 |
| Antigorite | 217 | 7,30 | 3,63 | 2,52 | 2,42 | 100 | 75 | 18 | 10 | ASTM-[7-417] |
| Gismondine | 243b | 7,30 | 3,24 | 2,73 | 4,19 | 100 | 100 | 100 | 80 | ASTM-[2-0096] |
| Eudialyte | 92 | 7,19 | 5,74 | 2,87 | 2,99 | 100 | 80 | 80 | 70 | CLARINGBULL, Br. Mus., 1930 |
| Nacrite | 222 | 7,18 | 4,36 | 3,59 | 4,13 | 100 | 80 | 80 | 70 | BAILEY, 1963 |
| Kaolinite | 220 | 7,17 | 1,49 | 3,58 | 4,37 | 100 | 90 | 80 | 60 | GOODYEAR & DUFFIN, 1961 |
| Dickite | 221 | 7,15 | 3,58 | 2,33 | 4,12 | 100 | 100 | 90 | 70 | NEWNHAM & BRINDLEY, 1957 |
| Euclase | 157a | 7,15 | 3,22 | 3,84 | 2,77 | 100 | 50 | 35 | 35 | MROSE & KNORRING, 1959 |
| Phillipsite | 243 | 7,15 | 3,20 | 4,12 | 3,24 | 100 | 100 | 80 | 80 | HOSS & ROY, 1960 |
| Chromium chlorite 6.47 weight-% $Cr_2O_3$ | 212a | 7,14 | 4,77 | 3,59 | 14,15 | 100 | 90 | 90 | 25 | LAPHAM, 1958 |
| Chromium Chlorite 8.31 weight-% $Cr_2O_3$ | 212a | 7,14 | 4,72 | 3,65 | 14,1 | 100 | 90 | 90 | 30 | KRAUSE, 1965 |
| Alumoserpentine group | 219 | 7,12 | 2,38 | 3,56 | 1,53 | 100 | 90 | 80 | 55 | BAILEY & TAYLOR, 1960 |
| Chromium Chlorite 3.08 weight-% $Cr_2O_3$ | 212a | 7,11 | 4,75 | 3,57 | 14,13 | 100 | 100 | 100 | 35 | LAPHAM, 1958 |
| Chromium Chlorite 1.16 weight-% $Cr_2O_3$ | 212a | 7,08 | 3,55 | 4,73 | 14,10 | 100 | 100 | 90 | 45 | LAPHAM, 1958 |
| Amesite (synth.) | 219a | 7,08 | 3,55 | 1,54 | 4,62 | 100 | 100 | 100 | 80 | NELSON & ROY, 1958 |
| Cronstedtite | 219b | 7,08 | 3,54 | 2,72 | 2,44 | 100 | 85 | 50 | 40 | MUS. Uni Oxford, 1964 |
| Alumoserpentine group | 219 | 7,08 | 2,49 | 3,55 | 4,60 | 100 | 80 | 70 | 40 | BAILEY & TAYLOR, 1960 |
| Amesite | 219a | 7,06 | 3,52 | 1,92 | 2,48 | 100 | 100 | 70 | 60 | BRINDLEY & al. 1958 |
| Vivianite | 121b | 6,80 | 2,97 | 2,71 | 3,20 | 100 | 67 | 67 | 53 | BARTH, 1937 |
| Natrolite | 235 | 6,44 | 5,81 | 4,32 | 3,18 | 100 | 100 | 100 | 100 | PENG, 1955 |
| Aegirine Augite group | 171 | 6,38 | 2,91 | 2,98 | 4,41 | 100 | 60 | 30 | 20 | ASTM [18-1221] |
| Sussexite | 122a | 6,32 | 2,74 | 2,48 | 3,43 | 100 | 80 | 80 | 40 | EPPRECHT & al. 1959 |
| Lepidocrocite | 107 | 6,30 | 3,30 | 1,94 | 2,48 | 100 | 70 | 60 | 50 | BERRY & THOMPSON, 1962 |
| Ascharite (-Szaibelyite) | 112a | 6,20 | 2,20 | 2,66 | 2,42 | 100 | 80 | 75 | 75 | TAKEUCHI, 1957 |
| Boehmite | 109a | 6,11 | 3,16 | 2,35 | 1,86 | 100 | 50 | 40 | 30 | ROOKSBY in BROWN, 1961 |
| Whewellite | 244 | 5,95 | 3,65 | 2,36 | 2,97 | 100 | 90 | 80 | 50 | PECORA & KERR, 1954 |
| Dumortierite (Fe-rich) | 193 | 5,85 | 5,06 | 2,09 | 3,43 | 100 | 90 | 80 | 60 | CLARINGBULL & HEY, 1958 |
| Hydromagnesite | 111 | 5,79 | 2,90 | 2,15 | 9,18 | 100 | 90 | 50 | 40 | MURDOCH, 1954 |

| Name | No. | d-spacings I | II | III | IV | intensity I | II | III | IV | Literature |
|---|---|---|---|---|---|---|---|---|---|---|
| Andalusite | 130 | 5,54 | 4,53 | 2,77 | 2,17 | 100 | 90 | 90 | 90 | SKINNER, CLARK & APPLEMAN, 1961 |
| Variscite | 121a | 5,36 | 4,26 | 3,05 | 4,82 | 100 | 100 | 90 | 40 | MURDOCH, [15-281] |
| Grandidierite | 155 | 5,16 | 5,01 | 5,51 | 2,76 | 100 | 90 | 60 | 50 | NEUMANN & BERGSOL, 1957 |
| Ludwigite | 112 | 5,12 | 2,55 | 2,52 | 2,03 | 100 | 70 | 70 | 50 | LEONARD & al. 1962 |
| Gibbsite | 109 | 4,85 | 4,37 | 2,39 | 4,32 | 100 | 16 | 8 | 7 | ASTM- [7-324] |
| Brucite | 84 | 4,82 | 2,37 | 1,80 | 1,38 | 100 | 100 | 80 | 40 | BERRY & THOMPSON, 1962 |
| Amblygonite group | 120 | 4,64 | 3,15 | 2,92 | 2,39 | 100 | 100 | 100 | 50 | ASTM- [8-102] |
| Viridine | 131 | 4,56 | 5,52 | 2,77 | 2,27 | 100 | 90 | 90 | 80 | HEINRICH & COREY, 1959 |
| Brewsterite | 243a | 4,53 | 6,15 | 2,89 | 3,21 | 100 | 90 | 90 | 80 | STRUNZ & TENNYSON, 1956 |
| Chloritoid trcl | 195 | 4,45 | 4,50 | 1,58 | 2,97 | 100 | 100 | 80 | 80 | HALFERDAHL, 1961 |
| Chloritoid mcl | 195 | 4,45 | 4,50 | 1,58 | 2,64 | 100 | 100 | 80 | 50 | HALFERDAHL, 1961 |
| Metahalloysite | 223 | 4,43 | 7,40 | 1,48 | 3,63 | 100 | 90 | 90 | 90 | BRINDLEY & ROBINSON, 1948 |
| Bertrandite | 157 | 4,38 | 3,19 | 2,54 | 2,28 | 100 | 90 | 80 | 60 | VERNON & WILLIAMS, 1960 |
| Tridymite S | 83 | 4,27 | 4,07 | 3,80 | 2,95 | 100 | 90 | 90 | 60 | FLÖRKE, 1961 |
| β-Carnegieite | 95a | 4,23 | 2,59 | 1,49 | 2,11 | 100 | 60 | 22 | 10 | SMITH & TUTTLE, 1957 |
| Microcline (low) | 230 | 4,22 | 3,26 | 3,25 | 3,29 | 100 | 80 | 80 | 60 | ASTM- [19-926] |
| Laumontite | 238 | 4,18 | 6,97 | 3,53 | 3,67 | 100 | 60 | 60 | 40 | COOMBS, 1952 |
| Goethite | 106 | 4,18 | 2,69 | 2,44 | 2,19 | 100 | 30 | 25 | 20 | BROWN, 1961 |
| α-Carnegieite | 95a | 4,17 | 2,55 | 3,03 | 3,61 | 100 | 40 | 20 | 15 | SMITH & TUTTLE, 1957 |
| Gmelinite | 100b | 4,10 | 12,0 | 2,96 | 7,69 | 100 | 90 | 80 | 60 | STRUNZ, 1956 |
| Tridymite - M | 83 | 4,10 | 4,32 | 3,82 | 2,96 | 100 | 33 | 20 | 18 | SATO, 1963 |
| Stilbite | 241 | 4,08 | 9,1 | 4,68 | 3,03 | 100 | 90 | 70 | 70 | MASON & al., 1954 |
| Cristobalite | 60 | 4,05 | 2,49 | 2,84 | 3,13 | 100 | 20 | 13 | 11 | ASTM- [11-695] |
| Levyne | 100c | 4,04 | 2,78 | 8,12 | 3,13 | 100 | 100 | 90 | 90 | STRUNZ, 1956 |
| Albite (low) | 233 | 4,03 | 3,22 | 3,66 | 3,20 | 100 | 65 | 60 | 55 | ASTM- [19-1184] |
| Diaspore | 108 | 3,99 | 2,32 | 2,13 | 2,08 | 100 | 56 | 52 | 49 | SWANSON & FUYAT, 1953 |
| Catapleite | 93 | 3,94 | 3,05 | 2,96 | 2,69 | 100 | 100 | 100 | 90 | SABINA & TRAILL, 1960 |
| Heulandite | 239 | 3,93 | 2,94 | 5,16 | 3,88 | 100 | 80 | 70 | 20 | STRUNZ & TENNYSON, 1956 |
| Epistilbite | 240 | 3,83 | 3,43 | 3,19 | 8,84 | 100 | 100 | 90 | 80 | STRUNZ & TENNYSON, 1956 |
| Petalite | 227 | 3,73 | 3,66 | 3,51 | 2,57 | 100 | 100 | 30 | 16 | VESASALO, 1959 |
| Hauyne | 63 | 3,72 | 2,63 | 6,45 | 2,88 | 100 | 50 | 30 | 30 | BARTH, 1932 |
| Bavenite | 157b | 3,71 | 3,35 | 3,22 | 3,12 | 100 | 90 | 80 | 80 | SWITZER & REICHEN, 1960 |
| Lasurite | 63 | 3,71 | 2,62 | 2,87 | 6,43 | 100 | 80 | 45 | 40 | MUS. Uni Oxford, 1965 |
| Nosean | 63 | 3,71 | 2,62 | 6,45 | 9,09 | 100 | 75 | 70 | 65 | BRAUN, 1964 |
| Davyne | 98 | 3,68 | 3,28 | 2,65 | 4,80 | 100 | 90 | 80 | 70 | NEUMANN, 1957 |
| Sodalite | 63 | 3,63 | 2,08 | 6,3 | 2,37 | 100 | 33 | 17 | 17 | ASTM - [3-0338] |
| Melanophlogite | 59 | 3,58 | 5,99 | 3,87 | 5,47 | 100 | 80 | 70 | 60 | SKINNER & APPLEMAN, 1963 |
| Edingtonite | 235 | 3,58 | 2,75 | 4,80 | 4,69 | 100 | 100 | 90 | 90 | CLARINGBULL, 1962 |
| Danburite | 224 | 3,57 | 2,65 | 2,74 | 2,96 | 100 | 75 | 70 | 70 | EUGSTER & WISE, 1963 |
| Anatase | 67 | 3,53 | 1,90 | 1,67 | 1,70 | 100 | 50 | 40 | 30 | BERRY & THOMPSON, 1962 |
| Sudoite | 216 | 3,52 | 2,39 | 2,50 | 4,75 | 100 | 86 | 83 | 73 | FRENZEL & SCHEMBRA, 1965 |
| Anhydrite | 113 | 3,50 | 2,85 | 2,33 | 2,21 | 100 | 33 | 22 | 20 | SWANSON, 1953 |

| Name | No. | d-spacings I | II | III | IV | intensity I | II | III | IV | Literature |
|---|---|---|---|---|---|---|---|---|---|---|
| Koenenite | 75 | 3,50 | 2,85 | 2,33 | 2,20 | 100 | 50 | 50 | 50 | SABINA & TRAILL, 1960 |
| Spodumene ( β ) | 175 | 3,49 | 3,92 | 1,88 | 1,87 | 100 | 60 | 60 | 60 | SKINNER & EVANS, 1960 |
| Brookite | 104 | 3,49 | 2,88 | 1,65 | 1,88 | 100 | 70 | 40 | 30 | BERRY & THOMPSON, 1962 |
| Mordenite | 243c | 3,48 | 3,22 | 9,10 | 6,61 | 100 | 100 | 90 | 90 | HARRYS & al., 1954 |
| Pseudobrookite | 103 | 3,48 | 2,75 | 1,54 | 1,86 | 100 | 80 | 60 | 40 | BERRY & THOMPSON, 1962 |
| Mizzonite | 74 | 3,46 | 3,07 | 3,82 | 3,02 | 100 | 70 | 60 | 55 | GIBBS & BLOSS, 1961 |
| Xenotime | 68 | 3,45 | 2,56 | 1,77 | 4,55 | 100 | 50 | 50 | 25 | CARRON & al., 1958 |
| Barite | 115 | 3,44 | 3,10 | 2,12 | 2,10 | 100 | 97 | 80 | 76 | SWANSON & FUYAT, 1953 |
| Marialite | 74 | 3,44 | 3,03 | 3,78 | 2,68 | 100 | 100 | 90 | 90 | ASTM- [2-0412] |
| Analcite | 61 | 3,43 | 5,61 | 2,93 | 1,74 | 100 | 80 | 80 | 60 | COOMBS, 1955 |
| Pollucite | 61a | 3,43 | 2,92 | 3,65 | 2,41 | 100 | 100 | 80 | 80 | KOPP, 1963 |
| Sillimanite (mullite-cont.) | 191 | 3,42 | 3,37 | 2,68 | 5,37 | 100 | 96 | 43 | 32 | ARAMAKI, 1961 |
| Aragonite | 110 | 3,40 | 1,98 | 3,27 | 2,70 | 100 | 65 | 52 | 46 | SWANSON & FUYAT, 1953 |
| Wairakite | 61b | 3,39 | 5,57 | 3,42 | 2,91 | 100 | 80 | 60 | 50 | COOMBS, 1955 |
| Mullite | 192 | 3,38 | 3,42 | 2,21 | 5,39 | 100 | 95 | 60 | 50 | US Bur. Stand., 1964 |
| Sillimanite (pure) | 191 | 3,36 | 2,20 | 3,41 | 2,53 | 100 | 100 | 90 | 90 | SCHOLZE, 1955 |
| Quartz | 81 | 3,34 | 4,25 | 1,38 | 1,82 | 100 | 60 | 60 | 50 | BERRY & THOMPSON, 1962 |
| Lepidolite (1 M) | 205 | 3,33 | 10,0 | 4,99 | 3,62 | 100 | 75 | 75 | 75 | LEVINSON, 1953 |
| Sanidine (synth.) | 228 | 3,33 | 3,79 | 3,22 | 4,24 | 100 | 80 | 80 | 50 | ASTM - [10-353] |
| Lepidolite (3 T) | 205 | 3,33 | 2,58 | 10,0 | 4,98 | 100 | 75 | 50 | 50 | LEVINSON, 1953 |
| Zircon | 69 | 3,30 | 4,43 | 2,52 | 1,72 | 100 | 45 | 45 | 40 | ASTM - [6-0266] |
| Vaterite | 85a | 3,30 | 3,59 | 2,74 | 2,06 | 100 | 100 | 100 | 100 | Mc CONNELL, 1960 |
| Milarite | 94a | 3,30 | 2,88 | 4,16 | 5,21 | 100 | 90 | 65 | 45 | BERRY, CDN, 1950 |
| Zinnwaldite | 205a | 3,29 | 9,8 | 1,98 | 3,09 | 100 | 80 | 55 | 40 | CUNDY & al., 1960 |
| Leuoite | 62 | 3,26 | 3,43 | 5,36 | 2,92 | 100 | 85 | 80 | 70 | FAUST, 1963 |
| Na-Sanidine $Or_{58}Ab_{41}An_{0.5}$ | 228 | 3,26 | 3,22 | 3,76 | 3,27 | 100 | 90 | 75 | 75 | ASTM - [19-1227] |
| Microcline (intermediate) | 230 | 3,24 | 3,29 | 4,22 | 2,16 | 100 | 50 | 45 | 30 | ASTM - [19-932] |
| Chrysoberyl | 101 | 3,24 | 2,09 | 1,62 | 4,01 | 100 | 90 | 80 | 50 | ASTM -[19-932] |
| K-Analbite (synth.) $Or_{30}Ab_{70}$ | 231 | 3,24 | 3,20 | 4,09 | 3,76 | 100 | 70 | 40 | 30 | ASTM - [10-361] |
| Rutile | 66 | 3,24 | 1,69 | 2,49 | 1,35 | 100 | 70 | 50 | 40 | BERRY & THOMPSON, 1962 |
| Sphene | 142 | 3,23 | 2,99 | 2,60 | 2,06 | 100 | 90 | 90 | 40 | ASTM - [11-142] |
| Xanthophyllite | 206 | 3,22 | 9,71 | 1,93 | 2,56 | 100 | 30 | 20 | 10 | HARADA & al., 1965 |
| Cancrinite (synth.) | 97 | 3,22 | 4,65 | 3,65 | 6,35 | 100 | 65 | 40 | 35 | EDGAR, 1964 |
| Hydroxy-Cancrinite (synth.) | 97 | 3,21 | 4,65 | 2,73 | 2,57 | 100 | 40 | 40 | 40 | EDGAR, 1964 |
| (K, Ca)-Analbite $Or_{32}Ab_{66}An_2$ | 231 | 3,21 | 3,24 | 4,11 | 2,16 | 100 | 90 | 15 | 15 | ASTM - [9-478]homogen. after heating |
| Bytownite (low) $An_{77}$ | 233 | 3,20 | 4,03 | 3,75 | 3,17 | 100 | 80 | 80 | 80 | ASTM - [9-467] |
| Oligoclase (high) $An_{30}$ | 233 | 3,20 | 4,02 | 3,74 | 3,17 | 100 | 80 | 80 | 80 | ASTM - [9-456] |
| Albite (low) | 233 | 3,20 | 3,78 | 6,39 | 3,68 | 100 | 25 | 20 | 20 | ASTM - [9-466] |
| Anorthite | 233 | 3,20 | 3,18 | 4,04 | 3,26 | 100 | 75 | 60 | 55 | ASTM - [12-301] |
| Clintonite | 206 | 3,20 | 2,54 | 2,10 | 1,51 | 100 | 90 | 70 | 70 | NEUMANN, 1957 |
| Labradorite (low) $An_{65}$ | 233 | 3,20 | 3,18 | 4,04 | 3,23 | 100 | 90 | 80 | 80 | ASTM - [9-465] |
| Cancrinite | 97 | 3,19 | 4,61 | 3,61 | 4,35 | 100 | 67 | 40 | 27 | ASTM - [3-0503] |
| Analbite (heated albite) | 231 | 3,18 | 3,75 | 3,21 | 4,04 | 100 | 30 | 30 | 30 | ASTM - [10-393] |

| Name | No. | d-spacings I | II | III | IV | intensity I | II | III | IV | Literature |
|---|---|---|---|---|---|---|---|---|---|---|
| Oligoclase (low) $An_{17}$ | 233 | 3,18 | 3,20 | 3,76 | 2,93 | 100 | 80 | 70 | 70 | ASTM - [9-457] |
| Baddeleyite | 105 | 3,18 | 2,84 | 3,69 | 3,63 | 100 | 65 | 18 | 14 | BERRY & THOMPSON, 1962 |
| Labradorite (intermed.) $An_{67}$ | 233 | 3,18 | 3,76 | 3,21 | 3,20 | 100 | 70 | 70 | 70 | ASTM - [18-1202] |
| Kyanite | 132 | 3,18 | 1,38 | 3,35 | 1,96 | 100 | 75 | 65 | 55 | ASTM - [11-46] |
| Enstatite | 164 | 3,17 | 2,87 | 2,49 | 2,94 | 100 | 87 | 51 | 44 | ASTM - [7-216] |
| Arfvedsonite | 183 | 3,16 | 2,73 | 8,51 | 3,42 | 100 | 80 | 70 | 45 | SADANAGA, 1962 |
| Hypersthene | 164 | 3,14 | 1,47 | 2,86 | 1,60 | 100 | 80 | 80 | 60 | CLAVAN, 1954 |
| "basaltic" Hornbl.(text. -free) | 185 | 3,13 | 8,46 | 2,16 | 2,80 | 100 | 85 | 60 | 55 | Miner. Inst. Freiburg, 1969 |
| Kaersutite | 185 | 3,13 | 8,42 | 3,29 | 2,81 | 100 | 69 | 32 | 21 | TOMITA, 1965 |
| Phenakite | 91a | 3,12 | 3,66 | 2,52 | 2,36 | 100 | 80 | 75 | 70 | ASTM - [9-431] |
| Kalsilite (synth.) | 96 | 3,12 | 2,58 | 3,97 | 2,17 | 100 | 50 | 45 | 17 | SMITH & TUTTLE, 1957 |
| Wurtzite | 44 | 3,12 | 1,91 | 1,63 | 3,25 | 100 | 90 | 70 | 50 | BERRY & THOMPSON, 1962 |
| Sphalerite | 44 | 3,12 | 1,91 | 1,63 | 1,11 | 100 | 80 | 70 | 50 | BERRY & THOMPSON, 1962 |
| Pargasite | 186 | 3,11 | 8,32 | 3,25 | 2,68 | 100 | 25 | 20 | 15 | Miner. Inst, Freiburg, 1969 |
| Datolite | 225 | 3,11 | 2,86 | 2,25 | 2,99 | 100 | 80 | 50 | 45 | EUGSTER & WISE, 1963 |
| Tschermakite group | 187 | 3,11 | 2,75 | 2,35 | 2,18 | vst | st | st | st | JAKOB & BRANDENBERGER, 1931 |
| Coesite | 81a | 3,10 | 3,43 | 2,77 | 2,68 | 100 | 50 | 15 | 15 | DACHILLE & ROY, 1959 |
| Gadolinite | 226 | 3,10 | 2,79 | 2,54 | 2,34 | 100 | 100 | 100 | 70 | ITO & MORI, 1953 |
| Monazite | 117 | 3,09 | 3,30 | 2,87 | 3,51 | 100 | 50 | 70 | 25 | CARRON, 1958 |
| Kaliophilite | 96 | 3,09 | 2,59 | 2,13 | 4,26 | 100 | 30 | 35 | 15 | SMITH & TUTTLE, 1957 |
| Grunerite | 178 | 3,08 | 2,77 | 8,38 | 2,64 | 100 | 100 | 80 | 80 | ASTM - [7-394] |
| Prehnite | 158 | 3,08 | 2,55 | 3,48 | 1,77 | 100 | 100 | 90 | 70 | NUFFIELD, 1943 |
| Tirodite | 178a | 3,07 | 8,30 | 9,03 | 3,25 | 100 | 90 | 80 | 80 | KLEIN, 1964 |
| Lazulite group | 119 | 3,07 | 3,14 | 6,15 | 3,23 | 100 | 95 | 75 | 75 | CAMPBELL, 1962 |
| Rinkite | 148 | 3,07 | 2,70 | 2,95 | 2,80 | 100 | 70 | 40 | 40 | SAHAMA & HYTÖNEN, 1957 |
| Anthophyllite | 176 | 3,06 | 8,33 | 3,23 | 4,49 | 100 | 70 | 50 | 35 | GREENWOOD, 1963 |
| Magnesiogedrite | 177 | 3,06 | 8,27 | 3,23 | 8,97 | 100 | 80 | 70 | 50 | MILTON & ITO, 1961 |
| Kaliophilite | 96 | 3,06 | 2,57 | 3,90 | 1,89 | 100 | 40 | 20 | 20 | NEUMANN, 1957 |
| Norbergite | 124-126a | 3,06 | 2,23 | 2,64 | 2,25 | 100 | 80 | 75 | 70 | ASTM - [11-686] |
| Pyrophyllite | 196 | 3,05 | 9,16 | 4,46 | 2,41 | 100 | 40 | 40 | 40 | BROWN, 1961 |
| Anthophyllite | 176 | 3,05 | 3,24 | 8,31 | 2,84 | 100 | 70 | 70 | 40 | NEUMANN, 1957 |
| Mosandrite | 149 | 3,05 | 2,69 | 2,93 | 2,01 | 100 | 20 | 15 | 15 | SAHAMA & HYTÖNEN, 1957 |
| Calcite | 85 | 3,05 | 1,92 | 1,88 | 2,29 | 100 | 70 | 60 | 50 | BERRY & THOMPSON, 1962 |
| Pandaite | 54 | 3,04 | 6,10 | 1,86 | 1,59 | 100 | 80 | 70 | 70 | JÄGER & al., 1959 |
| Apophyllite | 73 | 3,04 | 2,49 | 1,58 | 2,10 | 100 | 80 | 40 | 20 | DA SILVA, 1961 |
| Scawtite | 161a | 3,04 | 1,90 | 3,02 | 2,49 | 100 | 60 | 50 | 50 | MURDOCH, 1955 |
| Rankinite | 129a | 3,02 | 2,72 | 3,18 | 3,20 | 100 | 80 | 70 | 70 | ROY, 1958 |
| Staurolite | 134 | 3,01 | 3,56 | 2,69 | 2,37 | 100 | 60 | 50 | 50 | JUURINEN, 1956 |
| Tilleyite | 129b | 3,01 | 3,10 | 1,90 | 2,98 | 100 | 60 | 45 | 40 | HARKER, 1959 |
| Microlite | 54 | 3,01 | 2,97 | 1,83 | 1,57 | 100 | 100 | 100 | 100 | MURDOCH, 1955 |
| Basanite | 116a | 3,00 | 6,01 | 2,80 | 3,46 | 100 | 95 | 50 | 45 | ASTM - [14-453] |

| Name | No. | d-spacings I | II | III | IV | intensity I | II | III | IV | Literature |
|---|---|---|---|---|---|---|---|---|---|---|
| Nepheline | 95 | 3,00 | 4,18 | 3,27 | 2,88 | 100 | 70 | 70 | 70 | BARRER & WHITE, 1952 |
| Kornerupine | 153 | 3,00 | 2,61 | 3,34 | 1,49 | 100 | 100 | 90 | 90 | GIRAULT, 1952 |
| Johannsenite | 167a | 3,00 | 2,54 | 3,27 | 2,59 | 100 | 70 | 30 | 30 | ASTM - [14-374] |
| Pyrochlore | 54 | 3,00 | 1,84 | 1,57 | 5,98 | 100 | 60 | 50 | 25 | HOGARTH, 1961 |
| Alunite | 89 | 2,99 | 2,89 | 2,29 | 1,93 | 100 | 100 | 80 | 70 | PARKER, 1962 |
| Guarinite | 145 | 2,99 | 2,86 | 3,27 | | 100 | 100 | 50 | | NEUMANN, 1957 |
| Wagnerite | 118 | 2,99 | 2,83 | 3,15 | 3,32 | 100 | 100 | 90 | 80 | HENRIQUES, 1957 |
| Aegirine | 171 | 2,99 | 2,54 | 6,5 | 4,43 | 100 | 60 | 40 | 40 | ASTM - [3-0621] |
| Diopside | 167 | 2,99 | 2,53 | 2,89 | 2,52 | 100 | 40 | 30 | 30 | ASTM - [11-654] |
| Johannsenite | 167a | 2,99 | 2,54 | 1,07 | 1,64 | vst | st | m | w | SCHIAVINATO, 1953 |
| Augite group | 168 | 2,99 | 1,62 | 1,43 | 2,56 | 100 | 100 | 100 | 85 | ASTM - [3-0623] |
| Wollastonite | 161 | 2,98 | 3,83 | 3,52 | 3,31 | 100 | 80 | 80 | 80 | HELLER & TAYLOR, 1956 |
| Celestine | 114 | 2,97 | 3,30 | 2,73 | 3,17 | 100 | 98 | 63 | 59 | SWANSON & FUYAT, 1953 |
| Hedenbergite | 167 | 2,97 | 2,53 | 2,56 | 2,13 | 100 | 50 | 30 | 30 | NEUMANN, 1957 |
| Omphacite | 173 | 2,97 | 1,40 | 2,13 | 2,12 | 100 | 80 | 70 | 70 | WARNER, 1964 |
| Columbite group (Mn>Fe) | 102 | 2,96 | 3,65 | 1,73 | 1,48 | 100 | 60 | 60 | 50 | BERRY & THOMPSON, 1962 |
| Allanite | 138 | 2,96 | 3,50 | 2,67 | 1,63 | 100 | 80 | 80 | 60 | BERMANN, 1955 |
| Ureyite | 174a | 2,96 | 2,88 | 2,18 | 4,38 | 100 | 70 | 70 | 60 | FRONDEL & KLEIN, 1965 |
| Aegirine-Augite group | 171 | 2,95 | 6,45 | 3,00 | 3,23 | 100 | 60 | 60 | 55 | ASTM - [19-1] |
| Columbite group (Fe>Mn) | 102 | 2,95 | 3,62 | 1,73 | 1,71 | 100 | 60 | 50 | 50 | BERRY & THOMPSON, 1962 |
| Stishovite | 81b | 2,95 | 1,53 | 1,98 | 1,23 | 100 | 50 | 35 | 25 | CHAO & al., 1962 |
| Topaz | 133 | 2,94 | 3,20 | 3,69 | 2,36 | 100 | 65 | 60 | 45 | ASTM - [12-765] |
| Rosenbuschite | 147 | 2,94 | 3,06 | 1,89 | 3,96 | 100 | 80 | 60 | 40 | NEUMANN, 1962 |
| Gonnardite | 235 | 2,92 | 5,93 | 4,44 | 6,70 | 100 | 80 | 60 | 60 | MEIXNER & al., 1956 |
| Pectolite | 163 | 2,92 | 3,10 | 3,33 | 3,90 | 100 | 80 | 60 | 60 | HILDEBRAND, 1953 |
| Viséite | 61b | 2,92 | 1,74 | 3,46 | 5,68 | 100 | 60 | 50 | 40 | McCONNELL, 1952 |
| Chabasite | 100 | 2,91 | 4,29 | 9,31 | 5,47 | 100 | 90 | 80 | 70 | STRUNZ, 1956 |
| Allanite | 138 | 2,91 | 2,92 | 2,86 | 3,52 | 100 | 90 | 50 | 40 | BERMANN, 1955 |
| Spodumene | 175 | 2,91 | 2,78 | 4,19 | 2,45 | 100 | 80 | 60 | 60 | GABRIEL & al., 1942 |
| Allanite (heated above 700°C) | 138 | 2,91 | 1,64 | 3,47 | 1,47 | 100 | 80 | 60 | 60 | FARIA, 1964 |
| Låvenite | 144 | 2,90 | 2,81 | 3,22 | 6,79 | 100 | 90 | 50 | 30 | NEUMANN, 1957 |
| Pumpellyite | 141a | 2,90 | 2,74 | 3,79 | 2,45 | 100 | 50 | 50 | 40 | COOMBS, 1953 |
| Ankerite | 88 | 2,90 | 2,20 | 1,81 | 1,79 | 100 | 6 | 6 | 6 | HOWIE & BROADHURST, 1958 |
| Allanite | 138 | 2,90 | 1,64 | 2,70 | 1,88 | 100 | 90 | 70 | 60 | FARIA, 1964 |
| Chloromelanite group | 172 | 2,89 | 2,98 | 2,54 | | 100 | 90 | 50 | | FOSHAG, 1957 |
| Beckelite | 57a | 2,89 | 2,86 | 3,16 | 2,79 | st | m | m | m | GAY, 1957 |
| Dolomite | 88 | 2,89 | 2,19 | 1,79 | 2,01 | 100 | 30 | 30 | 15 | HOWIE & BROADHURST, 1958 |
| Clinoenstatite group | 165 | 2,88 | 2,98 | 3,17 | 2,46 | 100 | 95 | 60 | 60 | KUNO & HESS, 1953 |
| Clinozoisite | 136 | 2,88 | 2,38 | 2,67 | 2,58 | 100 | 65 | 60 | 60 | LAPHAM, 1957 |
| Beryl | 94 | 2,87 | 3,25 | 7,98 | 4,60 | 100 | 95 | 90 | 50 | ASTM - [9-430] |
| Bustamite | 162 | 2,87 | 3,20 | 2,98 | 2,22 | 100 | 70 | 60 | 50 | HUTTON, 1956 |

| Name | No. | d-spacings I | II | III | IV | intensity I | II | III | IV | Literature |
|---|---|---|---|---|---|---|---|---|---|---|
| Enstatite | 164 | 2,87 | 3,17 | 2,47 | 2,94 | 100 | 50 | 45 | 40 | GREENWOOD, 1963 |
| Triplite group | 118a | 2,87 | 3,02 | 3,25 | 1,63 | 100 | 90 | 80 | 70 | ASTM - [11-118] |
| Tephroite | 122a | 2,87 | 2,57 | 2,61 | 1,81 | 100 | 90 | 70 | 35 | HURLBUT, 1961 |
| Epidote | 136 | 2,87 | 1,63 | 2,39 | 1,87 | 100 | 85 | 80 | 70 | LAPHAM, 1957 |
| Mesolite | 235 | 2,86 | 5,79 | 4,35 | 6,44 | 100 | 70 | 50 | 40 | PENG, 1955 |
| Scolecite | 237 | 2,86 | 4,37 | 5,81 | 4,69 | 100 | 50 | 40 | 30 | PENG, 1955 |
| Thomsonite | 236 | 2,86 | 2,95 | 2,68 | 6,60 | 100 | 80 | 80 | 70 | TAYLOR, MEEK & JACKSON, 1933 |
| Melilite group | 71 | 2,85 | 1,75 | 2,43 | 2,41 | 100 | 100 | 70 | 70 | ASTM - [4-0690] |
| Woehlerite | 143 | 2,84 | 3,00 | 3,25 | 2,96 | 100 | 70 | 60 | 50 | NICKEL & al., 1958 |
| Jadeite | 174 | 2,83 | 2,92 | 2,53 | 2,50 | 100 | 70 | 40 | 30 | SEKI & al., 1960 |
| Fayalite | 122 | 2,83 | 2,50 | 2,57 | 3,56 | 100 | 70 | 50 | 30 | ASTM - [7-164] |
| Axinite | 152 | 2,81 | 3,16 | 3,46 | 6,30 | 100 | 90 | 80 | 70 | MILTON, 1953 |
| Apatite (O, OH) | 90 | 2,81 | 2,78 | 2,72 | 3,44 | 100 | 60 | 60 | 40 | ASTM - [9-432] |
| Larnite (synth.) | 127 | 2,80 | 2,74 | 2,78 | 2,61 | 100 | 95 | 90 | 65 | MIDGLEY, 1952 |
| Apatite (F) | 90 | 2,80 | 2,70 | 1,77 | 3,44 | 100 | 60 | 40 | 40 | ASTM - [15-876] |
| Siderite | 87 | 2,80 | 1,74 | 3,60 | 2,14 | 100 | 80 | 40 | 30 | BERRY & THOMPSON, 1962 |
| Apatite (Cl) (synth.) | 90 | 2,78 | 2,86 | 1,84 | 1,96 | 100 | 60 | 20 | 18 | CAROBBI & al., 1959 |
| Weddellite | 244 | 2,77 | 6,17 | 2,24 | 1,96 | 100 | 90 | 70 | 65 | HUTTON & TAFT, 1965 |
| Lamprophyllite | 150 | 2,77 | 3,43 | 2,13 | 3,73 | 100 | 55 | 45 | 40 | Uni-Museum, Oxford, LEWIS, 1965 |
| Apatite ($CO_3$) | 90 | 2,77 | 3,40 | 2,69 | 2,61 | 100 | 70 | 70 | 70 | AMES, 1959 |
| Rhodonite | 159 | 2,77 | 2,98 | 2,92 | 3,14 | 100 | 65 | 65 | 30 | ASTM - [13-138] |
| Forsterite | 122 | 2,77 | 2,51 | 2,46 | 3,88 | 100 | 100 | 80 | 60 | ASTM - [7-79] |
| Pyrophanite | 79 | 2,76 | 1,74 | 2,55 | 1,88 | 100 | 70 | 70 | 60 | LEE, 1955 |
| Cummingtonite | 178 | 2,75 | 3,07 | 8,38 | 3,87 | 100 | 80 | 70 | 60 | ASTM - [7-382] |
| Melinophane | 72 | 2,75 | 2,96 | 1,70 | 3,59 | 100 | 50 | 50 | 40 | NEUMANN & al., 1957 |
| Babingtonite | 160 | 2,75 | 2,87 | 3,12 | 2,95 | 100 | 80 | 70 | 60 | SABINA & TRAILL, 1960 |
| Vesuvianite | 70 | 2,75 | 2,59 | 1,62 | 2,45 | 100 | 80 | 60 | 50 | ASTM - [11-145] |
| Ilmenite | 78 | 2,75 | 1,73 | 2,54 | 1,87 | 100 | 80 | 70 | 50 | BERRY & THOMPSON, 1962 |
| Artinite | 111a | 2,74 | 5,34 | 3,69 | 2,21 | 100 | 65 | 48 | 42 | ASTM -[6-0484] |
| Magnesite | 86 | 2,74 | 2,10 | 1,70 | 2,50 | 100 | 45 | 35 | 17 | ASTM - [8-479] |
| Geikielite | 80 | 2,73 | 1,71 | 2,23 | 1,86 | 100 | 65 | 45 | 45 | WISE, 1959 |
| Chevkinite (unheated) | 140a | 2,72 | 3,18 | 1,97 | 2,17 | 100 | 80 | 80 | 60 | FARIA, 1962 |
| Hastingsite group $Fe^{2+}_{41}$ $Fe^{3+}_{73}$ | 184 | 2,72 | 3,14 | 8,50 | 2,80 | 100 | 80 | 65 | 50 | BRAUNER, 1961 |
| Hastingsite group $Fe^{2+}_{65}$ $Fe^{3+}_{65}$ | 184 | 2,72 | 3,12 | 8,44 | 2,75 | 100 | 75 | 60 | 40 | BRAUNER, 1961 |
| Perovskite | 52 | 2,72 | 1,93 | 1,57 | 0,91 | 100 | 60 | 50 | 50 | BERRY & THOMPSON, 1962 |
| Actinolite | 179 | 2,71 | 3,12 | 2,53 | 2,78 | 100 | 70 | 55 | 45 | BRAUNER, 1961 |
| Glaucophane | 181 | 2,71 | 3,12 | 2,50 | 3,41 | 100 | 90 | 80 | 70 | ERNST, 1961 |
| Ilvaite | 140 | 2,71 | 2,84 | 7,28 | 2,67 | 100 | 85 | 70 | 50 | LEONARD, 1962 |
| Perovskite | 52 | 2,71 | 1,91 | 1,56 | 1,35 | 100 | 100 | 100 | 70 | FRANK-KAMENETSKI, 1961 |

| Name | No. | d-spacings I | II | III | IV | intensity I | II | III | IV | Literature |
|---|---|---|---|---|---|---|---|---|---|---|
| Pseudozoisite | 139 | 2,70 | 4,07 | 2,02 | 2,87 | 100 | 58 | 46 | 40 | SEKI, 1959 |
| Meionite | 74 | 2,70 | 3,46 | 3,08 | 3,04 | 100 | 95 | 75 | 55 | GIBBS & BLOSS, 1961 |
| Chevkinite (heated) | 140a | 2,70 | 3,16 | 1,96 | 2,15 | 100 | 80 | 80 | 60 | FARIA, 1962 |
| "Basaltic" Hornblende | 185 | 2,70 | 3,13 | 2,56 | 2,16 | 100 | 85 | 85 | 55 | BRAUNER, 1961 |
| Karinthine | 186a | 2,70 | 3,11 | 2,54 | 2,78 | 100 | 75 | 60 | 40 | BRAUNER, 1961 |
| Andradite | 56 | 2,70 | 3,02 | 1,61 | 2,46 | 100 | 60 | 60 | 45 | ASTM -[10-288] |
| Zoisite | 139 | 2,70 | 2,87 | 4,03 | 8,09 | 100 | 65 | 50 | 40 | PISTORIUS, 1961 |
| Spurrite | 129 | 2,70 | 2,64 | 3,02 | 2,66 | 100 | 70 | 65 | 50 | GIBBS & SMITH, 1961 |
| Kaersutite | 185 | 2,69 | 3,11 | 8,38 | 3,36 | 100 | 80 | 65 | 65 | Uni-Mus. Oxford, 1965 |
| Hematite | 77 | 2,69 | 2,52 | 1,70 | 1,84 | 100 | 80 | 70 | 60 | BERRY & THOMPSON, 1962 |
| Hibschite | 58 | 2,68 | 3,00 | 1,61 | 1,95 | 100 | 80 | 80 | 60 | PABST, 1942 |
| Uvarovite | 56 | 2,68 | 3,00 | 1,60 | 2,45 | 100 | 70 | 60 | 55 | ASTM - [11-696] |
| Merwinite | 128 | 2,68 | 2,67 | 2,65 | 1,91 | 100 | 90 | 30 | 30 | ROY, 1956 |
| Monticellite | 123 | 2,67 | 3,64 | 2,59 | 4,19 | 100 | 40 | 40 | 35 | SAHAMA & HYTÖNEN, 1958 |
| Merwinite (synth.) | 128 | 2,66 | 1,90 | 1,53 | 1,87 | 100 | 70 | 70 | 60 | PHEMISTER, 1942 |
| Grossularite | 56 | 2,65 | 1,58 | 2,96 | 1,65 | 100 | 90 | 80 | 80 | ASTM - [3-0826] |
| Portlandite | 49 | 2,63 | 4,90 | 1,93 | 1,80 | 100 | 74 | 42 | 36 | SWANSON, 1950 |
| Lawsonite | 141 | 2,62 | 1,55 | 2,73 | 3,66 | 100 | 80 | 70 | 60 | DAVIS & PABST, 1960 |
| Spessartine | 56 | 2,60 | 1,55 | 1,08 | 1,61 | 100 | 80 | 70 | 60 | ASTM - [2-0992] |
| Carpholite | 194 | 2,59 | 5,66 | 3,33 | 3,02 | 100 | 90 | 80 | 80 | ASTM - [17-209] |
| Greenalite | 217a | 2,59 | 2,20 | 7,21 | 1,60 | 100 | 90 | 70 | 60 | YOUELL in BROWN, 1961 |
| Glauconite | 207 | 2,59 | 1,52 | 10,05 | 3,33 | 100 | 80 | 60 | 60 | VALETON, 1958 |
| Celadonite | 207 | 2,58 | 4,53 | 3,63 | 3,09 | 100 | 85 | 80 | 80 | WISE & EUGSTER, 1964 |
| Dravite | 91 | 2,58 | 3,99 | 2,96 | 4,22 | 100 | 85 | 85 | 65 | ASTM - [14-76] |
| Schorlite | 91 | 2,58 | 3,97 | 2,96 | 4,21 | 100 | 80 | 70 | 70 | NEUMANN & al., 1957 |
| Greenalite | 217a | 2,57 | 7,13 | 3,56 | 2,18 | 100 | 80 | 80 | 40 | GRUNER, 1936 |
| Almandine | 56 | 2,57 | 1,54 | 2,87 | 1,56 | 100 | 50 | 40 | 40 | ASTM - [9-427] |
| Ludwigite | 112 | 2,56 | 5,12 | 2,17 | 1,38 | 100 | 70 | 50 | 50 | THOMPSON & al., 1954 |
| Xanthophyllite | 206 | 2,56 | 3,23 | 1,50 | 1,48 | 100 | 80 | 70 | 70 | FORMAN, 1951 |
| Pyrope | 56 | 2,56 | 2,86 | 1,53 | 2,44 | 100 | 60 | 50 | 40 | Nat. Bur. Stand., 1965 |
| Dumortierite (Fe-poor) | 193 | 2,55 | 5,89 | 5,84 | 5,09 | 100 | 90 | 90 | 90 | ASTM - [7-71] |
| Triphyline | 120a | 2,54 | 4,29 | 3,51 | 3,03 | 100 | 90 | 90 | 90 | QUENSEL, 1957 |
| Elbaite | 91 | 2,54 | 3,96 | 2,91 | 3,43 | 100 | 90 | 90 | 80 | NEUMANN & al., 1957 |
| Rhoenite | 190 | 2,54 | 2,93 | 2,69 | 2,10 | 100 | 90 | 90 | 90 | WALENTA, 1965 |
| Paragonite (2 M) | 198 | 2,52 | 1,48 | 4,39 | 3,20 | 100 | 100 | 90 | 80 | HARDER, 1956 |
| Margarite | 200 | 2,51 | 3,20 | 4,40 | 3,39 | 100 | 90 | 80 | 80 | HARDER, 1956 |
| Chromite | 51 | 2,51 | 1,08 | 1,47 | 1,60 | 100 | 90 | 80 | 60 | BERRY & THOMPSON, 1962 |
| Galaxite | 50 | 2,47 | 2,90 | 1,58 | 1,45 | 100 | 60 | 60 | 60 | LEE, 1955 |
| Hercynite | 50 | 2,46 | 1,44 | 2,87 | 1,57 | 100 | 70 | 50 | 50 | BERRY & THOMPSON, 1962 |
| Gahnite | 50 | 2,45 | 2,87 | 1,56 | 1,44 | 100 | 80 | 70 | 70 | BERRY & THOMPSON, 1962 |
| Hoegbomite | 80a | 2,43 | 1,43 | 2,49 | 2,08 | 100 | 80 | 60 | 60 | McKIE, 1963 |

| Name | No. | d-spacings I | II | III | IV | intensity I | II | III | IV | Literature |
|---|---|---|---|---|---|---|---|---|---|---|
| Spinel | 50 | 2,43 | 1,43 | 2,01 | 1,55 | 100 | 70 | 60 | 60 | BERRY & THOMPSON, 1962 |
| Calciumoxide | 49 | 2,40 | 1,70 | 2,78 | 1,45 | 100 | 45 | 34 | 10 | SWANSON & TATGE, 1950 |
| Cookeite | 216a | 2,32 | 4,70 | 3,52 | 14,1 | 100 | 90 | 90 | 80 | BAILEY & BROWN, 1962 |
| Humite | 124-126a | 2,26 | 2,44 | 1,48 | 1,74 | 100 | 70 | 70 | 65 | ASTM - [12-755] |
| Chondrodite | 124-126a | 2,26 | 1,74 | 3,02 | 2,51 | 100 | 70 | 45 | 45 | ASTM - [12-527] |
| Clinohumite | 124-126a | 2,26 | 1,74 | 2,77 | 2,54 | 100 | 80 | 60 | 60 | VAN VALKENBURG, 1961 |
| Manganosite | 48 | 2,22 | 2,57 | 1,57 | 1,34 | 100 | 62 | 58 | 21 | SWANSON & al., 1955 |
| Manganosite | 48 | 2,22 | 1,57 | 0,99 | 2,57 | 100 | 70 | 50 | 40 | BERRY & THOMPSON, 1962 |
| Sinhalite | 122b | 2,14 | 1,62 | 3,24 | 2,62 | 100 | 90 | 70 | 70 | CLARINGBULL & HEY, 1952 |
| Periclase | 47 | 2,10 | 1,49 | 0,94 | 0,86 | 100 | 60 | 50 | 50 | BERRY & THOMPSON, 1962 |
| Corundum | 76 | 2,09 | 2,55 | 1,60 | 3,48 | 100 | 90 | 80 | 75 | ASTM - [10-173] |
| Boracite | 55 | 2,06 | 3,04 | 2,72 | 1,24 | 100 | 70 | 50 | 50 | ITO & al., ASTM - [5-0710] |
| Serendibite | 154 | 2,03 | 1,44 | 2,45 | 2,85 | 100 | 90 | 80 | 80 | SHABYNIN & al., 1956 |
| Sapphirine | 135 | 2,01 | 1,44 | 2,97 | 2,44 | 100 | 90 | 70 | 60 | FOSTER, 1950 |
| Cryolite | 46 | 1,94 | 2,75 | 3,88 | 2,32 | 100 | 95 | 60 | 50 | SMITH & al., ASTM - [12-257] |
| Fluorite | 45 | 1,93 | 3,16 | 1,12 | 1,65 | 100 | 60 | 50 | 40 | BERRY & THOMPSON, 1962 |
| Dysanalyte | 53 | 1,80 | 1,54 | 2,90 | 0,99 | 100 | 100 | 80 | 80 | GAERTNER, 1930 |
| Cassiterite | 65 | 1,76 | 3,36 | 2,64 | 1,41 | 100 | 80 | 70 | 50 | BERRY & THOMPSON, 1962 |
| Kaolinite | 220 | 1,48 | 2,33 | 4,34 | 2,55 | 100 | 95 | 85 | 85 | MELKA & SLANSKY, 1959 |

## b) Minerals soluble in water

| Name | No. | d-spacings I | II | III | IV | intensity I | II | III | IV | Literature |
|---|---|---|---|---|---|---|---|---|---|---|
| Ulexite | 26b | 12,2 | 7,75 | 6,00 | 4,16 | 100 | 80 | 30 | 30 | CLARK & CHRIST, 1959 |
| Koenenite | 10 | 7,8 | 3,84 | 2,59 | | 100 | 50 | 50 | | LOHSE, 1958 |
| Kernite | 26a | 7,41 | 6,63 | 2,47 | 3,70 | 100 | 80 | 40 | 30 | MUESSIG & ALLEN, 1957 |
| Kieserite | 16 | 4,79 | 3,40 | 3,31 | 2,53 | 100 | 100 | 80 | 40 | BRAITSCH, 1962 |
| Leonite | 19 | 4,76 | 3,42 | 3,30 | 3,04 | 100 | 85 | 80 | 45 | BRAITSCH, 1962 |
| Epsomite | 17 | 4,21 | 5,35 | 2,68 | 5,99 | 100 | 26 | 24 | 12 | ASTM - [8-467] |
| Bischofite | 6 | 4,10 | 2,65 | 2,88 | 2,72 | 100 | 75 | 50 | 44 | ASTM - [1-0431] |
| Schoenite | 21 | 4,07 | 3,70 | 7,16 | 4,16 | 100 | 100 | 85 | 55 | BRAITSCH, 1962 |
| Glauberite | 15 | 3,94 | 2,67 | 3,17 | 3,11 | 100 | 45 | 35 | 35 | BRAITSCH, 1962 |
| Ice | 1 | 3,90 | 3,67 | 3,44 | 2,25 | 100 | 75 | 55 | 40 | DOWELL & RINGRET, 1960 |
| Carnallite | 8 | 3,59 | 2,92 | 3,32 | 1,74 | 100 | 65 | 35 | 35 | BRAITSCH, 1962 |
| Koenenite | 10 | 3,50 | 2,85 | 2,34 | 2,20 | 100 | 50 | 50 | 50 | SABINA & TRAILL, 1960 |
| Loeweite | 20 | 3,47 | 3,19 | 2,70 | 4,31 | 100 | 100 | 100 | 90 | BRAITSCH, 1962 |
| Bloedite | 18 | 3,28 | 4,54 | 3,25 | 2,73 | 100 | 90 | 90 | 65 | BRAITSCH, 1962 |
| Syngenite | 23 | 3,16 | 9,5 | 2,89 | 4,75 | 100 | 65 | 25 | 20 | BRAITSCH, 1962 |
| Sylvine | 5 | 3,15 | 2,22 | 1,81 | 1,41 | 100 | 59 | 23 | 20 | SWANSON & TATGE, 1950 |
| Langbeinite | 14 | 3,14 | 4,04 | 2,65 | 2,99 | 100 | 30 | 30 | 16 | MOREY, 1950 |
| Colemanite | 26c | 3,13 | 5,64 | 3,85 | 2,55 | 100 | 50 | 50 | 50 | CHRIST, 1953 |
| Vanthoffite | 13 | 3,11 | 4,03 | 3,05 | 3,43 | 100 | 55 | 50 | 35 | BRAITSCH, 1962 |
| Kainite | 24 | 3,08 | 7,78 | 7,39 | 3,03 | 100 | 80 | 40 | 40 | BRAITSCH, 1962 |
| Soda Niter | 27 | 3,03 | 2,31 | 2,81 | 1,90 | 100 | 24 | 15 | 15 | SWANSON & al., 1956 |
| Polyhalite | 22 | 2,91 | 3,17 | 2,85 | 6,00 | 100 | 45 | 12 | 8 | BRAITSCH, 1961 |
| Glaserite | 12 | 2,83 | 3,13 | 2,03 | 3,64 | 100 | 75 | 40 | 30 | BRAITSCH, 1962 |
| Halite | 4 | 2,82 | 1,99 | 1,63 | 3,26 | 100 | 55 | 15 | 13 | SWANSON & FUYAT, 1953 |
| Thenardite | 11 | 2,78 | 4,66 | 3,18 | 3,10 | 100 | 73 | 51 | 47 | SWANSON & FUYAT, 1953 |
| Trona | 25 | 2,66 | 3,08 | 9,88 | 3,21 | 100 | 80 | 60 | 60 | PABST, 1959 |
| Trona | 25 | 2,65 | 3,07 | 9,80 | 4,92 | 100 | 85 | 50 | 35 | FAHEY, 1962 |
| Tachydrite | 9 | 2,60 | 3,80 | 3,08 | 2,87 | 100 | 45 | 43 | 40 | BRAITSCH, 1962 |
| Borax | 26 | 2,57 | 2,84 | 4,86 | 3,96 | 100 | 53 | 47 | 40 | ASTM - [1-1097] |
| Rinneite | 7 | 2,51 | 2,81 | 5,99 | 2,59 | 100 | 85 | 55 | 45 | BRAITSCH, 1962 |
| Calciumoxide | 2 | 2,40 | 1,70 | 2,78 | 1,45 | 100 | 45 | 34 | 10 | SWANSON & TATGE, 1950 |
| Villiaumite | 3 | 2,32 | 1,64 | 1,34 | 1,04 | 100 | 60 | 17 | 12 | SWANSON & TATGE, 1949 |

c) Minerals opaque or nearly opaque in thin section

| Name | No. | d-spacings I | II | III | IV | intensity I | II | III | IV | Literature |
|---|---|---|---|---|---|---|---|---|---|---|
| Aenigmatite | 42 | 8,11 | 3,14 | 2,70 | 2,54 | 100 | 100 | 80 | 80 | KELSEY & McKIE, 1964 |
| Ludwigite | 40 | 5,12 | 2,55 | 2,52 | 2,03 | 100 | 70 | 70 | 50 | LEONARD & al., 1962 |
| Graphite | 29 | 3,36 | 1,68 | 2,03 | 1,16 | 100 | 80 | 50 | 50 | BERRY & THOMPSON, 1962 |
| Columbite (Mn>Fe) | 39 | 2,96 | 3,65 | 1,73 | 1,48 | 100 | 60 | 60 | 50 | BERRY & THOMPSON, 1962 |
| Columbite (Fe>Mn) | 39 | 2,95 | 3,62 | 1,73 | 1,71 | 100 | 60 | 50 | 50 | BERRY & THOMPSON, 1962 |
| Pyrophanite | 36 | 2,76 | 1,74 | 2,55 | 1,88 | 100 | 70 | 70 | 60 | LEE, 1955 |
| Ilmenite | 35 | 2,75 | 1,73 | 2,54 | 1,87 | 100 | 80 | 70 | 50 | BERRY & THOMPSON, 1962 |
| Geikielite | 37 | 2,73 | 1,71 | 2,23 | 1,86 | 100 | 65 | 45 | 45 | WISE, 1959 |
| Ilvaite | 41 | 2,71 | 2,84 | 7,28 | 2,67 | 100 | 85 | 70 | 50 | LEONARD & al., 1962 |
| Melnikovite-Pyrite | 31 | 2,70 | 2,42 | 1,64 | 2,21 | 100 | 80 | 70 | 60 | VILLWOCK, 1961 |
| Marcasite | 31 | 2,70 | 1,75 | 3,43 | 2,41 | 100 | 90 | 60 | 60 | BERRY & THOMPSON, 1962 |
| Hematite | 34 | 2,69 | 2,52 | 1,70 | 1,84 | 100 | 80 | 70 | 60 | BERRY & THOMPSON, 1962 |
| Ludwigite | 40 | 2,56 | 5,12 | 2,17 | 1,38 | 100 | 70 | 50 | 50 | THOMPSON & al., 1954 |
| Rhoenite | 43 | 2,54 | 2,93 | 2,69 | 2,10 | 100 | 90 | 90 | 90 | WALENTA, 1965 |
| Magnetite | 32 | 2,53 | 1,48 | 1,61 | 2,09 | 100 | 90 | 80 | 70 | BERRY & THOMPSON, 1962 |
| Magnesioferrite | 32 | 2,53 | 1,48 | 1,61 | 1,09 | 100 | 80 | 70 | 60 | BERRY & THOMPSON, 1962 |
| Chromite | 33 | 2,51 | 1,08 | 1,47 | 1,60 | 100 | 90 | 80 | 60 | BERRY & THOMPSON, 1962 |
| Pyrrhotite | 30 | 2,08 | 2,65 | 1,73 | 3,00 | 100 | 60 | 50 | 40 | BERRY & THOMPSON, 1962 |
| Pyrite | 31 | 1,63 | 1,04 | 2,70 | 2,42 | 100 | 80 | 70 | 60 | BERRY & THOMPSON, 1962 |

# Index

All figures, except numbers of minerals, refer to numbers of pages.

| No. | Mineral | Tab. | Fig. | Diagr. | Appendix Dens. | d-spacings | Color etc. | other references |
|---|---|---|---|---|---|---|---|---|
| 110 | Aragonite | 44 | 47 | - | 155 | 167 | 158, 159 162, 163 | 31, 141 |
| 183 | Arfvedsonite | 84 | 89,93 | 90,91, 94 | 156 | 168 | 161, 162 | 19, 85, 87, 152 |
| 104 | Arkansite | 42 | - | - | | | | |
| 111a | Artinite | 44 | - | - | 154 | | 170 | 45 |
| - | "Asbestos" | - | - | - | | | | 84, 85, 113 |
| 112a | Ascharite | 44 | - | - | 154 | 165 | | 13, 27 |
| 18 | Astrakanite | 14 | - | - | | 173 | | 13, 15 |
| 151 | Astrophyllite | 64 | 69 | - | 156 | 164 | 160,161 163 | 43, 65 |
| 168 | Augite | 72 | 76,77 | 79,82 | 156 | 169 | 159,161 163 | 27,57,65,67, 73,85,87,103,152 |
| 152 | Axinite | 66 | 69 | 70,71 | 156 | 170 | 159,162 163 | |
| 181 | Bababuda-nite | - | - | - | 156 | | | |
| 160 | Babingtonite | 66 | 74 | 74 | 156 | 170 | 160-162 | |
| 105 | Baddeleyite | 42 | 46 | - | | 168 | 160-163 | 57, 65 |
| 115 | Barite | 44 | 47 | - | 157 | 167 | 158 | 45, 113 |
| 184a | Barkevikite | 86 | 89,93 | 91 | 156 | | | 87 |
| 187a | Barroisite | 86 | - | - | | | 161, 162 | 87 |
| 184a, 184b | Basaltic Hornblende | 86 | 89,93 | 91 | 156 | | | 87 |
| 116a | Basanite | - | - | - | | 168 | | |
| 218a | "Bastite" | 113 | - | - | | | | |
| 202a | Bauerite | 103 | - | - | | | | 103 |
| 214 | Bavalite | 110 | 114 | 117,118 | 155 | | 160,161 | |
| 157b | Bavenite | - | - | - | | 166 | | |
| 57a | Beckelite | - | - | - | | 169 | 160 | |
| 51 | Beresovs-kite | - | - | 22 | 157 | | | |
| 219 | Berthierine | 112 | 115 | 118 | | | | |
| 157 | Bertrandite | 66 | 69 | - | 154 | 166 | 160,165 | 49 |
| 94 | Beryl | 36 | 38 | 41 | 154 | 169 | 158-161, 163 | 37, 67, 103 |
| 202 | Biotite | 102 | 105 | 105,107 | 155 | 164 | 160,161 163 | 31, 65, 85, 87 101,102,103, 109,111 |
| 6 | Bischofite | 12 | - | - | | 173 | | 13, 15 |
| 18 | Bloedite | 14 | - | - | | 173 | | 13, 15 |
| 109a | Boehmite | 44 | - | - | 155 | 165 | | |
| 55 | Boracite | 18 | - | - | 155 | 172 | 158,161, 163 | |
| 26 | Borax | 14 | - | - | | 173 | | 15, 17 |
| 85-88 | Breunnerite | - | - | 39 | 155 | | | |
| 243a | Brewsterite | - | - | - | | 166 | | |
| 164 | Bronzite | 72 | - | 78 | 156 | | | 73 |
| 104 | Brookite | 42 | 46 | - | 157 | 167 | 159-163 | 19,27,30,43 |
| 87 | Brownspar = Siderite | 36 | 33 | 39 | 156 | 170 | 159, 160 162, 163 | 31,37,85,101, 113 |
| 84 | Brucite | 30 | 33 | - | 154 | 166 | 158, 162 163 | 13,19,27,37, 45,101 |
| 214 | Brunsvigite | 110 | 114 | 118 | | | | |
| 162 | Bustamite | 72 | 74 | 75 | 156 | 169 | 160, 161 | 49,57,67,73 |
| 233 | Bytownite | 120 | 125 | 126-137 | 154 | 167 | 158, 163 | |
| 85 | Calcite | 30 | 33 | 39 | 154 | 168 | 159, 162 163 | 31,37,45,67,85, 103,109,113,121 |
| 2, 49 | Calciumoxide = Lime | 12, 18 | - | - | 156 | 172,173 | | 19 |

| No. | Mineral | Tab. | Fig. | Diagr. | Appendix Dens. | d-spacings | Color etc. | other references |
|---|---|---|---|---|---|---|---|---|
| 105 | Caldasite | - | - | - | | | | 42 |
| 97 | Cancrinite | 36 | 38 | 41 | 154 | 167 | 158, 163 | 37 |
| 85-88 | Carbonates | 30,36 | 33 | 39 | 154 | 169, 170 | | |
| 8 | Carnallite | 12 | - | - | | 173 | | 13, 15 |
| 95a | Carnegieite | - | - | - | | 166 | | |
| 194 | Carpholite | 100 | 104 | - | 155 | 171 | 161, 163 | 85, 101 |
| 65 | Cassiterite | 20 | 29 | - | | 172 | 159-163 | 17,19,27,31, 37,43,49,109 |
| 93 | Catapleite | 36 | 38 | - | 154 | 166 | 160, 162 | 73 |
| 182 | Cats eye | 85 | - | - | | | | 85 |
| 207 | Celadonite | 109 | - | - | | 171 | | 109 |
| 114,81a | Celestite = Celestine | 44 | 47 | - | 156 | 169 | | 45 |
| 234 | Celsiane | 120 | 124 | 138 | 156 | | | 121 |
| 50 | Ceylanite | - | - | 22 | 156 | | | |
| 100 | Chabazite | 42 | 38 | 139 | 154 | 164, 169 | 162, 163 | 21,43,141 |
| 82 | Chalcedony | 30 | 33 | - | 154 | | 163 | 30,44,45,101 121 |
| - | Chalcopyrite | - | - | - | | | | 17 |
| 215,219 | Chamosite | 110, 112 | 115 | 118 | 155 | | 160, 161 | 109, 111 |
| 140a | Chevkinite =Tscheffkinite | 56 | - | - | 157 | 170, 171 | 160 | |
| 130 | Chiastolite | 50 | - | - | | | | 51 |
| 212-215 | Chlorite group | 110 | 114 | 117,118 | | | 161-163 | 85,101,103 |
| 195 | Chloritoid | 100 | 104 | - | 156 | 166 | 161-163 | 57,103,109 |
| 172 | Chloromelanite | 72 | - | - | 156 | 169 | 161 | 85 |
| 50 | Chlorospinel | 19 | - | - | 156 | | 161 | 19 |
| 124-126a | Chondrodite group | 50 | 53 | - | 155 | 172 | 160, 161 | 37,49,51,67 |
| 33,51 | Chromite | 16,18 | - | 22 | 157 | 171, 174 | 160, 163 | 17 |
| 219a | Chromium Amesite | - | - | - | | | | 113 |
| 212a | Chromium Chlorite | 110 | 114 | - | 154 | 165 | | |
| 50 | Chromium Picotite | - | - | 22 | 157 | | | |
| 50 | Chromium Pleonaste | - | - | - | 157 | | | |
| 101 | Chrysoberyl | 42 | 46 | - | 156 | 167 | 158 160-163 | 27 |
| 122 | Chrysolite | 48 | - | 54 | 156 | | 51 | |
| 218 | Chrysotile | 112 | 115 | - | 154 | 165 | 160, 161 | 85,109,111,113 |
| 233 | Cleavelandite | - | - | - | | | | 17,43 |
| 178-185 | Clinoamphibole | 84,86 | 80-90 93 | 90-98 | | | | 85 |
| 212 | Clinochlore | 110 | 114 | 118 | 154 | | 158, 159 163 | 111 |
| 165 | Clinoenstatite | 72 | 76,77 | 79,80 | | 169 | | 85 |
| 165 | Clinoferrosilite | 72 | 76 | 79,80 81 | | | | |
| 126 | Clinohumite | 50 | 53 | - | 155 | 172 | 160 | 51 |
| 165 | Clinohypersthene | 72 | 76 | 79 | | | 160, 161 | |
| 165-175 | Clinopyroxene | 72,84 | 76,77 | 79-83 | | | 163 | 73 |
| 136 | Clinozoisite | 56 | 58 | 62 | 156 | 169 | 158, 159 162,163 | 51, 57 |
| 206 | Clintonite | 108 | 105 | 106 | 155 | 167 | | |
| 28 | Coal | 16 | - | - | | | 163 | 17 |
| 105a | Coesite | 30,42 | - | - | 155 | 168 | | 21,27,31 |
| 26c | Colemanite | 16 | - | - | | 173 | | |

| No. | Mineral | Tab. | Fig. | Diagr. | Appendix Dens. | d-spacings | Color etc. | other references |
|---|---|---|---|---|---|---|---|---|
| 39,102 | Columbite | 16,42 | 46 | 60 | 157 | 169,174 | | 17,45,57 |
| 216a | Cookeite | 110 | 115 | - | 154 | 172 | 160, 161 | |
| 156 | Cordierite | 66 | 69 | 70 | 154 | 164 | 158 160, 162 | 31,49,51,57,67, 85,101,103,113 |
| 210 | Corrensite | 109 | - | - | | | | |
| 213 | Corundophi- lite | 111 | 114 | 118 | 155 | | | 111,113 |
| 76 | Corundum | 26 | 32 | - | 156 | 172 | 158-163 | 31,37,43,45,57, 67,101 |
| 42,189 | Cossyrite | 18,86 | - | - | | | | |
| 60 | Cristobalite | 20 | - | - | 154 | 166 | | 21,31 |
| 182 | Crocydolite | 85 | - | - | | | | 85 |
| 181 | Crossite | 84 | 88,93 | 90,91, 94 | 155 | | | |
| 46 | Cryolite | 18 | - | - | 155 | 172 | 162 | 13,19 |
| 205 | Cryophyllite | - | - | - | | 164 | | |
| 178 | Cummingto- nite | 84 | 88,93 | 90,91 92 | 156 | 170 | 160, 161 163 | 85 |
| 132 | Cyanite =Kyanite | 50 | 58 | 60 | 156 | 168 | 158, 162, 163 | 49,51,57,67, 73,85,101 |
| 224 | Danburite | 120 | 119 | - | 155 | 166 | | 67,121 |
| 214 | Daphnite =Bavalite | 110 | 114 | 117,118 | 155 | | 160, 161 | |
| 225 | Datolite | 120 | 119 | - | 155 | 168 | 163 | 67,121 |
| 98 | Davyne | 36 | 38 | 41 | 154 | 166 | | 37,43 |
| 215 | Delessite | 110 | 115 | - | 154 | | 160, 161 | 111 |
| 241 | Desmine | 140 | 139 | 139 | 154 | 166 | 163 | 141 |
| 213 | Diabantite | 110 | - | 118 | 155 | | | |
| 168 | "Diallage" | 72 | - | - | | | | |
| 108 | Diaspore | 42 | 47 | - | 156 | 166 | 158, 163 | 37,45,65, 103,113 |
| 156 | Dichroite | 66 | 69 | 70 | | | | |
| 221 | Dickite | 112 | 119 | - | 154 | 165 | | 109,113 |
| 167 | Diopside | 72 | 76,77 | 79,80 81,82 | 155 | 169 | 158, 163 | 67,73,85,87, 103,152 |
| 74 | Dipyre | 26 | 32 | 35 | 154 | | | 27 |
| 132 | Disthene = Kyanite | 50 | 58 | 60 | 156 | 168 | 158 161, 163 | 49,51,57,67, 73,85,101 |
| 88 | Dolomite | 36 | 33 | 39 | 155 | 169 | 159, 160, 162, 163 | 31 |
| 91 | Dravite | 36 | - | 39 | 155 | 171 | 160, 161 | 37,57 |
| 193 | Dumortie- rite | 100 | 104 | - | 156 | 165, 171 | 160, 161-163 | 49,57,67,85 |
| 53 | Dysanalyte | 18 | - | - | 157 | 172 | 161 | 19,31 |
| 202 | Eastonite | - | - | 107 | | | | |
| 183a | Eckermanite | 84,86 | 89,93 | 90,91 | 155 | | 161 | |
| 179-187 | Edenite | - | - | 90 | | | | |
| 235 | Edingtonite | - | - | - | | 166 | | |
| 91 | Elbaite | 36 | - | 39 | 155 | 171 | | 37 |
| 165-169 | Endiopside | - | - | 79 | | | | |
| 42,189 | Enigmatite =Aenigmatite | 18, 86 | 99 | 99 | 156 | 165, 174 | 160, 163 | 17,57,87 |
| 164 | Enstatite | 72 | 76 | 78 | 155 | 168, 170 | 160, 163 | 51,73,152 |
| 136 | Epidote | 56 | 58 | 62,63 | 156 | 170 | 158-163 | 17,27,49,57,65 67,85,101,109 152 |
| 240 | Epistilbite | 140 | 142, 143 | 139 | 154 | 166 | 163 | 141 |
| 17 | Epsomite | 14 | - | - | | 173 | | 15 |
| 100a | Erionite | - | - | - | | 164 | | |
| 157a | Euclase | 66 | - | - | | 165 | | |
| 92 | Eucolite | 36 | - | - | | | | 65 |
| 92 | Eudialyte | 36 | 38 | - | 155 | 165 | 160, 162 163 | 65 |

| No. | Mineral | Tab. | Fig. | Diagr. | Appendix Dens. | d-spacings | Coloretc. | other references |
|---|---|---|---|---|---|---|---|---|
| 164 | Eulite | 72 | - | 78 | 156 | | | |
| 169 | Fassaite | 72 | 76,77 | 79 | 155 | | 161, 162 | |
| 61c | Faujasite | 20 | - | - | | 164 | | |
| 122 | Fayalite | 48 | 53 | 54 | 157 | 170 | 163 | 49,57 |
| 228-234 | Feldspar group | 120 | 120, 124,125 | 123-138 | 154 | 166, 167 | 162 | 27,101,103, 113 |
| 184 | Femaghasting-site | - | - | 96 | | | | |
| 50 | Ferri-ceylanite | - | - | - | 157 | | | |
| 243d | Ferrierite | - | - | - | | 164 | | |
| 50 | Ferri-hercynite | - | - | - | 157 | | | |
| 199 | Ferri-muscovite | - | - | 106 | | | | |
| 50 | Ferripleo-naste | - | - | - | 157 | | | |
| 179 | Ferroacti-nolite | 84 | - | 91,93 | 156 | | | |
| 168 | Ferroaugite | 72 | - | 79 | | | | |
| 214 | Ferro-chlorite | 110 | 114 | 118 | | | | 113 |
| 85-88 | Ferro-dolomite | - | - | 39 | | | | |
| 176-178 | Ferro-gedrite | - | - | 90 | | 165 | | |
| 184 | Ferrohasting-site | - | - | 96 | | | | |
| 165-169 | Ferro-hedenbergite | - | - | 79 | | | | |
| 122 | Ferrohortono-lite | 48 | - | 54 | 157 | | | |
| 164 | Ferrohyper-sthene | - | - | 78 | 156 | | | |
| 119 | Ferrolazu-lite | - | - | 54 | | | | |
| 167 | Ferrosalite | 72 | 76 | 79,82 | 156 | | | |
| 164 | Ferrosilite | 72 | 76 | 78-80 | 156 | | | 152 |
| 211 | Ferrostilp-nomelane | 108 | - | 116 | | | | |
| 191 | Fibrolite = Sillimanite | 100 | 104 | - | 155 | 167 | 158, 163 | 43, 51, 67, 85,101 |
| 45 | Fluorite | 18 | - | - | 155 | 172 | 158,160, 161,163, | 13,19,21,65 |
| 133 | Fluortopaz | - | - | 61 | | | | |
| 183 | Fluotaramite | 84 | 89,93 | 90,91, 95 | 155 | | 161 | |
| 122 | Forsterite | 48 | 53 | 54 | 155 | 170 | 159,160, 163 | 152 |
| 199 | Fuchsite | 101 | - | - | | | | 101 |
| 226 | Gadolinite | 120 | 119 | - | 157 | 168 | 161 | 19,31,57 |
| 50 | Gahnite | 18 | - | 23 | 157 | 171 | 159 | 19 |
| 56 | Garnet group | 18 | - | 24,25 | | | 159,162, 163 | 19,27,37,49,51, 57,67,85,101, 109,111 |
| 181 | Gastaldite | - | - | 91 | | | | |
| 177 | Gedrite | 84 | 88 | 90,91, 92 | 156 | 168 | 160, 161 | 67,85 |
| 71 | Gehlenite | 26 | 32 | 34 | 155 | | 163 | 27,37,51 |
| 37,80 | Geikielite | 16,30 | 32 | - | 156 | 170,174 | 160 | 17,31,43 |
| 109 | Gibbsite = Hydrargillite | 44 | 47 | - | 154 | 166 | 162,163 | 31,45,49,101, 113,141 |

| No. | Mineral | Tab. | Fig. | Diagr. | Appendix Dens. | d-spacings | Color etc. | other references |
|---|---|---|---|---|---|---|---|---|
| 243b | Gismondite =Gismondine | 140 | - | 139 | | 165 | | |
| 12 | Glaserite =Aphthitalite | 12 | - | - | | 173 | | |
| 64 | Glass | 20 | - | - | 154 | | 160, 161 | 21,27 |
| 15 | Glauberite | 12 | - | - | | 173 | | 15 |
| 123 | Glaucochroite | - | - | 55 | | | | |
| 207 | Glauconite | 108 | 105 | 106 | 154 | 171 | 161 | 109,111 |
| 181 | Glaucophane | 84 | 88,93 | 90,91 94 | 155 | 170 | 160-162 | 49,57,73,85, 101,109 |
| 100b | Gmelinite | 42 | - | 139 | | 166 | | |
| 22 | Goergeyite | - | - | - | | | | 15 |
| 106 | Goethite | 42 | 46 | - | 157 | 166 | 159-161, 163 | 17,27,30,31, 43,57 |
| 235 | Gonnardite | - | - | - | | 169 | | |
| 179 | Grammatite =Tremolite | 84 | 93 | 91,93 | 155 | 165 | 158, 161, 163 | 49,57,67,85, 101,152 |
| 155 | Grandidierite | 66 | 69 | - | 155 | 166 | 161 | 49,67,101 |
| 29 | Graphite | 16 | - | - | 154 | 174 | 163 | 17 |
| 217a | Greenalite | 112 | - | 118 | 156 | 171 | 161 | 101 |
| 210a | Griffithite | 108 | - | - | | | | |
| 212 | Grochauite | 110 | - | 117,118 | 154 | | | 111 |
| 56 | Grossularite | 18 | - | 24,25 | 156 | 171 | 159,161, 162 | 19,20,27,67, 109 |
| 142 | "Grothite" | 64 | - | 63 | | | 158 | |
| 178 | Grunerite | 84 | 88,93 | 71,90 91,92 | 156 | 165,168 | | |
| 145 | Guarinite | 64 | 68 | - | 155 | 164,169 | 160,163 | 65 |
| 199b | Guembelite | 103 | - | - | | | | 103 |
| 116 | Gypsum | 44 | 47 | - | 154 | 165 | 158 | 13,15,31,45 |
| 146 | Hainite | 64 | 68 | - | 155 | | 160 | 65 |
| 4 | Halite | 12 | - | - | | 173 | 159 | 13,15,17,45 |
| 223 | Halloysite | 112 | 119 | - | 154 | 164 | | 113 |
| 242 | Harmotome | 140 | 142,143 | - | 154 | 165 | 163 | 141 |
| 184 | Hastingsite | 86 | 89,93 | 91 | 156 | 170 | | 87 |
| 63 | Hauyne | 20 | - | - | 154 | 166 | 162, 163 | 43 |
| 167 | Hedenbergite | 72 | 76,77 | 79-83 | 156 | 169 | 159,161 | 17,57,152 |
| 34,77 | Hematite | 16,30 | 32 | - | 157 | 171,174 | 159-163 | 13,17,27,30, 31,43,49,113 |
| 50 | Hercynite | 18 | - | 22,23 | 157 | 171 | 161 | 19 |
| 239 | Heulandite | 140 | 142, 143 | 139 | 154 | 166 | 163 | 141 |
| - | Hexahydrite | - | - | - | | | | 15 |
| 57 | Hibschite | 20 | - | - | 155 | 171 | 160 | |
| 231 | "High Albite" =Analbite | 120 | 124 | 123,124 | 154 | 167 | | 121 |
| 156 | High Cordi-erite | 67 | - | - | | | | 67 |
| 145 | Hjortdahlite | 64 | 68 | 71 | | | 161 | |
| 80a | Hoegbomite | 30 | - | - | 156 | 171 | | 121 |
| 177a | Holmquistite | - | - | 91 | | 165 | | |
| 184,186 187 | Hornblende | 86 | 89 | 90,91 98 | 155 | 165,168, 171 | 160, 161 163 | 17,19,57,85,87, 109,152 |
| 122 | Hortonolite | 48 | - | 54 | 156 | | | 49,152 |
| 125 | Humite | 50 | 53 | - | 155 | 172 | 160,162 163 | 49,51 |
| 234 | Hyalophane | 120 | 124 | 138 | 154 | | | 121 |
| 122 | Hyalosiderite | 48 | - | 54 | 156 | | | 152 |
| 109 | Hydrargillite =Gibbsite | 44 | 47 | - | 154 | 166 | 162,163 | 31,45,49,101, 113,141 |
| 202a | Hydrobiotite | 102 | - | - | | 164 | 160 | 103,109 |
| 111 | Hydromagne-site | 44 | 47 | - | 154 | 165 | 162,163 | 31,45,49 |
| 199b | Hydromica | 102 | - | - | 154 | 164 | | 103,111 |

| No. | Mineral | Tab. | Fig. | Diagr. | Appendix Dens. | d-spacings | Color etc. | other references |
|---|---|---|---|---|---|---|---|---|
| 199b | Hydromusco-vite | 102 | - | - | 154 | 164 | | 103,111 |
| 99 | Hydronephe-lite | 36 | - | - | | | 163 | |
| 199b | Hydrophen-gite | 103 | - | - | | | | 103 |
| 201 | Hydrophlogo-pite | - | - | - | | | | 103 |
| 133 | Hydroxyl-topaz | - | - | 61 | 156 | | | |
| 164 | Hypersthene | 72 | 76 | 78 | 156 | 168 | 160,161 163 | 51,152 |
| 1 | Ice | 12 | - | - | | 173 | 158 | |
| 122 | Iddingsite | - | - | - | 154 | | 160,161 | 49 |
| 70 | Idocrase =Vesuvianite | 26 | - | - | | | | |
| 199b | Illite | 103 | - | - | | | | 103,109,113 |
| 35,78 | Ilmenite | 16,30 | 32 | - | 157 | 170,174 | 160,162 163 | 17,27,31,43, 73,87 |
| (38) | (Ilmeno-rutile) | - | - | - | 157 | | | |
| 41,140 | Ilvaite =Lievrite | 16,56 | 59 | - | 156 | 170,174 | 160,161, 163 | 17,43,45,57, 73 |
| 180 | Imerinite | - | - | 91 | | | | |
| - | Inyoite | - | - | - | | | | 15 |
| 156 | Iron Cordierite | 66 | - | 70 | 154 | | | |
| 230 | "Iso-Micro-cline" | - | - | - | | | | 121 |
| 229 | "Iso-Ortho-clase" | - | - | - | | | | 121 |
| 174 | Jadeite | 84 | 77 | 82 | 156 | 170 | 160,161 | 65,73,85 |
| 217 | Jenkinsite | 111 | - | - | | | 160,161 | 73 |
| 167a | Johannsenite | 72 | - | - | | 169 | | |
| 149 | Johnstrupite | 64 | - | - | | | | |
| 212a | Kaemmererite | 110 | 114 | - | | | 160 | 57 |
| 185 | Kaersutite | 86 | 89,93 | 91,97 | 155 | 168,171 | 160 | 87,152 |
| 24 | Kainite | 14 | - | - | | 173 | | 13,15 |
| 96 | Kaliophilite | 36 | 38 | - | 154 | 168 | | 37 |
| 96 | Kalsilite | 37 | - | - | | 168 | | 37 |
| 220 | Kaolinite | 112 | 119 | - | 154 | 165,172 | 163 | 45,101,103, 109,111,113 |
| 186a | Karinthine | 86 | - | - | 155 | 171 | 160 | 87 |
| 188 | Katophorite | 86 | 93 | - | 156 | | 160 | |
| 142 | "Keilhauite" =Yttrotitanite | 64 | - | 63 | | | | |
| 26a | Kernite | 14 | - | - | | 173 | | 15 |
| 228-230 | K-feldspar | 120 | 122,124 | 123,124 | | | | 121,152 |
| 16 | Kieserite | 14 | - | - | | 173 | | 13,15 |
| 215 | Klementite | 111 | - | - | | | | |
| 10,75 | Koenenite | 12,26 | 32 | - | | 167,173 | 160 | 13,15,45 |
| 54 | Koppite | 18 | - | - | 157 | | | 19 |
| 153 | Kornerupine | 66 | 69 | 71 | 156 | 169 | 160,161 | |
| 212a | Kotchubeite | 110 | 114 | - | | | | |
| 205 | Kryophyllite =Cryophyllite | - | - | - | | 164 | | |
| 175 | Kunzite | - | - | - | | | 162 | |
| 178 | "Kupfferite" | - | - | - | 155 | | | |
| - | Kurnakovite | - | - | - | | | | 15 |
| 132 | Kyanite =Cyanite | 50 | 58 | 60 | 156 | 168 | 158, 162,163 | 49,51,57,67,73, 85,101 |

| No. | Mineral | Tab. | Fig. | Diagr. | Appendix Dens. | d-spacings | Color etc. | other references |
|---|---|---|---|---|---|---|---|---|
| 233 | Labradorite | 120 | 125 | 126-137 | 154 | 167,168 | 158,163 | |
| 150 | Lamprophyllite | 64 | 69 | - | 156 | 170 | 160,161 | 65 |
| 14 | Langbeinite | 12 | - | - | | 173 | | 13,15 |
| 127 | Larnite | 50 | - | - | 156 | 170 | | 51 |
| 63 | Lasurite | - | - | - | | 166 | | |
| 238 | Laumontite | 140 | 142, 143 | 139 | 154 | 166 | 162,163 | 141 |
| 144 | Låvenite | 64 | 68 | - | 156 | 169 | 160-163 | 43,65 |
| 141 | Lawsonite | 56 | 59 | - | 155 | 171 | 162 | 51,65,67 |
| 119 | Lazulite | 48 | 52 | 54 | 155 | 168 | 162,163 | 49,57,67 |
| 58a | Lechatelierite | 20 | - | - | 154 | | | 31,43 |
| - | Leonhardite | - | - | - | | | | 15 |
| 19 | Leonite | 14 | - | - | | 173 | 162 | 13,15 |
| 107 | Lepidocrocite | 42 | 46 | - | 157 | 165 | 160 | 17,27,31,43 |
| 205 | Lepidolite | 102 | 105 | 106,107 | 155 | 167 | 160,162 | 21,45,49,85,101, 109,111,121 |
| 202 | Lepidomelane | 102 | 105 | 106,107 | 155 | 164 | | |
| 215 | Leptochlorite | 110 | 114 | - | | | | 111 |
| 212 | Leuchtenbergite | 111 | - | 118 | 154 | | | |
| 62 | Leucite | 20 | - | - | 154 | 167 | 158,162 163 | 21,31 |
| (67),78a (142) | Leucoxene | 30 | - | - | | | | 17,19,27,31,65 |
| 100c | Levynite | 42 | - | - | | 166 | | |
| 41,140 | Lievrite | 16,56 | 59 | - | 156 | 170,174 | 160,161, 163 | 17,43,45,57,73 |
| 2,49 | Lime = Calciumoxide | 12,18 | - | - | 156 | 172,173 | | 19 |
| 106,107 | Limonite | 43 | - | - | | | | 43,85,101 |
| 204 | Lithionite | 102 | 105 | 107 | 155 | | 160,163 | 37,103 |
| 204 | Lithium-Siderophyllite | 102 | 105 | 107 | | | | |
| 218a | Lizardite | 112 | 115 | - | 154 | 165 | | |
| 20 | Loeweite = Loewite | 14 | - | - | | 173 | 160 | 13,15 |
| 40,112 | Ludwigite | 16,44 | 47 | - | 156 | 166,171 174 | 160,161 | 17,57 |
| 60 | Lussatine = Lussatite | 20 | - | - | | | | 31 |
| 183 | Magnesio-arfvedsonite | 84 | 89,93 | 91 | 156 | | | |
| 50 | Magnesiochromite | - | - | 22 | | | | |
| 50 | Magnesioferrite | - | - | 22 | 157 | 174 | | |
| 184 | Magnesiohastingsite | 86 | 93 | 90,91,96 | 155 | | | |
| 50 | Magnesiomagnetite | - | - | 22 | 157 | | | |
| 181 | Magnesioriebeckite | 84 | 88,93 | 91,94 | | | | |
| 119 | Magnesioskorzalite | - | - | 54 | | | | |
| 86 | Magnesite | 30 | - | 39 | 155 | 170 | 160,163 | |
| 212 | Magnesium-chlorites | 110 | 114 | 118 | | | | 111,113 |
| 32 | Magnetite | 16 | - | 22 | 157 | 174 | 163 | 17,19,43,67, 87,101,109 |

| No. | Mineral | Tab. | Fig. | Diagr. | Appendix Dens. | d-spacings | Coloretc. | other references |
|---|---|---|---|---|---|---|---|---|
| 180 | Magnophorite | - | - | 91 | | | | |
| 152 | Mangan-axinite | - | - | 70 | | | | |
| - | Mangano-phyllite | - | - | - | | | | 103 |
| 85-88 | Mangano-siderite | - | - | 39 | | | | |
| 48 | Manganosite | 18 | - | - | 157 | 172 | | 19 |
| 85-88 | Mangano-spherite | - | - | 39 | | | | |
| 31 | Marcasite | - | - | - | | 174 | | 17 |
| 200 | Margarite | 102 | 105 | 106,117 | 155 | 171 | 163 | 101 |
| 74 | Marialite | 26 | 32 | 35 | 154 | 167 | | 27 |
| 74 | Meionite = Mejonite | 26 | 32 | 35 | 154 | 171 | 163 | 27 |
| 56 | Melanite | 18 | - | - | 156 | | 160-162 | 17,19 |
| 59 | Melano-phlogite | 20 | - | - | 154 | 166 | | |
| 71 | Melilite | 26 | 32 | 34 | 155 | 170 | 158,160-163 | 19,27,37 |
| 72 | Melinophane | 26 | 32 | - | 155 | 170 | 160, 161 | 49,65 |
| 72 | Meliphanite | 26 | 32 | - | | | | |
| 202 | Meroxene | 102 | 105 | 106,107 | 155 | 164 | | |
| 128 | Merwinite | 50 | 53 | - | 155 | 171 | 162 | 43,49,51 |
| 85-88 | Mesitite = Mesitinspar | - | - | 39 | 156 | | | |
| 92 | Mesodialite | 36 | - | - | | | | |
| 235 | Mesolite | - | 139 | 139 | | 170 | 163 | |
| 223 | Metahalloy-site | 113 | - | - | | 166 | | 112 |
| 198-207 | Mica group | 100,102 108 | 104, 105 | 107,117 | | 164 | 162,163 | 27,51,57,111, 113, 152 |
| 230 | Microcline | 120 | 124 | 123,124 | 154 | 166,167 | 158, 163 | 121 |
| 94a | Milarite | - | - | - | | 167 | | |
| 197a | Minnesotaite | 100 | - | - | 155 | 164 | 160,161 | |
| 74 | Mizzonite | 26 | 32 | 35 | 154 | 167 | | 27 |
| - | Molybdenite | - | - | - | | | | 17 |
| 117 | Monazite | 48 | 52 | - | 157 | 168 | 159,160, 161-163 | 27,57,65 |
| 120 | Montebrasite | 48 | 52 | 55 | 155 | | | |
| 123 | Monticellite | 48 | 53 | 55 | 155 | 171 | 158,163 | 19,49,51,109 |
| 208 | Montmorillo-nite | 108 | 114 | - | 154 | 164 | | 109,113 |
| 243c | Mordenite | 140 | 139 | 116,139 | | 167 | | |
| 149 | Mosandrite | 64 | 69 | - | 155 | 168 | 160,162 163 | 65 |
| 192 | Mullite | 100 | 104 | - | 155 | 167 | 159,160 | |
| 199 | Muscovite | 100 | 104 | 106,107 | 155 | 164 | 158,160, 161,163 | 31,37,67,101, 102,103,109, 121 |
| 222 | Nacrite | 112 | 119 | - | 154 | 165 | | 109,113 |
| 228 | Na-Sanidine | 120 | 122 | 123 | 154 | 167 | 158,162 | 31,37,103,121 |
| 98 | Natrodavyne | - | - | 41 | | | | |
| 235 | Natrolite | 140 | 139, 142 | 139 | 154 | 165 | 158,163 | 31,37,141 |
| 95 | Nepheline = Nephelite | 36 | 38 | 40 | 154 | 169 | 158,162, 163 | 21,31,37,121, 141,152 |
| (179) | Nephrite | 85 | - | - | | | | 85 |
| 102 | Niobite | 16,42 | 46 | 60 | | | 160 | 43 |
| - | Nitrocalite | - | - | - | | | | 17 |
| 27 | Nitronatrite | 16 | - | - | | 173 | | |
| 209 | Nontronite | 108 | 114 | 116 | 154 | 164 | 160,161 | 109,111 |
| 124-126a | Norbergite | - | - | - | | 168 | | |

| No. | Mineral | Tab. | Fig. | Diagr. | Appendix Dens. | d-spacings | Color etc. | other references |
|---|---|---|---|---|---|---|---|---|
| 63 | Nosean | 20 | - | - | 154 | 166 | 160,161 163 | 43 |
| 233 | Oligoclase | 120 | 125 | 126-137 | 154 | 167,168 | 158,163 | 31 |
| 85-88 | Oligonite = Oligonspar | - | - | 39 | | | | |
| 122 | Olivine | 48 | 53 | 54 | | | 158-163 | 31,49,51,73,101, 109,111,113, 152 |
| 173 | Omphacite | 84 | 77 | 82 | 156 | 169 | 161 | 73,85 |
| 58 | Opal | 20 | - | 28 | 154 | | 163 | 19,21,27,31 |
| 138 | Orthite = Allanite | 56 | 59 | 62,63 | 156 | 169 | 160-163 | 21,31,49,57,65, 121 |
| 176,177 | Ortho-amphibole | 84 | 88 | 90-92 | | | | 85 |
| 212-214 | Ortho-chlorite | 110 | 114 | 117,118 | | | | 111 |
| 229 | Orthoclase | 120 | 122 | 123 | 154 | | | 121 |
| 164 | Orthoferro-silite | 72 | - | 78 | | | | |
| 164 | Ortho-pyroxene | 72 | 76 | 78 | | | | 73,113 |
| 182 | Osannite | 84 | 89,93 | 91 | 156 | 164 | 160, 162 | |
| 195 | Ottrelite | 101 | - | - | | | 163 | |
| 203a | Oxybiotite | 102 | - | 106 | 155 | | 160 | 103 |
| 184b | Oxyhorn-blende | 86 | - | 91 | | | | 87 |
| 184b | Oxykaer-sutite | 86 | - | 91 | | | | |
| 198 | Paragonite | 100 | 104 | 106,117 | 155 | 164,171 | | 101,103 |
| 161 | Parawollas-tonite | - | - | - | | | | 67 |
| 186 | Pargasite | 86 | 89,93 | 90,91, 97 | 155 | 168 | 161 | 49,67,87 |
| 211 | Parsetten-site | 109 | - | - | | | | 109 |
| 204 | Paucili-thionite | - | - | 107 | | | | |
| 163 | Pectolite | 72 | 74 | 75 | 155 | 169 | 163 | 37,65,67,87 |
| 212 | Pennine = Penninite | 110 | 114 | 118 | 154 | | 160-163 | 111 |
| - | Pentahydrite | - | - | - | | | | 15 |
| 47 | Periclase | 18 | - | - | 156 | 172 | 159,160, 163 | 19,31 |
| 52 | Perovskite | 18 | - | - | 156 | 170 | 160,162, 163 | 19,27,31 |
| 227 | Petalite | 120 | 122 | - | 154 | 166 | | 85,111 |
| 91a | Phenakite | 36 | - | - | | 168 | | 37,57 |
| 199a | Phengite | 102 | 104 | 106 | 154 | | | 101 |
| 243 | Phillipsite | 140 | 142, 143 | 139 | 154 | 165 | 163 | 141 |
| 201 | Phlogopite | 102 | 105 | 106,107 | 154 | 164 | 160,163 | 51,87,103,109, 111 |
| 90 | Phosphorite | 36 | - | - | | | | |
| 198-207 | Phyllosi-licates | 100,102 108 | 104, 105 | 107,117 | | 164 | 162 | 27,51,57,111, 113,152 |
| 50 | Picotite | 18 | - | 22 | 157 | | 160,161 | 17,19 |
| 21 | Picromerite | 14 | - | - | | 173 | | 13,15,37 |
| 137 | Piemontite | 56 | 58 | 62,63 | 156 | | 160,161 | 51,73 |
| 166 | Pigeonite | 72 | 76,77 | 79,80 | 156 | | 160,161 | 51,73 |
| 165-169 | Pigeonitic Augite | 72 | - | 79 | | | | |
| 199a | Pinite | 103 | - | - | | | | 67 |

| No. | Mineral | Tab. | Fig. | Diagr. | Appendix Dens. | d-spacings | Color etc. | other references |
|---|---|---|---|---|---|---|---|---|
| 136 | Pistacite see Epidote | 56 | 58 | 62,63 | 156 | 170 | 158-163 | 17,27,49,57,65,67, 85,101,109,152 |
| 85-88 | Pistomesite | - | - | 39 | 156 | | | |
| 233 | Plagioclase | 120 | 125 | 125, 126-137 | | 167,168 | 163 | 21,51,67,101, 141,152 |
| 50 | Pleonaste | 18 | - | 22 | 157 | | 161,162 | 19,67,103 |
| 61a | Pollucite | 20 | - | - | 155 | 167 | | |
| 22 | Polyhalite | 14 | - | - | | 173 | 162 | 13,15,45 |
| 205 | Polylithi-onite | 102 | 105 | 106,107 | 155 | | | |
| 228-230 | Potassic Feldspar | 120 | 122, 124 | 123,124 | | | | 121,152 |
| 85-88 | Ponite | - | - | 39 | | | | |
| 158 | Prehnite | 66 | 69 | 71 | 155 | 168 | 162,163 | 49,57,67,73,121 |
| 153 | Prismatine | (66) | - | - | | | 163 | 51 |
| 213 | Prochlorite = Rhipidolite | 110 | 114 | 117,118 | | | | |
| 204 | Proto-lithionite | 102 | 105 | 107 | 155 | | 160,163 | 37,103 |
| 103 | Pseudo-brookite | 42 | 46 | - | | 167 | 160,163 | 17,43,57 |
| 62 | Pseudoleucite | - | - | - | | | | 21 |
| 139 | Pseudo-zoisite | 56 | 59 | - | | 171 | | |
| 243c | Ptilolite | - | - | 139 | | | | |
| 141a | Pumpellyite | 64 | 59 | 63 | 155 | 169 | 161 | 57 |
| 213 | Pycnochlo-rite | 110 | - | 118 | | | | |
| 31 | Pyrite | 16 | - | - | 157 | 174 | 163 | 17,113 |
| 54 | Pyrochlore | 18 | - | - | 157 | 168,169 | 160 | 19 |
| 56 | Pyrope | 18 | - | 24,25 | 156 | 171 | 158-160 | 19 |
| 36,79 | Pyrophanite | 16,30 | 32 | - | 157 | 170,174 | 160 | 17,31,43 |
| 196 | Pyrophyllite | 100 | 104 | - | 154 | 168 | | 43,101,113 |
| 164-175 | Pyroxene | 72,84 | 76,77 | 78-83 | | | 159,162 | 49,57,67,73,85, 109,111 |
| 54 | Pyrrhite | 18 | - | - | 157 | | | 19 |
| 30 | Pyrrhotine = Pyrrhotite | 16 | - | - | 157 | 174 | 163 | 17 |
| 81 | Quartz | 30 | 33 | - | 154 | 167 | 158 | 21,27,30,31,37,49, 57,67,85,101,103, 109,113,121 |
| 82 | Quartzine | 30 | 33 | - | | | | 30 |
| 129a | Rankinite | 50 | - | - | 155 | 168 | | 49,51 |
| 213 | Rhipidolite = Ripidolite | 110 | 114 | 117,118 | 155 | | 161 | 111 |
| 85-88 | Rhodochro-site | - | - | 39 | | | | 67 |
| 159 | Rhodonite | 66 | 74 | 74,75 | 156 | 170 | 160,161 | 49,51,73,85 |
| 181 | Rhodusite | 85 | - | - | 156 | | | 85 |
| 43,190 | Rhoenite | 18,86 | 99 | 99 | | 171,174 | 160, 163 | 17,19,43,57,87, 103 |
| 180 | Richterite | 84 | 88,93 | 90,91 | 155 | 164 | 160,161 | 85 |
| 182 | Riebeckite | 84 | 89,93 | 90,91 | 156 | 164 | 162 | 37,85,101 |
| 148 | Rinkite | 64 | 68 | - | 156 | 168 | 159,160 162,163 | 65 |
| - | Rinkolite | - | - | - | | | | 65 |
| 7 | Rinneite | 12 | - | - | | 173 | 160,162 | |
| 147 | Rosenbu-schite | 64 | 68 | - | 156 | 169 | 160,163 | 65,73 |
| 66 | Rutile | 26 | 29 | - | 157 | 167 | 159-163 | 17,19,21,30, 31,43,49,51, 57,85,101 |

| No. | Mineral | Tab. | Fig. | Diagr. | Appendix | | | other references |
|---|---|---|---|---|---|---|---|---|
| | | | | | Dens. | d-spacings | Coloretc. | |
| 66 | Sagenite | 26 | - | - | | | | |
| 167 | Salite | 72 | 76 | 79,82,83 | 156 | | | 51 |
| 228 | Sanidine | 120 | 122 | 123 | 154 | 167 | 158,162 | 31,37,103,121 |
| 210a | Saponite group | 108 | - | - | 154 | 164 | | |
| 135 | Sapphirine | 56 | 58 | - | 156 | 172 | 160-163 | 27,51,67,101 |
| (136) | Saussurite | - | - | - | | | | 57 |
| 74 | Scapolite = Skapolite | 26 | 32 | 35 | | | 158-160 | 51 |
| 161a | Scawtite | 72 | - | - | | 168 | | |
| (168) | Schefferite | - | - | - | | | | 85 |
| 21 | Schoenite | 14 | - | - | | 173 | | 13,15,37 |
| 91 | Schorlite | 36 | 38 | 39 | 155 | 171 | 160-163 | 73 |
| 237 | Scolecite | 140 | 142, 143 | 139 | 154 | 170 | 163 | 141 |
| 119 | Scorzalite | - | - | 54 | | | | |
| 207 | Seladonite = Celadonite | 109 | - | - | | 171 | | 109 |
| 154 | Serendi- bite | 66 | 69 | - | 156 | 172 | 161-163 | 49,57,67 |
| 199 | Sericite | 101, 103 | - | - | | | | 51,101,103, 111,113 |
| 217-219 | Serpentine group | 110, 112 | 115 | 118 | | | 163 | 49,51,152 |
| 212 | Sheridanite | 111 | - | 117,118 | 154 | | | 31 |
| 87 | Siderite | 36 | 33 | 39 | 156 | 170 | 159,160, 162,163 | 31,37,85,101, 113 |
| 202 | Sidero- phyllite | 102 | 105 | 106,107 | 156 | 164 | | |
| 85-88 | Sidero- plesite | - | - | 39 | 156 | | | |
| 191 | Sillimanite | 100 | 104 | - | 155 | 167 | 158,163 | 43,51,67,85, 101 |
| 122b | Sinhalite | 48 | - | - | 156 | 172 | 160,161 | 49 |
| 179a | Smaragdite | 84 | - | - | | | | 85 |
| 197 | Soapstone = Talc | 100 | 104 | - | 154 | 164 | 163 | 13,27,31,45,85, 101,103,109, 111 |
| 63 | Sodalite | 20 | - | - | 154 | 166 | 158,163 | 19,21,141 |
| 180-184 | Sodic Hornblende | 84, 86 | 88, 89 | 93- 96 | 155,156 | | 161,162 | |
| 56 | Spessartite | 18 | - | 24,25 | 157 | 171 | 158-160 | 19,43,57,67,73 |
| 44 | Sphalerite | 18 | - | - | 157 | 168 | 159,160, 163 | 113 |
| 142 | Sphene | 64 | 59 | 63 | 156 | 167 | 159,160, 161-163 | 27,30,31,37, 49,57,65,87 |
| 50 | Spinel | 18 | - | 22,23 | 156 | 172 | 158, 162 163 | 19,31,49,51,57, 67,73,85,109, 121 |
| 175 | Spodumene | 84 | 77 | - | 155 | 167,169 | 158,160 | 21,49,103,111, 121 |
| 129 | Spurrite | 50, | 53 | - | 155 | 171 | 162 | 51 |
| 55 | Stassfurtite | 18 | - | - | 155 | | | 45 |
| 134 | Staurolite | 56 | 58 | 61 | 156, | 168 | 158-163 | 51,65,67,101 |
| 197 | Steatite = Talc | 100 | 104 | - | 154 | 164 | 163 | 13,27,31,45,85, 101,103,109,111, 113 |
| 241 | Stilbite | 140 | 142, 143 | 139 | 154 | 166 | 163 | 141 |
| 211 | Stilpnome- lane | 108 | 114 | 116 | 154 | 164 | 160,161 | 101,103,109 |
| 67a, 81b | Stishovite | 26,30 | - | - | 157 | 169 | | 31,43 |
| 216 | Sudoite | 110 | 115 | - | 154 | 166 | 161 | |
| 97 | Sulfatic Cancrinite | - | - | 41 | | | | |

| No. | Mineral | Tab. | Fig. | Diagr. | Appendix Dens. | d-spacings | Color etc. | other references |
|---|---|---|---|---|---|---|---|---|
| 56 | Uvarovite = Uwarowite | - | - | - | 156 | 171 | 159 | 19,57 |
| 91 | Uvite | 36 | 38 | 39 | | | | |
| 13 | Vanthoffite | 12 | - | - | | 173 | | 13,15 |
| 121a | Variscite | 48 | - | - | 154 | 166 | | |
| 85a | Vaterite | - | - | - | | 167 | | |
| 210 | Vermiculite | 108 | 114 | - | 154 | 164 | 161 | 103,109 |
| 70 | Vesuvianite = Idocrase | 26 | 29 | 35 | 156 | 170 | 160-163 | 27,37,43,57, 109 |
| 3 | Villiaumite | 12 | - | - | 154 | 173 | 160 | 19 |
| 131 | Viridine | 50 | 58 | - | 155 | 166 | 160,161 | 57 |
| 61b | Viséite | - | - | - | | 169 | | |
| 121b | Vivianite | 48 | - | - | 154 | 165 | 161,162 | |
| 64 | Volcanic Glass | 20 | - | - | 154 | | 160,161 | 21, 27 |
| 118 | Wagnerite | 48 | 52 | - | 155 | 169 | 163 | |
| 61b | Wairakite | 20 | - | - | | 167 | | |
| 121 | Wavellite | 48 | 52 | - | 154 | 165 | 158,163 | 45 |
| 244 | Weddellite | - | - | - | | 170 | | |
| 244 | Whewellite | 140 | 143 | - | 154 | 165 | | |
| 70 | Wiluite | 26 | - | 35 | | | | |
| 180 | Winchite | - | - | 91 | | | | |
| 97 | Wischnewite | 36 | - | 41 | | | | 37 |
| 203 | Wodanite | 102 | - | - | | | 160,161 | |
| 143 | Woehlerite | 64 | 68 | - | 156 | 170 | 160,162 | 65 |
| - | Wolfeite | - | - | - | | | | 49 |
| - | Wolframite | - | - | - | | | | 17,43,49,103, 109 |
| 161 | Wollastonite | 66 | 74 | 74,75 | 155 | 169 | 163 | 51,57,73,85,87 |
| 206 | Xantho-phyllite | 108 | 105 | 106 | 155 | 167,171 | 160,161 163 | 101,103,109 |
| 68 | Xenotime | 26 | 29 | - | 157 | 167 | 159-163 | 27,49 |
| 142 | Yttro-titanite | 64 | - | - | 156 | | | 65 |
| 235-243 | Zeolite group | 140 | 142, 143 | 139 | | | 162 | 21,27,31,43, 49,109,121, 141,152 |
| 44 | Zincblende = Sphalerite | 18 | - | - | 157 | 168 | 159,160, 163 | 113 |
| 205a | Zinnwaldite | 108 | 105 | 106 | 155 | 167 | 160 | 103 |
| 69 | Zircon | 26 | 29 | - | 157 | 167 | 159-163 | 27,49,65 |
| 139 | Zoisite | 56 | 59 | - | 156 | 171 | 162,163 | 27,37,49,57, 65,67,73,85, 101 |